AF889076

Encyclopaedia of Mathematical Sciences

Volume 23

Editor-in-Chief: R.V. Gamkrelidze

I. R. Shafarevich (Ed.)

Algebraic Geometry I

Algebraic Curves
Algebraic Manifolds and Schemes

With 49 Figures

Springer-Verlag
Berlin Heidelberg New York
London Paris Tokyo
Hong Kong Barcelona
Budapest

Consulting Editors of the Series:
A. A. Agrachev, A. A. Gonchar, E. F. Mishchenko,
N. M. Ostianu, V. P. Sakharova, A. B. Zhishchenko

Title of the Russian edition:
Itogi nauki i tekhniki, Sovremennye problemy matematiki,
Fundamental'nye napravleniya, Vol. 23, Algebraicheskaya geometriya 1
Publisher VINITI, Moscow 1988

Mathematics Subject Classification (1991):
14Axx, 14Cxx, 14Exx, 14Fxx, 14Hxx, 14Kxx, 14Mxx, 14Nxx, 30Fxx

ISBN 3-540-51995-5 Springer-Verlag Berlin Heidelberg New York
ISBN 0-387-51995-5 Springer-Verlag New York Berlin Heidelberg

Library of Congress Cataloging-in-Publication Data
Algebraicheskaia geometriia 1. English.
Algebraic geometry I: algebraic curves, algebraic manifolds and schemes / I. R. Shafarevich (ed.).
p. cm. — (Encyclopaedia of mathematical sciences; v. 23)
Includes bibliographical references and index.
ISBN 0-387-51995-5
1. Curves, Algebraic. 2. Manifolds (Mathematics) 3. Schemes (Algebraic geometry)
I. Shafarevich, I. R. (Igor' Rostislavovich), 1923- . II. Title. III. Series.
QA565.A4413 1994 516.3'5—dc20 93-4995

This work is subject to copyright. All rights are reserved, whether the whole or part of the material is concerned, specifically the rights of translation, reprinting, reuse of illustrations, recitation, broadcasting, reproduction on microfilm or in any other way, and storage in data banks. Duplication of this publication or parts thereof is permitted only under the provisions of the German Copyright Law of September 9, 1965, in its current version, and permission for use must always be obtained from Springer-Verlag. Violations are liable for prosecution under the German Copyright Law.
© Springer-Verlag Berlin Heidelberg 1994
Printed in the United States of America
Typesetting: Springer $\TeX$ in-house system
41/3140 - 5 4 3 2 1 0 - Printed on acid-free paper

List of Editors, Authors and Translators

Editor-in-Chief

R.V. Gamkrelidze, Russian Academy of Sciences, Steklov Mathematical Institute, ul. Vavilova 42, 117966 Moscow, Institute for Scientific Information (VINITI), ul. Usievicha 20a, 125219 Moscow, Russia

Consulting Editor

I. R. Shafarevich, Steklov Mathematical Institute, ul. Vavilova 42, 117966 Moscow, Russia

Authors

V. I. Danilov, Central Economic and Mathematical Institute of the Russian Academy of Sciences, ul. Krasikova 32, 117418 Moscow, Russia
V. V. Shokurov, Department of Mathematics, The Johns Hopkins University, Baltimore, Maryland 21218, USA

Translators

D. Coray, Section de Mathématiques, Université de Genève, 2-4, rue du Lièvre, Case postale 240, 1211 Genève 24, Switzerland
V. N. Shokurov †

Contents

I. Riemann Surfaces and Algebraic Curves

V. V. Shokurov

Translated from the Russian
by V. N. Shokurov

Contents

Introduction[1]

The name 'Riemann surface' is a rare case of a designation which is fully justified historically: all fundamental ideas connected with this notion belong to Riemann. Central among them is the idea that an analytic function of a complex variable defines some natural set on which it has to be studied. This need not coincide with the domain of the complex plane where the function was initially given. Usually, this natural set of definition does not fit into the complex plane $\mathbb{C}$, but is a more complicated surface, which must be specially constructed from the function: this is what we call the Riemann surface of the function. One can get a complete picture of the function only by considering it on the whole of its Riemann surface. This surface has a nontrivial geometry, which determines some of the essential characters of the function.

The extended complex plane, obtained by adjoining a point at infinity, can be perceived as an embryonic form of this approach. Topologically, the extended plane is a two-dimensional sphere, also known as the Riemann sphere. This example already displays some features which are characteristic of the general notion of a Riemann surface:

1) The Riemann sphere $\mathbb{CP}^1$ can be defined by gluing together two disks (i.e., circles) of the complex plane; for instance, the disks $|z| < 2$ and $|w| < 2$, in which the annuli $\frac{1}{2} < |z| < 2$ and $\frac{1}{2} < |w| < 2$ are identified by means of the correspondence $w = z^{-1}$. (This yields the shaded area in Fig. 1.)

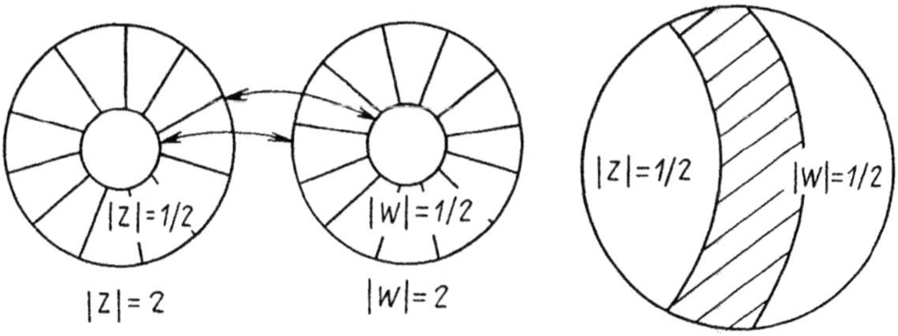

Fig. 1

2) The relation $w = z^{-1}$, which defines the gluing, is a one-to-one and analytic (conformal) correspondence of the domains it identifies. For that reason the property of being analytic at some point agrees in both circles, $|z| < 2$ and $|w| < 2$, on the identified regions. This leads to a unified notion

[1] The author expresses his profound gratitude to Professor I. R. Shafarevich for numerous remarks and suggestions, which have contributed to the improvement of the text, and for writing this introduction, which provides a fascinating bird's-eye view of the charming world of algebraic geometry.

of analytic function on the Riemann sphere glued from them. It is therefore possible to state and prove such theorems as: 'a function which is holomorphic on the whole Riemann sphere is constant', or: 'a function on the Riemann sphere which has only poles for singularities, is a rational function'.

The same principles underlie the general notion of a Riemann surface. We shall deal only with compact Riemann surfaces. By definition, this is a closed (compact) surface S glued from a finite number of disks $U_1, \dots, U_m$ in the complex plane: for any two disks, U_i and U_j, some domains, $V_{ij} \subset U_i$ and $V_{ji} \subset U_j$, are identified by means of a correspondence $\varphi_{ij}\colon V_{ij} \to V_{ji}$, which is one-to-one and analytic.

In other words, a Riemann surface is a union of sets $U_1, \dots, U_N$, each of which is endowed with a coordinate function z_i $(i = 1, \dots, N)$. This is a one-to-one mapping of U_i onto a disk in the complex plane. Further, in an intersection $V_{ij} = U_i \cap U_j$, the coordinate z_j is expressed in terms of z_i as an analytic function, and similarly z_i in terms of z_j.

Thus, just as in the case of the Riemann sphere, there is a well-defined notion of analyticity for a continuous complex-valued function, given in a neighbourhood of some point $p \in S$. Further, we can carry over to functions given on the surface S such notions as a pole, the property of being meromorphic, and so forth. Hence a Riemann surface is a set on which it makes sense to say that a function is analytic, and locally (in a sufficiently small domain) this amounts to the ordinary concept of analyticity in some domain of the complex plane. This definition is explained in detail in § 1 of Chapter 1.

So, with the notion of a Riemann surface, we run into an entity of a new mathematical nature. It must be rated on a par with such notions as a Riemannian manifold in geometry, or a field in algebra. Just as some metric concepts are defined in a Riemannian manifold, and algebraic operations in a field, so is the notion of analytic function on a Riemann surface. In particular, it is now possible to formulate and prove the theorem stating that a function which is holomorphic on an entire (compact) Riemann surface is constant.

That the concept of Riemann surface is nontrivial, is manifest from its connection with the theory of multivalued analytic functions. In fact, for every such function one can construct a Riemann surface on which it becomes single-valued. We restrict ourselves to algebraic functions, so the corresponding Riemann surfaces are compact.

The simplest case, represented by the function $w = \sqrt[n]{z}$, does not yet necessitate any new type of surface. Indeed we have $z = w^n$; so, even though w is a multivalued function of z, the function $z(w)$ is single-valued. Therefore we can regard w as an independent variable, running over the Riemann sphere S, which is just the Riemann surface of the function w. The relation $z = w^n$ defines a mapping of the w-sphere S onto the z-sphere $\mathbb{CP}^1$. One can think of the sphere S as lying 'above' $\mathbb{CP}^1$ (in some larger space), in such a way that above each point $z = z_0$ we find the points which are mapped into it. Then for $z_0 \neq 0, \infty$ the inverse image on S of a disk $U\colon |z - z_0| < \varepsilon$, for sufficiently small ε, is made up of n disjoint domains W_i, $i = 1, \dots, n$:

$$w = w_i g(t), \quad |t| < \frac{\varepsilon}{|z_0|}, \quad g(t) = \sqrt[n]{1+t}, \quad g(0) = 1, \quad t = \frac{z}{z_0} - 1,$$

where the w_i are the distinct values of $\sqrt[n]{z_0}$ (Fig. 2a). But, in a neighbourhood of the point 0 (respectively, of ∞), the inverse image of a disk $|z| < \varepsilon$ (respectively, $|t| < \varepsilon$, with $t = z^{-1}$) is constituted by a single circle W: $|w| < \sqrt[n]{\varepsilon}$, which lies above the disk in the form of a 'helix' (see Fig. 2b, where $n = 2$).

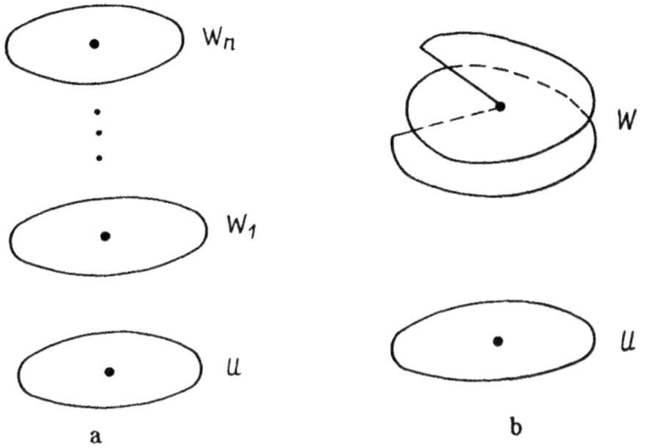

Fig. 2

In the general case, an algebraic function is defined by an equation $f(z, w) = 0$, where $f(z, w)$ is a polynomial $f(z, w) = a_0(z)\, w^n + \ldots + a_n(z)$, and the $a_i(z)$ are polynomials in z. As a first, rough approximation to the Riemann surface of the function w, we shall look at the set $\widetilde{S}$ of all solutions (z, w) of $f(z, w) = 0$. On this set, w is tautologically the function that takes on the value w_0 at (z_0, w_0). However, this definition must be made more precise. We shall assume that $\widetilde{S} \subset \mathbb{C}^2$, where $\mathbb{C}^2$ is the plane of the two complex variables z, w, and where the topology of $\widetilde{S}$ is inherited from $\mathbb{C}^2$. In other words, $\widetilde{S}$ is a complex algebraic curve lying in the plane $\mathbb{C}^2$.

To start with, suppose z_0 is such that $f(z_0, w) = 0$ has n distinct roots $w_1, \ldots, w_n$. This means that $a_0(z_0) \neq 0$ and $f'_w(z_0, w_i) \neq 0$. Then, by the implicit function theorem, w is an analytic function $g_i(z)$ of z in some neighbourhood $|z - z_0| < \varepsilon$ of z_0. More precisely, all solutions of $f(z, w) = 0$ close to (z_0, w_i) can be represented in the form $(z, g_i(z))$, $i = 1, \ldots, n$. That is to say, the solutions with $|z - z_0| < \varepsilon$ fall into n disks W_i, $i = 1, \ldots, n$:

$$|z - z_0| < \varepsilon, \quad w = g_i(z),$$

exactly as in Fig. 2a. We call them disks because the function z maps them in a one-to-one manner onto the disk U: $|z - z_0| < \varepsilon$.

It remains to consider the cases we have omitted, in which the number of solutions of $f(z_0, w) = 0$ is less than n, and also the case where $z_0 = \infty$ on the Riemann sphere $\mathbb{CP}^1$. In all these cases there exists a disk $U: |z - z_0| < \varepsilon$ (respectively, $|t| < \varepsilon$, $t = z^{-1}$, if $z_0 = \infty$) with the property that, for all points $z \in U$, $z \neq z_0$, we are in the case previously considered. We denote by $\widetilde{U}$ the associated punctured disk: $|z - z_0| < \varepsilon$, $z \neq z_0$, and by $\widetilde{W}$ its inverse image in $\widetilde{S}$. The set $\widetilde{W}$ may turn out to be disconnected.

Trivially, if $f(z_0, w) = 0$ has two distinct solutions, w_i and w_j, then two small neighbourhoods in $\widetilde{S}$ do not meet and give rise to different connected components of $\widetilde{W}$, like the sets $W_1, \ldots, W_n$ in Fig. 2*a*. But there are less trivial cases in which various connected components of $\widetilde{W}$ converge to the same point of $\widetilde{S}$. The idea is that in reality these components must define distinct points of the Riemann surface S of w: they must be 'separated' in S. If, for instance, $w^2 = z^2 + z^3$ then $w = z\sqrt{1+z}$. Now the function $\sqrt{1+z}$ has two branches, $g_1(z)$ and $g_2(z) = -g_1(z)$, in a neighbourhood of $z_0 = 0$. So $\widetilde{W}$ consists of two components: $\widetilde{W}_1 = \{|z| < \varepsilon,\ z \neq 0,\ w = zg_1(z)\}$ and $\widetilde{W}_2 = \{|z| < \varepsilon,\ z \neq 0,\ w = zg_2(z)\}$, which merge as $z \to 0$ (Fig. 3*a*).

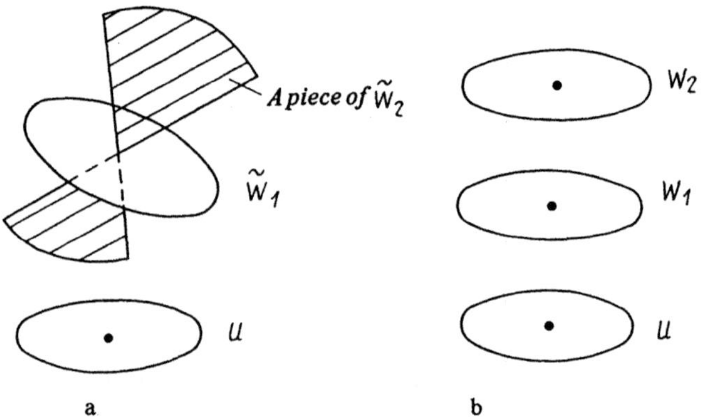

Fig. 3

In the general case, we denote by $\widetilde{W}_1, \ldots, \widetilde{W}_r$ the connected components of $\widetilde{W}$. The Riemann surface S is defined in such a way that in it the $\widetilde{W}_i$ are, so to speak, 'isolated' from each other: their closures do not meet as $z \to z_0$. Set-theoretically, S differs from $\widetilde{S}$ in that now there are r distinct points above z_0, each corresponding to its own component $\widetilde{W}_i$. More precisely, each $\widetilde{W}_i$ is a connected unramified covering of the punctured disk $\widetilde{U}$: above every point $z \in \widetilde{U}$, we find the same number n_i (say) of points in $\widetilde{W}_i$, and $n_1 + \ldots + n_r = n$. It is easy to prove that a function w_i can be defined on each $\widetilde{W}_i$ in such a way that $\widetilde{W}_i$ is given as the punctured disk $|w_i| < \varepsilon^{1/n_i}$, $w_i \neq 0$, and the mapping $\widetilde{W}_i \to \widetilde{U}$ is defined as $z - z_0 = w_i^{n_i}$. We can then look at the unpunctured disk W_i: $|w_i| < \varepsilon^{1/n_i}$. The various disks W_i are regarded as disjoint sets in the Riemann surface S (cf. Fig. 3*b*). Each of them is mapped by the function w_i onto a disk of the complex plane, and they lie above the Riemann z-sphere as in Fig. 2*b*.

From all the disks W_i we have constructed, above the various points $z_0 \in \mathbb{CP}^1$ (including $z_0 = \infty$), we can select a finite number, $W_1, \dots, W_N$, whose union already contains all the others. From the analyticity of all the mappings we have encountered, it is easy to deduce that the variety obtained by gluing the disks $W_1, \dots, W_N$ verifies the condition occurring in the definition of a Riemann surface. Thus, S is indeed a Riemann surface. For a detailed justification of this construction, see Chapter 1, § 2.

An arbitrary Riemann surface carries with it a large amount of geometric information. In particular, the Riemann surface of an algebraic function reveals some important characteristics of that function. Since the gluings φ_{ij} are conformal, and hence orientation-preserving, transformations, any Riemann surface is orientable. So, from a topological point of view it has a unique invariant: the genus. In Fig. 4 are depicted surfaces of genus $g = 0, 1, 2, 3, 4$.

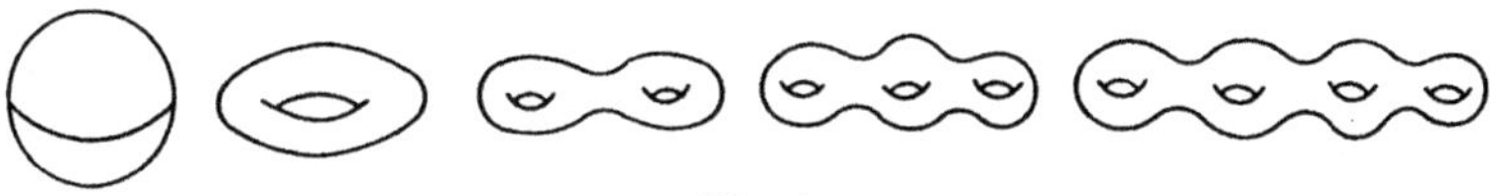

Fig. 4

If, for example, a polynomial $f(z)$ (of degree $2n$ or $2n-1$, say) has no multiple roots, then the Riemann surface of the function $w = \sqrt{f(z)}$ is of genus $n-1$. But, in addition, one can define on a Riemann surface all the notions which are invariant under conformal transformations: it has a 'conformal geometry'. Among such notions are the Laplace operator and harmonic functions. In particular, the real and imaginary parts of a function which is analytic in some domain of a Riemann surface, are harmonic. This enables us to study functions on a Riemann surface by applying the apparatus of elliptic differential operators and even some physical intuition. A harmonic function on a Riemann surface can be conceived as a description of a stationary state of some physical system: a distribution of temperatures, for instance, in case the Riemann surface is a homogeneous heat conductor. Klein (following Riemann) had a very concrete picture in his mind:

"This is easily done by covering the Riemann surface with tin foil ... Suppose the poles of a galvanic battery of a given voltage are placed at the points A_1 and A_2. A current arises, whose potential u is single-valued, continuous, and satisfies the equation $\Delta u = 0$ across the entire surface, except for the points A_1 and A_2, which are discontinuity points of the function."

[*Vorlesungen über die Entwicklung der Mathematik im 19. Jahrhundert,* p. 260]

The existence of functions, which is suggested by such physical considerations, is established on the basis of the theory of elliptic partial differential equations. This provides an absolutely new method of constructing analytic functions on a Riemann surface: once a harmonic function u has been constructed, we select its conjugate function v, so that $u + iv$ is analytic.

In particular, this enables one to describe the stock of all meromorphic functions on any Riemann surface S. If S is the Riemann surface of an algebraic function w given by $f(z, w) = 0$, then both w and z are meromorphic functions on S. Therefore any rational function of w and z is meromorphic. It can easily be proved that this is the way all meromorphic functions on S are obtained. This is a generalization of the theorem saying that a meromorphic function on the Riemann sphere is a rational function of z. For an arbitrary Riemann surface, however, it is by no means obvious that there is even one nonconstant meromorphic function. Such a function is constructed, as we have just said, by using methods from the theory of elliptic partial differential equations. Furthermore, one can construct along the same lines two meromorphic functions w and z on S, connected by a relation of the form $f(z, w) = 0$, where f is a polynomial, and with the property that S is just the Riemann surface of the algebraic function w defined by the equation $f = 0$. This result is known as 'Riemann's existence theorem'.

Hence the abstract notion of a (compact) Riemann surface reduces to that of Riemann surface for an algebraic function. This is a highly nontrivial result, with powerful applications. Indeed, in a number of particular situations, what arises is an 'abstract' Riemann surface. Then the preceding theorem provides a very explicit realization of such a surface. The simplest example of such a situation is when S is the quotient group $\mathbb{C}/\Lambda$ of the complex plane $\mathbb{C}$ modulo a lattice $\Lambda = \{\omega_1 n_1 + \omega_2 n_2 \mid n_1, n_2 \in \mathbb{Z}\}$, spanned by two complex numbers ω_1 and ω_2. Let U be any sufficiently small disk, so that no two of its points differ by a vector from Λ. Then the coordinate z on $\mathbb{C}$ is a one-to-one mapping of U onto a domain in $S = \mathbb{C}/\Lambda$ (Fig. 5). Further, these disks form a covering of S. Topologically S is a torus: it is of genus 1. In this situation, Riemann's existence theorem shows that S is the Riemann surface of an algebraic function $w = \sqrt{z^3 + az + b}$, where a and b are some complex numbers and the polynomial $z^3 + az + b$ has no multiple roots. It can be shown that every Riemann surface of genus 1 can be obtained in this way. The meromorphic functions on S are interpreted as being all meromorphic functions of z which are invariant under translations by vectors of the lattice Λ, that is, elliptic functions. In this case, Riemann's existence theorem furnishes a very explicit description of an elliptic function field.

Such a description is possible for Riemann surfaces of genus $g > 1$ as well. One has to consider discrete groups of linear fractional transformations acting in the disk $|z| < 1$. Two points are identified if they are sent to each other by an element of such a group Γ. Thus the Riemann surface is represented as a quotient $\Gamma \backslash \mathbb{D}$, where $\mathbb{D}$ is the unit disk. Just like the plane $\mathbb{C}$ (for surfaces of genus 1), the unit disk, for genus $g > 1$, is the universal covering of the Riemann surface S. For $g = 0$, S is nothing else than the Riemann sphere and is its own universal covering. In the plane $\mathbb{C}$ the Euclidean metric

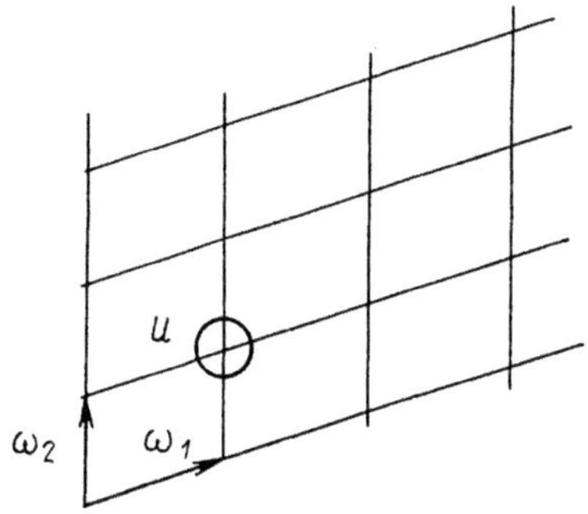

Fig. 5

$ds^2 = |dz|^2$ is invariant under transformations of the group Λ and specifies a metric of zero-curvature on the surface S. Likewise, in the unit circle the metric $ds^2 = |dz|^2/(1-|z|^2)$ defines a Lobachevskian geometry of constant negative curvature, and hence a similar metric on the surface $S = \Gamma\backslash\mathbb{D}$ as well. Finally, there is a metric of constant positive curvature on the sphere $\mathbb{CP}^1$. In all three cases, these metrics provide the Riemann surface S with a 'conformal geometry'. Hence the properties of Riemann surfaces depending on their topology can be summarized in the following Table:

Genus	Type of universal covering	Metric of constant curvature K
0	Riemann sphere $\mathbb{CP}^1$	$K > 0$
1	$\mathbb{C}$	$K = 0$
> 1	$\mathbb{D} = \{z,\ |z| < 1\}$	$K < 0$

One sees from this table that on any Riemann surface S one can define a metric ds^2 of constant curvature K which provides the surface with a conformal geometry. The converse is also true: any metric $ds^2 = Edx^2 + 2Fdxdy + Gdy^2$ on a compact orientable surface S defines on it a Riemann surface structure. Namely, it can be proved that, in a neighbourhood U of any point on the surface, any such metric can be written in some coordinate system as $ds^2 = \lambda(dx^2 + dy^2)$ (x and y are called isothermal coordinates). Setting $z = x + iy$ we may write the metric as $ds^2 = \lambda\, dz\, d\bar{z}$. If, similarly, $ds^2 = \mu\, dw\, d\bar{w}$ in another domain V, one checks easily (in view of the orientability of the surface) that $dw = \varphi dz$. It follows that w is an analytic function of z. Thus the domains U, together with their coordinates z, define a Riemann surface structure on S. Two metrics define the same Riemann surface if they differ by a factor ψ, where ψ is an everywhere positive, real function on S. Multiplication by such a function is called a gauge transformation. Thus a Riemann surface is the same as a surface with a metric of differential geometry, which is considered up to gauge transformations. This is how Riemann surfaces arise in quantized field theory, in the so-called boson string theory.

Starting from an algebraic curve $\widetilde{S} \subset \mathbb{C}^2$, given by an equation $F(z,w) = 0$, we showed above how to construct a Riemann surface S. This Riemann surface can then be used for studying the algebraic curve. On the one hand, every polynomial $G(z,w)$ may be regarded as defining a new algebraic curve C, with equation $G(z,w) = 0$; further, the common solutions of $F(z,w) = 0$ and $G(z,w) = 0$ can be viewed as the intersection points of the curves $\widetilde{S}$ and C. On the other hand, $G(z,w)$ is a meromorphic function on the Riemann surface S; its zeros correspond to the intersection points of $\widetilde{S}$ and C, where the order of contact is defined by the multiplicity of the corresponding zero, etc. To give a geometrical meaning also to the points of S where z or w go to infinity, one has to consider algebraic curves in the projective plane $\mathbb{CP}^2 \supset \mathbb{C}^2$.

If, for example, the curve $\widetilde{S}$ is of degree 3, then it is easy to see that, in some coordinate system, its equation can be written as $w^2 = z^3 + az + b$. As stated above, the associated Riemann surface is of the form $\mathbb{C}/\Lambda$ (cf. Fig. 5). Thus, each point $p \in \widetilde{S}$ corresponds to a point on the surface $S = \mathbb{C}/\Lambda$, that is, to a complex number z, given up to an element $w \in \Lambda$. The following 'dictionary' shows how the geometric properties of points $p \in \widetilde{S}$ are reflected by the corresponding complex numbers. The proof is omitted.

Three points $p_1, p_2, p_3 \in \widetilde{S}$ are collinear	$z_1 + z_2 + z_3 \in \Lambda$
The points $p_1, \ldots, p_m$ form the intersection of $\widetilde{S}$ with a curve C of degree n	$m = 3n$, $\sum_1^m z_i \in \Lambda$

Here, for example, $p_1 = p_2$ if there is a contact at p_1; and $p_1 = p_2 = p_3$ in the case of a contact of order 2 (an inflection point), etc.

For instance, suppose we wish to find the tangents to $\widetilde{S}$ that pass through a point $p \in \widetilde{S}$. Then a complex number z' corresponding to a point of contact p' satisfies the relation $2z' + z \in \Lambda$ (where z corresponds to the point p). In other words,

$$2z' + z = n_1\omega_1 + n_2\omega_2, \quad n_i \in \mathbb{Z}$$

$$z' = -\tfrac{1}{2}z + \frac{\nu_1\omega_1 + \nu_2\omega_2}{2} + \omega, \quad \nu_i = 0, 1; \quad \omega \in \Lambda.$$

Hence there are four such tangents ($\nu_1 = 0, 1$; $\nu_2 = 0, 1$), as in Fig. 6.

In exactly the same way, the inflection points of $\widetilde{S}$ correspond to the complex numbers z such that $3z \in \Lambda$. Therefore

$$z = \frac{\nu_1\omega_1 + \nu_2\omega_2}{3} + \omega, \quad \nu_i = 0, 1, 2; \quad \omega \in \Lambda.$$

So, there are nine of them. As a further consequence, if a straight line passes through two inflection points, its third intersection with $\widetilde{S}$ is again a flex. The

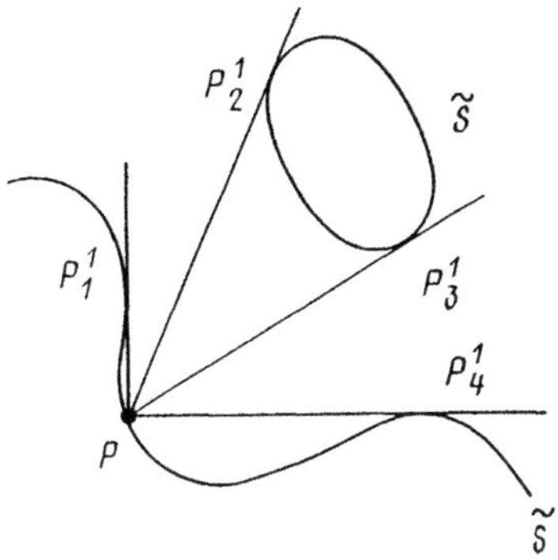

Fig. 6

more general problem of inflection points 'of order n' can be dealt with in much the same way. These points $p \in \widetilde{S}$ are such that some curve of degree n meets $\widetilde{S}$ in p and in no other point. They correspond to the complex numbers z such that $3nz \in \Lambda$; so, their number is $9n^2$. For the other curves, which are attached to Riemann surfaces of genus $g > 1$, these surfaces can also be employed to yield an avalanche of new geometrical properties of the curves.

Riemann surfaces are not simply one of the methods for investigating the properties of algebraic curves: the two theories are in fact 'isomorphic'. They can be regarded as two languages for describing the same system of logical relations. The choice of the language is far from being unimportant, however, for it implies its own intuition and its own way of formulating problems. In particular, it is possible to take algebraic curves as a starting point, rather than multivalued algebraic functions or abstract Riemann surfaces. This gives rise to a branch of 'synthetic' geometry which, in spirit, is a direct continuation of the theory of conic sections, while remaining wholly compatible with Riemann surface theory. In particular, the genus of the Riemann surface corresponding to an algebraic function, determined (say) by an equation $f(z, w) = 0$, can be defined in a purely geometric way as an invariant of the algebraic curve with the same equation $f(z, w) = 0$. Thus there is a notion of genus for an algebraic curve. As special cases, straight lines and conic sections are curves of genus 0, cubics being of genus 1. Conformal equivalence of Riemann surfaces corresponds to a relation between algebraic curves which can be defined geometrically, namely birational equivalence. Perhaps the most striking thing is that even results associated with integration on Riemann surfaces ('abelian integrals') have an algebro-geometric equivalent.

A noteworthy feature of this point of view is that all geometric constructions can be described as algebraic operations on the coordinates of the points involved and on the coefficients of the equations of lines. In view of this we may assume that these quantities are in an arbitrary field and not necessarily in the field of complex numbers $\mathbb{C}$. When an algebraic geometer writes:

'... consider an algebraic curve defined over an arbitrary field k,' he thereby declares that he will keep within the framework of a synthetic, purely geometric, study of curves. If k happens to be the field of complex numbers, then Riemann's existence theorem guarantees that this is precisely equivalent to the theory of compact Riemann surfaces. Using other types of fields opens up entirely new possibilities of applying the theory of algebraic curves. For example, to explore algebraic surfaces given by $f(x, y, z) = 0$, where f is a polynomial, we may consider x as a parameter and adjoin it to the coefficients. Then f is a polynomial in y and z with coefficients in the field $\mathbb{C}(x)$ of rational functions, and an algebraic surface is an algebraic curve over the field $\mathbb{C}(x)$. This approach to the study of algebraic surfaces has proved very fruitful.

Another example is when $k = \mathbb{R}$ is the field of real numbers. In this case, an algebraic curve is situated in the real plane and is just a standard object of study in analytic geometry. As a rule, it is not connected, but consists of several pieces, called 'ovals'. (Cf. Fig. 6, which represents the curve $y^2 = x^3 + ax + b$ in the case where the polynomial $x^3 + ax + b$ has three real roots). The number and the relative position of the ovals raise a lot of questions. Some of them can be investigated with methods from the theory of algebraic curves, while the answers to others are unknown. One of the best known results is that a curve of genus g decomposes into at most $g + 1$ ovals. The curve is considered in projective plane, so that, for example, the two branches of a hyperbola make up a single oval. Figure 6 illustrates the case where $g = 1$.

If $k = \mathbb{F}_p$ is the field with p elements (the residue class field modulo a prime p), then the equation of an algebraic curve $f(x, y) = 0$ becomes a congruence $f(x, y) \equiv 0 \pmod{p}$. The application of methods from the theory of algebraic curves has produced the most profound results of number theory on the subject of congruences. This is a case in which the number of points on the curve (in the field $\mathbb{F}_p$) is finite, and questions about the topology of this set of points are replaced by questions about their number. Let N be the number of points on the curve C with equation $f(x, y) = 0$, including the points at infinity (again, the curve must be examined in projective plane). This number can be compared with the number of points on a line, which is $p + 1$ (including the point at infinity). The assertion, for a curve of genus g, is as follows:

$$|N - (p + 1)| \leq 2g\sqrt{p}.$$

Of particular interest for number theory is the case in which k is the field $\mathbb{Q}$ of rational numbers. Then the points lying on an algebraic curve correspond to the solutions in rational numbers of an indeterminate equation $f(x, y) = 0$. If, for instance, $f(x, y) = x^n + y^n - 1$, we are concerned with 'Fermat's Last Theorem'. Here, too, everything hinges on the genus of the curve. For curves of genus 0, there exists an explicit rational parametrization of the set solutions, which is analogous to the classical parametrization

$$x = \frac{1-t^2}{1+t^2}, \qquad y = \frac{2t}{1+t^2}$$

of the solutions of $x^2 + y^2 = 1$. In the case of curves of genus 1, the 'dictionary' on page 13 enables one to extend addition in the group $\mathbb{C}/\Lambda$ to the points on the curve. This operation is defined in a purely geometric way. So, in particular, if the coefficients of its equation are rational then the rational points on the curve form a group. A fundamental theorem asserts that this group is finitely generated. Finally, for curves of genus $g > 1$ it is an essential theorem that such a curve has finitely many rational points. These results are also valid if k is a finite extension of the field $\mathbb{Q}$. Thus we can add yet another column to the Table of page 11, one characterizing algebraic curves from the point of view of their arithmetic.

Genus	Set of points with coordinates in a finite extension k of $\mathbb{Q}$
0	Explicit rational parametrization
1	Finitely generated group
> 1	Finite set

It is interesting to note how – in all cases – the genus of an algebraic curve appears as the main characteristic of its set of points. In the case of curves over the field of complex numbers, it characterizes the topology of that set, the type of its universal covering, and its properties pertaining to differential geometry. In the case of the field $\mathbb{R}$, it provides an estimate of the number of connected components, or 'ovals', of that set (just as the degree of a real polynomial merely gives a bound and does not indicate the exact number of its real roots). In the case of a finite field, it characterizes the deviation of the number of points from an 'average value'. In the case of the field of rational numbers, the genus determines the 'type' of the set of rational points.

Chapter 1
Riemann Surfaces

... dans l'application de l'algèbre à la géométrie,
l'imagination est le coefficient du calcul,
et les mathématiques deviennent poésie.

Victor Hugo, *William Shakespeare*

This chapter is a survey of the basic notions and main results of the theory of Riemann surfaces. Attention is centred on the compact case, as it is directly related to the theory of algebraic curves. A detailed exposition, and proofs, can be found in Ahlfors-Sario [1960], Forster [1977], Springer [1957], and Weyl [1923].

§ 1. Basic Notions

Currently, Riemann surfaces are most conveniently regarded as special complex analytic manifolds. This section therefore begins with basic definitions from complex analytic geometry. This is also justified by the fact that many important notions and results of the theory of Riemann surfaces are difficult to explain without resorting to some more general complex analytic manifolds. A more detailed discussion of the basic theory of these manifolds can be found in Griffiths-Harris [1978], Narasimhan [1968], and Wells [1973].

1.1. Complex Chart; Complex Coordinates. Consider a topological space M. By a *complex chart* on M we mean a homeomorphism $\varphi\colon U \to \mathbb{C}^n$ of an open subset $U \subset M$ onto an open subset $\varphi(U) \subset \mathbb{C}^n$. The coordinates of the complex vector space $\mathbb{C}^n$ determine continuous complex-valued functions $z_1, \ldots, z_n$ on U, which are called *complex coordinates* on U. Every point $p \in U$ is uniquely determined by the ordered set of its coordinates $(z_1(p), \ldots, z_n(p))$, and the chart φ has the following coordinate representation: $\varphi(p) = (z_1(p), \ldots, z_n(p))$. Conversely, given an ordered set $(z_1, \ldots, z_n)$ of continuous complex-valued functions on U, it is a complex coordinate system on U if the mapping $\varphi\colon U \to \mathbb{C}^n$ defined by the above representation is a chart, that is, a homeomorphism onto an open subset of $\mathbb{C}^n$.

Example. Let $\mathbb{CP}^n$ be n-dimensional complex projective space, with the usual topology. Consider some homogeneous coordinate system $(x_0 : \ldots : x_n)$ on it. Then we have the following complex chart:

$$\varphi\colon U \to \mathbb{C}^n$$
$$(x_0 : \ldots : x_n) \mapsto (x_0/x_n, \ldots, x_{n-1}/x_n)$$

with coordinates $z_1 = x_0/x_n, \ldots, z_n = x_{n-1}/x_n$, where

$$U = \{(x_0 : \ldots : x_n) \mid x_n \neq 0\}.$$

Such *charts,* and the corresponding *coordinates,* are said to be *affine.* Every affine chart is defined on an open subset which is the complement of a hyperplane in $\mathbb{CP}^n$, and its image is the whole of $\mathbb{C}^n$. In particular, the homogeneous coordinates $(x_0 : x_1)$ on the complex projective line $\mathbb{CP}^1$ determine a single affine coordinate $z = x_0/x_1$, which is undefined only at the point $(1 : 0)$. The symbol $\infty = 1/0$ is naturally viewed as the z-coordinate of this point.

1.2. Complex Analytic Atlas. Let $f = (f_1, \ldots, f_m)$ be a mapping from an open subset $U \subset \mathbb{C}^n$ to an open subset $V \subset \mathbb{C}^m$. We say that f is holomorphic (or analytic) if so are its components $f_i(z_1, \ldots, z_n)$, in the sense of the theory of functions of several complex variables (cf. Shabat [1969] and Hörmander [1966]).

A *complex atlas* on a topological space M is a (possibly infinite) collection of complex charts $\Phi = \{\varphi_i \colon U_i \to \mathbb{C}^{n_i} \mid i \in I\}$, whose domains of definition U_i cover the entire space M. We say that Φ is an *analytic atlas* if the transition maps

$$\varphi_j \circ \varphi_i^{-1} \colon \varphi_i(U_i \cap U_j) \to \varphi_j(U_i \cap U_j)$$

are holomorphic for all $i, j \in I$. The components of $\varphi_j \circ \varphi_i^{-1}$ are the transition functions from the coordinates of φ_i to those of the chart φ_j in their common domain of definition, $U_i \cap U_j$.

Example. An atlas on $\mathbb{CP}^n$ consisting of affine charts is always analytic. In particular, there is an atlas on $\mathbb{CP}^1$ with only two charts: $\mathbb{CP}^1 - \{\infty\} \xrightarrow{z} \mathbb{C}$ and $\mathbb{CP}^1 - \{0\} \xrightarrow{1/z} \mathbb{C}$. The transition function $1/z$ on $\mathbb{C}^\times = \mathbb{C} - \{0\}$ is obviously holomorphic.

1.3. Complex Analytic Manifolds. A Hausdorff space M, equipped with a complex analytic atlas Φ, is called a *complex analytic*, or simply a *complex, manifold.* It is customary to use the same notation, M, for this complex manifold and for its set of points, the topology and the atlas being assumed to be fixed.

It is convenient to provide oneself from the very outset with as many coordinate systems on M as possible. A *complex chart* $\varphi \colon U \to \mathbb{C}^n$ on a complex manifold M is said to be *analytic* if it can be added to Φ without destroying the analyticity of this atlas. This means that the transition functions between the coordinates of φ and those of any chart of Φ are analytic. Any atlas made up of complex analytic charts on M is analytic. Moreover, the atlas consisting of all complex analytic charts on M is a maximal analytic atlas: the adjunction of any new chart makes it non-analytic. This maximal analytic atlas is called a *complex analytic structure* on M. Further on, unless stated to the contrary, a local coordinate system on M will always mean a complex coordinate system corresponding to a complex analytic chart on M.

Remark. Complex manifolds are defined in a way that reflects their local structure, which is the same as that of a ball in $\mathbb{C}^n$. In particular, the use of coordinates locally reduces the study of manifolds to the theory of analytic functions of n variables. For Riemann surfaces, functions of one variable normally suffice.

Example 1. The space $\mathbb{C}^n$ can be provided with an atlas consisting of only one chart, $\mathbb{C}^n \xrightarrow{(z_1,\ldots,z_n)} \mathbb{C}^n$. The corresponding analytic structure consists of biholomorphic homeomorphisms $\varphi\colon U \to \mathbb{C}^n$ of an open subset $U \subset \mathbb{C}^n$ onto an open subset $V \subset \mathbb{C}^n$. It is always assumed that the complex manifold $\mathbb{C}^n$ is equipped with just this analytic structure.

Example 2. We shall assume that the space $\mathbb{CP}^n$ is provided with the analytic structure that corresponds to an atlas consisting of affine charts.

Example 3. Let $\Lambda \subset \mathbb{C}^n$ be a discrete lattice. Then the quotient space $\mathbb{C}^n/\Lambda$ carries a complex manifold structure, defined by the quotient mapping $\pi\colon \mathbb{C}^n \to \mathbb{C}^n/\Lambda$. As an atlas on $\mathbb{C}^n/\Lambda$, one can take the set of local sections of π, that is, continuous mappings $s\colon U \to \mathbb{C}^n$ of an open subset $U \subset \mathbb{C}^n/\Lambda$, such that $\pi \circ s(p) = p$ for every $p \in U$. This manifold is compact if and only if Λ has maximal rank, $2n$. In this case, $\mathbb{C}^n/\Lambda$ is called a *complex torus*.

Example 4. The *product* $M \times N$ of two complex manifolds has a natural complex manifold structure. As an atlas on $M \times N$, one can take all charts $(\varphi, \psi) : U \times V \to \mathbb{C}^m \times \mathbb{C}^n$, where φ and ψ are complex analytic charts on M and N, respectively.

Example 5. Let $U \subset M$ be an open subset. The complex analytic charts on M whose domain of definition is contained in U, define a natural analytic structure on U. The manifold U, with this structure, is called an *open submanifold* of M. In what follows, any open subset in M is considered to be a manifold in this sense.

Example 6. More generally, a subset N of a complex manifold M is said to be a *submanifold* if it is defined locally by a system of equations $f_1 = \ldots = f_n = 0$, where $f_1, \ldots, f_n$ are holomorphic functions of the coordinates $z_1, \ldots, z_m$ and the matrix $\left(\dfrac{\partial f_i}{\partial z_j}\right)$ is of rank n. In view of the complex analytic version of the inverse mapping theorem, this is equivalent to N being locally flat in M, in the complex analytic sense. This means that every point $p \in N$ has an open neighbourhood U in M, with coordinates $z_1, \ldots, z_m$, such that $U \cap N = \{p \in U \mid z_1(p) = \ldots = z_n(p) = 0\}$. Thus N carries a natural complex manifold structure, and we can take as coordinates on $U \cap N$ the functions $z_{n+1}, \ldots, z_m$. Any complex projective subspace of $\mathbb{CP}^n$ is a closed submanifold.

1.4. Mappings of Complex Manifolds. A mapping $f\colon M \to N$ of complex manifolds is said to be *holomorphic* if in local coordinates it is given by holomorphic functions. This means that the functions $w_i = f_i(z_1, \dots, z_m)$, which define f in local coordinates $z_1, \dots, z_m$ on M and $w_1, \dots, w_n$ on N, are holomorphic in their domain of definition. We observe that it is not possible to check that f is holomorphic for all coordinate representations of this mapping, but only for a certain set of representations, whose domains of definition include all points of M.

An invertible mapping of complex manifolds whose inverse is holomorphic, is called an *isomorphism*. An *automorphism* is an isomorphism of a manifold onto itself. Clearly, complex manifolds, together with holomorphic mappings, form a category (cf. Shafarevich [1986]) with the isomorphisms and automorphisms just defined. Holomorphic mappings of the form $f\colon M \to \mathbb{C}$ are called *holomorphic functions* on the complex manifold M. If $f\colon M \to N$ is a holomorphic mapping, and $g\colon N \to \mathbb{C}$ a holomorphic function, then we say that the holomorphic function $f^*(g) \stackrel{\text{def}}{=} g \circ f\colon M \to \mathbb{C}$ is the *pull-back* of g by f.

Example 1. Any complex analytic chart $\varphi\colon U \to \mathbb{C}^n$ on M is holomorphic, and its coordinates are holomorphic functions on U.

Example 2. A *complex Lie group* is a complex manifold G with a group structure, such that the group law $G \times G \xrightarrow{g_1 \cdot g_2} G$ and the inverse map $G \xrightarrow{g^{-1}} G$ are holomorphic. $\mathbb{C}^n$ and the quotient manifolds $\mathbb{C}^n/\Lambda$ are additive complex Lie groups, and $\pi\colon \mathbb{C}^n \to \mathbb{C}^n/\Lambda$ is a holomorphic homomorphism of these groups.

Example 3. Let $\mathbb{CP}^n$ be projective space, with homogeneous coordinates $(x_0 : \ldots : x_n)$, and let $M = (m_{ij})$ be an invertible $(n+1) \times (n+1)$-matrix. Then

$$\begin{gathered}\mathbb{CP}^n \to \mathbb{CP}^n \\ (x_0 : \ldots : x_n) \mapsto (m_{00}x_0 + \ldots + m_{0n}x_n : \ldots : m_{n0}x_0 + \ldots + m_{nn}x_n)\end{gathered}$$

is a holomorphic automorphism. Such mappings are called *linear fractional*, since in affine coordinates they are given by linear fractional functions:

$$(z_1, \dots, z_n) \mapsto \left(\frac{m_{00}z_1 + \ldots + m_{0\,n-1}z_n + m_{0n}}{m_{n0}z_1 + \ldots + m_{n\,n-1}z_n + m_{nn}}, \dots, \right. \\ \left. \frac{m_{n-1\,0}z_1 + \ldots + m_{n-1\,n-1}z_n + m_{n-1\,n}}{m_{n0}z_1 + \ldots + m_{n\,n-1}z_n + m_{nn}} \right)$$

Example 4. Let $p \in \mathbb{CP}^2$ be a point, and let $\mathbb{CP}^1 \subset \mathbb{CP}^2$ be a line not passing through p. Then the *projection map*

$$\begin{gathered}\pi\colon \mathbb{CP}^2 - \{p\} \to \mathbb{CP}^1 \\ q \mapsto \overline{pq} \cap \mathbb{CP}^1\end{gathered}$$

is holomorphic. Here $\overline{pq}$ denotes the complex line through p and q.

1.5. Dimension of a Complex Manifold. The *dimension* of a chart $\varphi\colon U \to \mathbb{C}^n$ is the number n, that is, the number of its coordinates. For a *connected* complex manifold M, this number is independent of the choice of φ on M and is called the *dimension* of M. In what follows, all manifolds will be assumed connected. The dimension of M is denoted by $\dim_{\mathbb{C}} M$ or, simply, $\dim M$.

Example. $\dim \mathbb{C}^n = \dim \mathbb{CP}^n = \dim \mathbb{C}^n/\Lambda = n$. In particular, the complex dimension of $\mathbb{C}^n$ and of $\mathbb{CP}^n$, as linear spaces, is the same as their complex analytic dimension.

1.6. Riemann Surfaces

Definition. A *Riemann surface* is a connected complex analytic manifold of dimension one.

Example 1.

$\mathbb{CP}^1$ is the *Riemann sphere,*
$\mathbb{C}$ is the *Gaussian plane,*
$\mathbb{H} = \{\operatorname{Im} z > 0\}$ is the *upper half-plane,*
$\mathbb{D} = \{|z| < 1\}$ is the *unit disk,* and
$\mathbb{D}^\times = \{0 < |z| < 1\}$ is the *punctured unit disk.*

The upper half-plane is isomorphic to the unit disk. The isomorphism can be given by a linear fractional function:

$$\begin{aligned}\mathbb{H} &\to \mathbb{D}\\ z &\mapsto \frac{z-a}{z-\bar{a}},\end{aligned}$$

with $a \in \mathbb{H}$.

Example 2. A one-dimensional complex torus is called a *complex elliptic curve.*

Example 3. Let $\mathbb{CP}^2$ be projective plane, with homogeneous coordinates $(x : y : z)$, and let $f(x, y, z)$ be a nontrivial homogeneous polynomial. The set of zeros of this polynomial,

$$C = \{(x : y : z) \mid f(x, y, z) = 0\},$$

is called a *complex algebraic plane curve.* Any such curve, with the subspace topology, is compact and connected (see Corollary 5 of Sect. 2.11 in the irreducible case, and Corollary 2 of Chap. 2, Sect. 2.3 in the general case). This curve C is said to be *nonsingular* if it is a complex submanifold of $\mathbb{CP}^2$. This submanifold, which is one-dimensional, is called the *Riemann surface of* (or: *associated with*) C. To check that C is nonsingular, the following criterion is convenient. If, for every $p \in C$, we have

$$\left(\frac{\partial f}{\partial x}(p) : \frac{\partial f}{\partial y}(p) : \frac{\partial f}{\partial z}(p)\right) \neq (0:0:0),$$

then C is nonsingular. This criterion is easily derived from the complex analytic version of the inverse mapping theorem.

Example 4. A homogeneous quadratic form $ax^2 + bxy + cy^2 + \ldots + dz^2$ of rank 3 defines on $\mathbb{CP}^2$ a nonsingular algebraic curve, called a *conic*. The projection of a conic from any of its points to the Riemann sphere $\mathbb{CP}^1$ (see Fig. 7) extends by continuity to an isomorphism of the Riemann surface of the conic onto $\mathbb{CP}^1$.

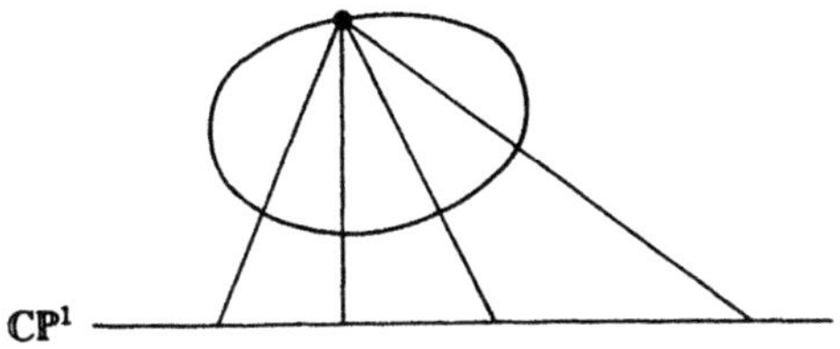

Fig. 7

Example 5. The equation $x^d + y^d = z^d$ defines in $\mathbb{CP}^2$ a nonsingular algebraic curve, which is called the *Fermat curve* (of degree d).

Example 6. A typical example of a Riemann surface is the Riemann surface of an algebraic function. For simplicity, we assume that $F(z)$ is an *algebraic function* on $\mathbb{CP}^1$. In other words, F is a multivalued function satisfying an equation of the form $f(z, F) = 0$, where f is a complex polynomial in two variables (of degree n in F). Suppose, also, that f is irreducible. Then one can assert the existence of a Riemann surface S on which F becomes single-valued. More precisely, there is a holomorphic mapping $g\colon S \to \mathbb{CP}^1$ and a meromorphic function φ on S (see Sect. 2.2), such that $F = \varphi \circ g^{-1}$. This Riemann surface is called the *Riemann surface of the algebraic function* F. We sketch its construction. For almost[2] all $z \in \mathbb{C}$, the algebraic equation $f(z, F) = 0$ for F has the same number n of roots $F_1, \ldots, F_n$. The corresponding pairs (z, F_i), with $f(z, F_i) = 0$, constitute an open subset $U \subset S$ (complementary to a finite subset of S). To close values of z correspond close roots F_i. This induces a topology on U. Further, the mapping g is given on U by the projection $(z, F_i) \mapsto z$. Obviously, g is a finite topological covering over U. This means that any point $z \in g(U) \subset \mathbb{C}$ has a neighbourhood V whose inverse image, $g^{-1}(V)$, can be written as a union of open sets $V_1, \ldots, V_n$ in such a way that the maps $g\colon V_i \to V \subset \mathbb{C}$ are homeomorphisms. Now we introduce an analytic structure on U by considering these isomorphisms as charts on U. Then the projection g turns into a holomorphic mapping (cf. the Proposition in Sect. 2.9). To construct S, it remains to complete U above $\mathbb{CP}^1 - g(U)$ by a finite collection of points and to extend g by continuity. In fact, let $z_0 \in \mathbb{CP}^1 - g(U)$, and consider a punctured disk $D^{\times}_{\varepsilon, z_0} = \{0 < |z - z_0| < \varepsilon\}$, above which the surface S is already defined. It

[2] i.e., with finitely many exceptions

is not hard to check that each connected component $V \subset g^{-1}(D^{\times}_{\varepsilon,z_0})$ determines a mapping $g_V : V \to D^{\times}_{\varepsilon,z_0}$, which is isomorphic to the standard map $D^{\times}_{\sqrt[m]{\varepsilon},0} \xrightarrow{z^m} D^{\times}_{\varepsilon,0}$. Here m is the number of points in a fibre $g_V^{-1}(z)$, $z \in D^{\times}_{\varepsilon,z_0}$. Hence we obtain S by adding one point to each such component V. The proof of isomorphism is based on the following fact: if we watch the points on the fibre $g^{-1}(z)$, as z moves around z_0, we notice that at the end of a revolution they are permuted. This is called *monodromy* (see also Sect. 3.6). It turns out that the points of the fibre $g_V^{-1}(z)$, which lie in the connected component V, are permuted cyclically under monodromy. The same is true for z^m. This allows us to construct the required isomorphism topologically, and then check that it is holomorphic (see Example 3 in Sect. 2.9). The function φ (on U) is defined by the rule $\varphi(z, F_i) = F_i$. As in Example 3, the main difficulty lies in proving the connectedness of S or, equivalently, that of U. If the surface U were not connected, then the decomposition of U into connected components would correspond to a decomposition of f into factors, which is impossible since f is irreducible. This is treated at greater length in Sect. 2.11, where a more general construction is presented.

Remark. Now let $f(x, y, z)$ be a real homogeneous polynomial of degree d. Its set of zeros in the real projective plane $\mathbb{RP}^2$ is called a *real algebraic plane curve.* This curve is not always connected. Harnack's theorem says that the number of connected components of the curve is at most $\frac{1}{2}(d-1)(d-2)+1$ (cf. Chebotarev [1948]).

1.7. Differentiable Manifolds. On replacing the space $\mathbb{C}^n$ by $\mathbb{R}^n$, and requiring that the transition maps be differentiable, instead of holomorphic, we define *differentiable manifolds* and local *differentiable coordinate systems* on them. Coordinates allow us to introduce the notion of a *differentiable mapping* of differentiable manifolds. More about these manifolds can be found in the books by Dubrovin, Novikov & Fomenko [1984], Hirsch [1976], and Narasimhan [1968]. Any complex manifold M can be regarded as a differentiable manifold with the same atlas. It suffices to replace the space $\mathbb{C}^n$ by the corresponding real space $\mathbb{R}^{2n}$ and, from the coordinate point of view, each complex coordinate $z_i = x_i + \sqrt{-1}\, y_i$ by the two *real* ones, x_i and y_i. Since the transition mappings are holomorphic, those of the corresponding real atlas are differentiable. Clearly,

$$\dim_{\mathbb{R}} M = 2 \dim_{\mathbb{C}} M,$$

where $\dim_{\mathbb{R}} M$ is the *real dimension* of M, that is, the dimension of M regarded as a differentiable manifold. Differentiable manifolds of dimension two are called *surfaces.* A Riemann surface is a surface exactly in this sense, or in a still weaker topological sense.

§ 2. Mappings of Riemann Surfaces

This section begins with a discussion of meromorphic functions: we give the simplest examples and mention some famous problems concerning the existence of meromorphic functions with assigned properties. However, the main problem of the theory of Riemann surfaces – that of finding at least one nonconstant meromorphic function on a Riemann surface – is put off until § 4 (see Sect. 4.14). The subsections 2.1, 2.4, and 2.5 deal with the elementary topological properties of holomorphic mappings of Riemann surfaces and some consequences. From an algebraic point of view, the most interesting mappings of Riemann surfaces are the so-called finite mappings. They are discussed in the remaining part of the present section, starting from Sect. 2.7. Their interest stems from the algebraic fact that the corresponding extension of meromorphic function fields is finite (see Sect. 2.11). Moreover, the Riemann surface of an algebraic function – which is one of the most important examples of a Riemann surface – is constructed as a finite covering on which the function becomes single-valued.

All mappings of Riemann surfaces will be assumed holomorphic.

2.1. Nonconstant Mappings of Riemann Surfaces are Discrete. A mapping $f\colon X \to Y$ of topological spaces is said to be *discrete* if the inverse image $f^{-1}(p)$ of each point $p \in Y$ is a discrete subset of X.

Uniqueness theorem. *Suppose $f_1, f_2\colon S_1 \to S_2$ are two mappings of Riemann surfaces which coincide on some nondiscrete subset of S_1. Then they coincide on the whole of S_1.*

This is a simple generalization of the uniqueness theorem for holomorphic functions of one complex variable.

Corollary. *Any nonconstant mapping of Riemann surfaces is discrete.*

2.2. Meromorphic Functions on a Riemann Surface. Let S be a Riemann surface.

Definition. A *meromorphic function* on S is a partially defined function f on S which, locally, is meromorphic in the usual sense (cf. Shabat [1969]). Thus a meromorphic function f on S is holomorphic on some open subset U, whose complement $S - U$ is discrete in S and consists of poles of f.

A *pole* $p \in S - U$ is defined in one of the following equivalent ways:

(a) $\lim_{q \to p} f(q) = \infty$;

(b) $f(z)$ can be written locally as a Laurent series

$$f(z) = \sum_{i=-n}^{+\infty} a_i z^i,$$

with $a_{-n} \neq 0$ and $n > 0$, where z is a *local parameter* at p (i.e., a local coordinate such that $z(p) = 0$);

(c) locally $f = g/h$, where g and h are holomorphic functions in some neighbourhood of p; further, $g(p) \neq 0$ and $h(p) = 0$.

The set of all meromorphic functions on S is denoted by $\mathcal{M}(S)$.

A meromorphic function f can be continued to a mapping $f\colon S \to \mathbb{CP}^1$: $f(p) = \infty$ at each pole p. The fact that f is holomorphic follows from Riemann's removable singularity theorem. Conversely, if $f\colon S \to \mathbb{CP}^1$ is a mapping of Riemann surfaces and z an affine coordinate on $\mathbb{CP}^1$ such that $f(z) \not\equiv \infty$, then $f^*(z)$ is a meromorphic function on S, whose poles form the set $f^{-1}(\infty)$. Hence the meromorphic functions $f \in \mathcal{M}(S)$ can be identified with the holomorphic mappings $f\colon S \to \mathbb{CP}^1$.

Example 1. Every polynomial $f(z)$ defines a meromorphic function on $\mathbb{CP}^1$, which has a unique pole at ∞, provided the degree of the polynomial is ≥ 1. In fact, every rational function $f(z)$ defines a meromorphic function on $\mathbb{CP}^1$, which fails to have a pole at ∞ precisely if $f(z)$ is a proper fraction. Conversely, every meromorphic function f on $\mathbb{CP}^1$ is rational. This can be proved (cf. Forster [1977]) by selecting principal parts (see Sect. 2.3) and using the ordinary Maximum Principle (cf. Shabat [1969] and Proposition 3 of Sect. 2.5) or Liouville's theorem. In this way we also show the existence and uniqueness of the expansion of a proper fraction into partial fractions in the complex sense. Thus $\mathcal{M}(\mathbb{CP}^1) \simeq \mathbb{C}(z)$, where $\mathbb{C}(z)$ is the field of rational functions of one variable z. Hence meromorphic functions constitute a natural generalization of rational functions. Furthermore, these notions coincide for any compact Riemann surface, once the rationality of a function is suitably defined (see Sect. 6.5).

Example 2. Let $f\colon S_1 \to S_2$ be a nonconstant mapping of Riemann surfaces. The pull-back $f^*(g)$ of any meromorphic function $g \in \mathcal{M}(S_2)$ is meromorphic on S_1.

Example 3. Consider a holomorphic mapping $f\colon S \to \mathbb{CP}^2$ of a Riemann surface S into complex projective plane $\mathbb{CP}^2$, with affine coordinates z_1, z_2. Suppose that some rational function $g(z_1, z_2)$ is defined at least at one point of $f(S)$. This means that it can be written as a ratio of homogeneous polynomials of the same degree:

$$g(z_1, z_2) = \frac{p(x_0, x_1, x_2)}{q(x_0, x_1, x_2)},$$

where $(x_0 : x_1 : x_2)$ are homogeneous coordinates on $\mathbb{CP}^2$ corresponding to z_1, z_2, and the set of zeros of the polynomial $q(x_0, x_1, x_2)$ does not contain $f(S)$. Then the pull-back of this rational function is meromorphic on S.

Example 4. The *Weierstrass $\wp$-function* is a meromorphic function on the Gaussian plane $\mathbb{C}$, which is defined by the convergent series

$$\wp(z) \stackrel{\text{def}}{=} \wp(z,\Lambda) \stackrel{\text{def}}{=} \frac{1}{z^2} + \sum_{\lambda\in\Lambda-\{0\}} \left[\frac{1}{(z-\lambda)^2} - \frac{1}{\lambda^2}\right].$$

Here $\Lambda \subset \mathbb{C}$ denotes a lattice of maximal rank. This function is
(a) even: $\wp(-z) = \wp(z)$;
(b) periodic: $\wp(z+\lambda) = \wp(z)$ for all $\lambda \in \Lambda$;
(c) further it has no other poles than the points $\lambda \in \Lambda$ of the lattice.

The $\wp$-function, being periodic, induces a well-defined meromorphic function on the elliptic curve $\mathbb{C}/\Lambda$, whose unique pole is at the origin. In particular, there is a nonconstant meromorphic function on any such curve.

Lemma 1. *The meromorphic functions on a Riemann surface S form a field $\mathcal{M}(S)$, with the natural addition and multiplication operations.*

Lemma 2. *A nonconstant mapping $f\colon S_1 \to S_2$ of Riemann surfaces defines a field extension, that is, a homomorphic embedding $f^*\colon \mathcal{M}(S_2) \hookrightarrow \mathcal{M}(S_1)$ of the fields of meromorphic functions.*

Remark. We shall identify constant functions with complex numbers. Obviously, $f^*\colon \mathcal{M}(S_2) \hookrightarrow \mathcal{M}(S_1)$ is a $\mathbb{C}$-extension, in the following sense: $f^*(c) = c$ for all $c \in \mathbb{C}$.

2.3. Meromorphic Functions with Prescribed Behaviour at Poles. The general question of the structure of the meromorphic function field $\mathcal{M}(S)$ for a Riemann surface S, and particularly the proof that $\mathcal{M}(S) \neq \mathbb{C}$, is quite important. It will be discussed at a number of places in this survey (see Sections 2.11, 4.8, and 4.14). Nevertheless, the first really difficult results concerning the existence of nontrivial meromorphic functions and differentials will only come up at the end of § 4. At this point, we give only the statement of one famous problem on the existence of a meromorphic function with prescribed principal parts.

Let p be a point on a Riemann surface S. We fix a local parameter z at p. If $f = \sum_{i=-n}^{+\infty} a_i z^i$ is the Laurent expansion of a meromorphic function in some neighbourhood of p, then the initial segment of this series, $\sum_{i=-n}^{-1} a_i z^i$, is called the *principal part of the function* f. Note that, up to a summand which is holomorphic in a neighbourhood of p, the principal part is independent of the choice of the local parameter z.

Mittag-Leffler's problem for meromorphic functions. *Let $\left\{\sum_{i=-n}^{-1} a_i z^i\right\}$ be a set of principal parts, defined on a discrete set of points p of a Riemann surface S. It is required to find a meromorphic function $f \in \mathcal{M}(S)$ having poles only at these points, and with the given principal parts.*

As Mittag-Leffler has established, the problem is always solvable on the Gaussian plane $\mathbb{C}$ (cf. Shabat [1969]). The same is true in the general case for noncompact Riemann surfaces (see Forster [1977]). In the compact case, Mittag-Leffler's problem is solvable if the coefficients a_i of the principal parts satisfy a finite set of linear relations, which depend on the topology of the Riemann surface (see Remark 1 of Sect. 4.9 and Theorem 1 of Sect. 6.3).

2.4. Multiplicity of a Mapping; Order of a Function. Let $f\colon S_1 \to S_2$ be a nonconstant mapping of Riemann surfaces. We choose a local parameter z at a point $p \in S_1$, and w at $f(p) \in S_2$. In these coordinates f can be written as

$$w = z^n g(z), \tag{1}$$

where n is some integer and the function $g(z)$ is holomorphic in a neighbourhood of the origin, with $g(0) \neq 0$. In fact there is a more precise statement.

Lemma (on normal form). *One can find local parameters z and w at the points $p \in S_1$, respectively $f(p) \in S_2$, such that the mapping f takes the form $w = z^n$.*

Definition 1. The number n in relation (1) is called the *multiplicity* of f at p and is denoted by $\operatorname{mult}_p f$.

Definition 2. The number $r_p(f) \stackrel{\text{def}}{=} \operatorname{mult}_p f - 1$ is called the *ramification index* of f at p. A point $p \in S_1$ is said to be a *ramification point* if $r_p(f) \geq 1$.

Definition 3. The *order* at $p \in S$ of a meromorphic function $f\colon S \to \mathbb{CP}^1$ is defined as follows:

$$\operatorname{ord}_p f = \begin{cases} \operatorname{mult}_p f & \text{if } f(p) = 0, \text{ i.e., if } p \text{ is a zero of } f; \\ -\operatorname{mult}_p f & \text{if } f(p) = \infty, \text{ i.e., if } p \text{ is a pole of } f; \\ 0 & \text{otherwise.} \end{cases}$$

That $\operatorname{mult}_p f$ is well-defined, follows from the geometric interpretation of multiplicity. In the full inverse image $f^{-1}(q)$ of a point $q \in S_2 - f(p)$ close to $f(p)$, one finds precisely $\operatorname{mult}_p f$ points close to p.

Example 1. $\mathbb{D} \xrightarrow{z^n} \mathbb{D}$ has 0 as a single ramification point, with index $n-1$.

Example 2. $\mathbb{C} \xrightarrow{\exp} \mathbb{C}$ has no ramification points.

Example 3. If $f(z) \in \mathcal{M}(\mathbb{CP}^1)$ is a polynomial of degree d then $\operatorname{ord}_\infty f = -d$, and $\operatorname{ord}_p f$ is equal to the multiplicity of p as a root of f if $f(p) = 0$.

Example 4. The Weierstrass $\wp$-function (see Example 4 of Sect. 2.2) has second-order poles at the lattice points: $\operatorname{ord}_\lambda \wp = -2$ for all $\lambda \in \Lambda$ (see Hurwitz [1922,1964]). More generally one says that p is a *pole of order* n if $\operatorname{ord}_p f = -n$.

Remark 1. The ramification points of $f\colon S_1 \to S_2$ form a discrete set on S_1. Indeed, if $w = f(z)$ is a local description of f then the ramification points are just the zeros of the derivative $f'(z)$.

Remark 2. If $f \equiv c$ is a constant function then $\operatorname{ord}_p f = 0$ for $c \neq 0$; it is convenient to consider that $\operatorname{ord}_p 0 = +\infty$.

2.5. Topological Properties of Mappings of Riemann Surfaces. All propositions in this subsection have a local nature. Hence they are easy to derive from the corresponding facts belonging to the theory of analytic functions of one variable. (For instance, the first two propositions follow in an obvious way from the Lemma on normal form stated in the preceding subsection).

Proposition 1. *Any nonconstant mapping of Riemann surfaces is open.*

Corollary 1. *Let $f\colon S_1 \to S_2$ be a nonconstant mapping of Riemann surfaces, where S_1 is compact. Then S_2 is also compact, and f is surjective.*

Proposition 2. *If a mapping of Riemann surfaces is injective then it is an open immersion, that is, an isomorphism onto an open subset.*

Proposition 3 (the Maximum Principle). *Let $f\colon S \to \mathbb{C}$ be a nonconstant holomorphic function on a Riemann surface S. Then $|f|$ does not attain any maximum value on S.*

Corollary 2. *On a compact Riemann surface, any holomorphic function is constant.*

This partly explains why meromorphic functions should be introduced, especially in the compact case.

Remark. Most of the above statements have higher-dimensional generalizations (see Gunning-Rossi [1965]). So, for instance, Corollary 2 holds for a compact complex manifold of any dimension.

2.6. Divisors on Riemann Surfaces. When we investigate ramification points together with their multiplicities, or attempt to formalize the problem of finding a function with prescribed zeros and poles, and in many other questions of Riemann surface theory, we are led naturally to the notion of divisor.

Definition 1. A *divisor* D on a Riemann surface S is a locally finite, formal linear combination

$$D = \sum a_i p_i,$$

where $a_i \in \mathbb{Z}$ and $p_i \in S$. 'Locally finite' means that the *support* of D,

$$\operatorname{Supp} D \stackrel{\text{def}}{=} \{p_i \mid a_i \neq 0\},$$

is a discrete subset of S.

Definition 2. The divisors on a Riemann surface S form an additive group $\operatorname{Div} S$, called the *divisor group*.

Definition 3. A divisor $D = \sum a_i p_i$ is said to be *effective* if all $a_i \geq 0$. The usual notation is: $D \geq 0$.

Definition 4. For finite divisors $D = \sum a_i p_i$, that is, for divisors whose support is finite, there is a notion of *degree*:

$$\deg D \stackrel{\text{def}}{=} \sum a_i.$$

Example 1. On a compact Riemann surface S, every divisor is finite; and we have the degree epimorphism $\deg\colon \operatorname{Div} S \twoheadrightarrow \mathbb{Z}$.

Example 2. Let $f\colon S_1 \to S_2$ be a nonconstant mapping of Riemann surfaces. Each point $p \in S_2$ determines an effective divisor

$$f^*(p) = \sum_{q \in f^{-1}(p)} \operatorname{mult}_q f \cdot q,$$

whose support is the *fibre* $f^{-1}(p)$. By additivity, this defines a homomorphism

$$f^*\colon \operatorname{Div} S_2 \to \operatorname{Div} S_1,$$
$$\sum a_i p_i \mapsto \sum a_i f^*(p_i).$$

The divisor $R \stackrel{\text{def}}{=} \sum r_p(f) \cdot p \in \operatorname{Div} S_1$, where $r_p(f)$ is the ramification index of f at p, is called the *ramification divisor* of f.

Example 3. Now let f be a nonconstant meromorphic function on a Riemann surface S. The effective divisors

$$(f)_0 \stackrel{\text{def}}{=} \sum_{f(p)=0} \operatorname{ord}_p f \cdot p \qquad \text{and} \qquad (f)_\infty \stackrel{\text{def}}{=} \sum_{f(p)=\infty} \operatorname{ord}_p f \cdot p$$

are called the *divisor of zeros* and the *divisor of poles* of f, respectively. The divisor

$$(f) \stackrel{\text{def}}{=} \sum \operatorname{ord}_p f \cdot p = (f)_0 - (f)_\infty$$

is the *divisor of the function* f. This notion enables us to define a homomorphism

$$(\ )\colon \mathcal{M}(S)^\times \to \operatorname{Div} S$$
$$f \mapsto (f)$$

A divisor in the image of this map, that is, one of the form (f), is said to be *principal.* The kernel of this homomorphism consists of the holomorphic functions on S that are everywhere nonzero. For a compact surface S, in particular, this kernel consists of all nonzero constant functions. It follows that, on a compact Riemann surface, a function is uniquely determined by its divisor, up to multiplication by a constant.

Example 4. A divisor $D = \sum a_i p_i$ on the Riemann sphere $\mathbb{CP}^1$ is principal if and only if its degree is equal to zero. Indeed, if $\deg D = 0$ then $D = (f)$, where $f = \prod(z - p_i)^{a_i}$ and the product is taken over all $p_i \neq \infty$. The converse is obtained from the fundamental theorem of algebra (cf. Corollary 3 of Sect. 2.7), since any meromorphic function on $\mathbb{CP}^1$ is rational (see Example 1 of Sect. 2.2). According to the next subsection, a principal divisor on a compact Riemann surface is always of degree 0. But if the surface is not isomorphic to $\mathbb{CP}^1$, not every such divisor is principal. Abel's theorem (see Chap. 3, Sect. 2.1) specifies which divisors of degree 0 on a compact Riemann surface are actually principal.

Definition 5. Two divisors D_1, D_2 on a Riemann surface S are said to be *linearly equivalent*, and we write $D_1 \sim D_2$, if they differ by a principal divisor: $D_1 - D_2 = (f)$, with $f \in \mathcal{M}^\times(S)$. It is easy to check that this is indeed an equivalence relation.

Example 4 (*continued*). Two divisors D_1, D_2 on $\mathbb{CP}^1$ are linearly equivalent if and only if they have the same degree: $\deg D_1 = \deg D_2$.

2.7. Finite Mappings of Riemann Surfaces. A *mapping* of topological spaces $f\colon X \to Y$ is said to be *proper* if the inverse image of any compact subset is compact. For example, this is always the case if X is compact.

Definition. A mapping of Riemann surfaces is said to be *finite* if it is nonconstant and proper.

Example 1. $\mathbb{D} \xrightarrow{z^n} \mathbb{D}$ is a finite mapping.

Example 2. Let $f\colon S_1 \to S_2$ be a finite mapping of Riemann surfaces and let U be an open subset of S_2. Then, for any connected component S of $f^{-1}(U)$, the mapping $f\colon S \to f(S)$ is finite.

Example 3. Even though all the fibres of an inclusion $\mathbb{C} \hookrightarrow \mathbb{CP}^1$ are finite, this map is not finite in the sense of the definition.

Example 4. Similarly, the mapping $\mathbb{C} - \{1\} \xrightarrow{z^2} \mathbb{C}$ is not finite either.

Lemma (numerical criterion of finiteness). *A mapping of Riemann surfaces $f\colon S_1 \to S_2$ is finite if and only if all its fibres are finite and the divisors $f^*(p)$ have one and the same degree for all $p \in S_2$.*

If f is a finite mapping of Riemann surfaces, the degree of any one of its fibres $f^*(p)$ is called the *degree* of f and is denoted by $\deg f$.

Corollary 1. *Any finite mapping of Riemann surfaces is surjective. In particular, on a compact Riemann surface a nonconstant meromorphic function takes on all complex values and ∞.*

Corollary 2. *Every principal divisor on a compact Riemann surface S is of degree 0, that is, $\deg(f) = 0$ for all $f \in \mathcal{M}(S)^\times$.*

Corollary 3 (Fundamental Theorem of Algebra). *A complex polynomial of degree n has n complex roots, counted with multiplicities.*

We shall give below the basic methods for constructing finite mappings. This will also be the occasion for discussing our first nontrivial statements about Riemann surfaces and meromorphic functions on them.

2.8. Unramified Coverings of Riemann Surfaces. A mapping of topological spaces $f\colon Y \to X$ is said to be an *unramified covering* if each point $p \in X$ has an open neighbourhood U such that $f^{-1}(U) = \bigcup U_i$, where the U_i are pairwise disjoint open subsets of Y and all $f\colon U_i \to U$ are homeomorphisms.

Definition 1. A nonconstant mapping of Riemann surfaces is said to be *unramified* if it has no ramification points.

Definition 2. A mapping of Riemann surfaces is said to be an *unramified covering* if it is so topologically. Then it is also unramified in the sense of Definition 1.

Example. A finite, unramified mapping of Riemann surfaces is obviously an unramified covering. Up to a discrete subset, every finite mapping of Riemann surfaces has this property of 'looking locally like a pack of cards' (see the beginning of Sect. 2.10).

2.9. The Universal Covering. In what follows we assume that the reader is familiar with the simplest notions and results concerning the fundamental group $\pi(X)$ of a topological space X (namely, the group of *loops* (closed paths) up to continuous deformations) and its universal covering $\widetilde{X}$. By definition, $\widetilde{X}$ is equipped with an unramified covering $\widetilde{X} \to X$ having the following universal property: for every unramified covering $Y \to X$ there exists a continuous mapping $\widetilde{X} \to Y$ such that the triangle

$$\begin{array}{ccc} \widetilde{X} & \longrightarrow & Y \\ & \searrow \quad \swarrow & \\ & X & \end{array}$$

is commutative. A connected (complex or differentiable) manifold X has a universal covering $\widetilde{X}$, which is connected and simply connected. The fundamental group $\pi(X)$ acts freely and discretely on $\widetilde{X}$, and $X \simeq \widetilde{X}/\pi(X)$. Moreover, there is a one-to-one correspondence

$$\widetilde{X}/\Gamma \to \widetilde{X}/\pi(X) \simeq X \leftrightarrow \Gamma$$

between the set of connected unramified coverings $Y \to X$ (up to isomorphism) and that of all subgroups $\Gamma \subset \pi(X)$ (up to conjugation). In the case of an *n-sheeted covering* $Y \to X$ (which means that the inverse image of a point $p \in X$ consists of n points), we have $n = (\pi(X) : \Gamma)$. To specialize these results to Riemann surfaces, we need the following

Proposition. *Let S be a Riemann surface, and let $f\colon M \to S$ be a connected unramified covering of topological spaces. There exists a unique complex analytic structure on M which makes f into an unramified covering of Riemann surfaces.*

As charts on M, one can take all compositions $\varphi \circ f\colon U \to \mathbb{C}$, where $f\colon U \to V$ is a homeomorphism onto an open subset $V \subset S$ and $\varphi\colon V \to \mathbb{C}$ is a chart on S.

Corollary 1. *There is a one-to-one correspondence between the unramified coverings $S_1 \to S$ of a Riemann surface S and the subgroups of its fundamental group $\pi(S)$. An n-sheeted covering corresponds to a subgroup of index n.*

Corollary 2. *The universal covering surface $\widetilde{S}$ of a Riemann surface S is a Riemann surface, on which the fundamental group $\pi(S)$ acts by holomorphic automorphisms.*

Thus, to describe all Riemann surfaces, it suffices to describe those which are simply connected, together with groups of automorphisms acting freely and discretely on them. This idea is further developed in § 5.

Example 1. The unramified covering $\mathbb{C} \to \mathbb{C}/\Lambda$, where $\Lambda \subset \mathbb{C}$ is any discrete lattice, is universal and $\pi(\mathbb{C}/\Lambda) \simeq \Lambda$.

Example 2. As a special case, the covering $\mathbb{C} \xrightarrow{\exp} \mathbb{C}^\times$ is universal, and $\pi(\mathbb{C}^\times) \simeq \mathbb{Z}$, the action on $\mathbb{C}$ being given by $z \mapsto z + 2\pi\sqrt{-1}\,n$, for $n \in \mathbb{Z}$.

Example 3. In a similar way, the covering $\mathbb{H} \xrightarrow{\exp(\sqrt{-1}\,z)} \mathbb{D}^\times$ is universal, and $\pi(\mathbb{D}^\times) \simeq \mathbb{Z}$, the action on $\mathbb{H}$ being given by $z \mapsto z + 2\pi n$, for $n \in \mathbb{Z}$. Hence, for every $n > 0$ there exists a unique unramified n-sheeted covering $S \to \mathbb{D}^\times$. It is isomorphic to $\mathbb{D}^\times \xrightarrow{z^n} \mathbb{D}^\times$.

2.10. Continuation of Mappings. The fundamental group is also useful for the description of finite mappings. If $f\colon S_1 \to S_2$ is a finite mapping of Riemann surfaces then we have a finite unramified covering $f\colon S_1 - f^{-1}(\Delta) \to S_2 - \Delta$, where the *branch locus* $\Delta \subset S_2$ is the discrete subset above which the ramification points lie. Conversely:

Proposition. *Let $\Delta \subset S_2$ be a discrete subset. A finite unramified covering $U \to S_2 - \Delta$ has a unique continuation to a (possibly ramified) finite mapping $S_1 \to S_2$, where $S_1 \supset U$.*

This statement is obviously local. Hence we may assume that $S = \mathbb{D}$ and $\Delta = \{0\}$. Now, every connected unramified covering $U \to \mathbb{D}^\times$ is isomorphic to $\mathbb{D}^\times \xrightarrow{z^n} \mathbb{D}^\times$ (see Example 3 above), which can be continued to the finite mapping $\mathbb{D} \xrightarrow{z^n} \mathbb{D}$.

Corollary 1. *There is a one-to-one correspondence between the finite mappings $S_1 \to S_2$ of degree n that are ramified only over $\Delta \subset S_2$, and the subgroups of index n in $\pi(S_2 - \Delta)$.*

If S_2 is a compact surface then Δ is finite and the fundamental group $\pi(S_2 - \Delta)$ has an explicit finite presentation (see Sect. 3.4).

Definition. A mapping of Riemann surfaces $f\colon S_1 \to S_2$ is said to be *normal*, or *Galois*, if its automorphism group

$$\operatorname{Aut} f \stackrel{\text{def}}{=} \{g \in \operatorname{Aut} S_1 \mid f \circ g = f\}$$

acts transitively on the fibres $f^{-1}(p)$, $p \in S_2$.

Corollary 2. *A normal finite mapping $f\colon S_1 \to S_2$ corresponds to a normal subgroup $\Gamma \triangleleft \pi(S_2 - \Delta)$, and $\operatorname{Aut} f \simeq \pi(S_2 - \Delta)/\Gamma$.*

2.11. The Riemann Surface of an Algebraic Function

Proposition 1. *Let*

$$P(T) = T^n + c_1 T^{n-1} + \ldots + c_n \in \mathcal{M}(S_2)[T]$$

be an irreducible polynomial. Then there exists a finite mapping of Riemann surfaces $f\colon S_1 \to S_2$, of degree n, and a meromorphic function $F \in \mathcal{M}(S_1)$ that satisfies the equation

$$F^n + f^*(c_1)\, F^{n-1} + \ldots + f^*(c_n) = 0. \tag{2}$$

The function F is algebraic over the field $\mathcal{M}(S_2)$ and it can be regarded as an n-valued function on S_2. Its values form the points of a surface S_1, which is therefore called the *Riemann surface of the algebraic function F*. More precisely, let $\Delta \subset S_2$ be a discrete subset which contains the poles of all the functions $c_1, \ldots, c_n$ and also the points $p \in S_2$ where the polynomial

$$P_p(T) \stackrel{\text{def}}{=} T^n + c_1(p)T^{n-1} + \ldots + c_n(p) \in \mathbb{C}[T]$$

has multiple roots. The last points are the zeros of the discriminant of P. The submanifold

$$U = \{(p, z) \in (S_2 - \Delta) \times \mathbb{C} \mid P_p(z) = 0\} \subset (S_2 - \Delta) \times \mathbb{C}$$

is a Riemann surface. The connectedness of U is a nontrivial fact, which follows from the converse to Proposition 1.

Proposition 2. *Let $f\colon S_1 \to S_2$ be a finite mapping of Riemann surfaces. Then every function $F \in \mathcal{M}(S_1)$ is algebraic over $\mathcal{M}(S_2)$ and verifies some equation (2), of degree $n \le \deg f$.*

On removing the branch points and the poles of f, we obtain an unramified covering of $S_2 - \Delta$. Let $V \subset S_2 - \Delta$ be a (connected) open set such that $f^{-1}(V) = \bigcup V_i$, where $V_i \cap V_j = \varnothing$ for $i \neq j$, and the maps $f\colon V_i \to V$ are isomorphisms. Set $F_i = \tau_i^*(F)$ and $c_i = (-1)^i s_i$, where $\tau_i = f^{-1}\colon V \to V_i$ and the s_i are the elementary symmetric functions of $F_1, \ldots, F_n$. The functions c_i are well-defined and they are holomorphic on $S_2 - \Delta$. Further F satisfies equation (2). By Riemann's removable singularity theorem, the coefficients c_i have meromorphic continuations on S_2.

We return to the construction of the Riemann surface of the algebraic function F. Note that this function is holomorphic on U and is given by the projection map $(p, z) \mapsto z$. The other projection $(p, z) \mapsto p$ defines an unramified covering $U \to S_2 - \Delta$. Again, by Riemann's removable singularity theorem and by the Proposition of Sect. 2.10, we obtain the required mapping $f\colon S_1 \to S_2$, $U \subset S_1$, and a meromorphic function F on S_1. But according to Proposition 2 the decomposition of S_1 into connected components defines a factorization of P, whence S_1 (and U) are connected.

Example. Let $f(z) = (z - a_1) \ldots (z - a_n)$ be a polynomial with pairwise distinct roots $a_1, \ldots, a_n \in \mathbb{C}$. The polynomial $P(T) = T^2 - f$, which is irreducible over $\mathcal{M}(\mathbb{CP}^1) = \mathbb{C}(z)$, defines the algebraic function $\sqrt{f}$. Its Riemann surface S is called *hyperelliptic*. It is compact. The corresponding *mapping* $\gamma\colon S \to \mathbb{CP}^1$, of degree 2, and the *involution* $j\colon S \to S$, which permutes the points in the fibres of γ, are also called *hyperelliptic*. The ramification points of γ are the fixed points of j. They lie above the points

$$a_1, \ldots, a_n, \infty \quad \text{if } n \text{ is odd, and}$$
$$a_1, \ldots, a_n \quad \text{if } n \text{ is even.}$$

It would be more convenient to present the hyperelliptic surface S as being the plane curve $y^2 = f(z)$. But, for $n \geq 4$ this has a singular point at infinity. Hence S may be viewed as its desingularization (see Corollary 4 below).

In view of the primitive element theorem (cf. Shafarevich [1986]), we obtain from the above Propositions:

Theorem 1. *If $S_1 \to S_2$ is a finite mapping of Riemann surfaces then the field extension $f^*\colon \mathcal{M}(S_2) \hookrightarrow \mathcal{M}(S_1)$ is finite, and its degree is $\leq \deg f$.*

Theorem 2. *Let S_2 be a Riemann surface and let $\varphi\colon \mathcal{M}(S_2) \hookrightarrow K$ be a finite $\mathbb{C}$-extension of degree n. Then there exists a finite mapping of Riemann surfaces $f\colon S_1 \to S_2$, of degree n, whose associated field extension $f^*\colon \mathcal{M}(S_2) \hookrightarrow \mathcal{M}(S_1)$ is isomorphic to φ. Such a mapping f is unique up to isomorphism.*

Corollary 1. *The field of meromorphic functions $\mathcal{M}(S)$ of a compact Riemann surface S is finitely generated over $\mathbb{C}$ and of transcendence degree ≤ 1.*

Indeed, if there are no nonconstant meromorphic functions on S then $\mathcal{M}(S) = \mathbb{C}$ and $\operatorname{tr.deg} \mathcal{M}(S)/\mathbb{C} = 0$. Otherwise, applying Theorem 1 to some nonconstant meromorphic function $f\colon S \to \mathbb{CP}^1$, we obtain:

$$\text{tr. deg}\, \mathcal{M}(S)/\mathbb{C} = \text{tr. deg}\, \mathbb{C}(z)/\mathbb{C} = 1.$$

Remark 1. As a matter of fact, equality holds both in Theorem 1 and in Corollary 1 (see Corollaries 3 and 7 in Sect. 4.14).

Conversely, from Theorem 2 we deduce:

Corollary 2. *Any field K of transcendence degree 1 which is finitely generated over $\mathbb{C}$, is isomorphic to the field of meromorphic functions $\mathcal{M}(S)$ of some compact Riemann surface S.*

Such a surface is called a *model* of the field K. For example, the Riemann sphere $\mathbb{CP}^1$ is a model of the purely transcendental extension $\mathbb{C}(z)$ of $\mathbb{C}$.

Remark 2. About the uniqueness of the model, see Corollary 12 in Sect. 4.14 (cf. the theorem on the model for curves in Sect. 1.7 of Chap. 2).

An *algebraic curve* given in $\mathbb{CP}^2$ by an irreducible homogeneous polynomial $F(x_0, x_1, x_2)$ is called *irreducible* (cf. Chap. 2, Sect. 1.4).

Corollary 3. *Let $f\colon S \to \mathbb{CP}^2$ be a nonconstant holomorphic mapping of a compact Riemann surface S into $\mathbb{CP}^2$. Then $f(S)$ is an irreducible algebraic curve.*

The equation of $f(S)$ is obtained as follows. Let $(x_0 : x_1 : x_2)$ be homogeneous coordinates in $\mathbb{CP}^2$. By Corollary 1, the meromorphic functions $f^*(x_0/x_2)$ and $f^*(x_1/x_2) \in \mathcal{M}(S)$ are algebraically dependent over $\mathbb{C}$. Let $F(z_1, z_2)$ be a complex polynomial of minimal degree d defining an algebraic relation $F(f^*(x_0/x_2), f^*(x_1/x_2)) = 0$. Then the polynomial $x_2^d F(x_0/x_2, x_1/x_2)$ is irreducible and defines the curve $f(S)$.

Conversely, we have:

Corollary 4. *Let $C \subset \mathbb{CP}^2$ be an irreducible algebraic curve. Then there exists a compact Riemann surface S and a holomorphic mapping $f\colon S \to \mathbb{CP}^2$, whose image is identical with C. For a suitable choice of S, the mapping f is generically injective. Such a mapping f is called a* desingularization *of the curve C* (see Fig. 8 and cf. Sect. 1.7 of Chap. 2).

Let $F(x_0, x_1, x_2)$ be an irreducible homogeneous polynomial defining C. Take for S a model of the quotient field of the integral domain $\mathbb{C}[x_0, x_1, x_2]/(F(x_0, x_1, x_2)\,,\, x_2 - 1)$, where $(F(x_0, x_1, x_2)\,,\, x_2 - 1)$ is the ideal generated by the polynomials $F(x_0, x_1, x_2)$ and $x_2 - 1$. The mapping $f(p) = (x_0(p) : x_1(p) : 1)$, where $p \in S$, extends by continuity to a desingularization of the curve C. As a special case, if C is nonsingular then S is the Riemann surface of C and $f = id$. If, on the other hand, C has some singularities then f resolves them, for it is an isomorphism over the nonsingular points.

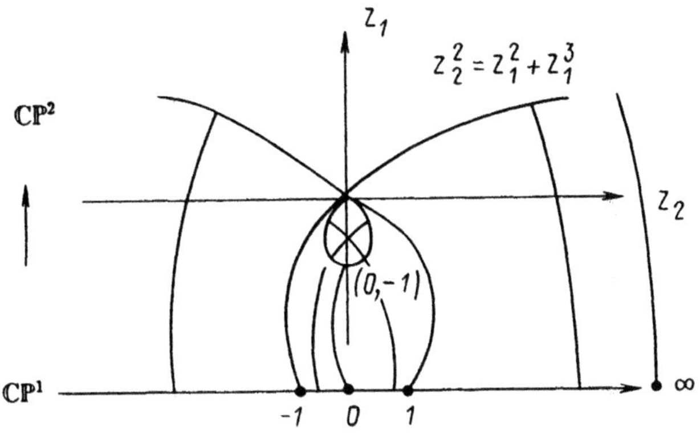

Fig. 8. Desingularization of Descartes' folium: The curve $x_2 x_1^2 = x_0^2 x_2 + x_0^3$ is represented in the affine coordinates $z_1 = x_0/x_2$, $z_2 = x_1/x_2$. Its desingularization is given by the parametrization $z_1 = z^2 - 1$, $z_2 = z(z^2 - 1)$; $\infty \mapsto (0:1:0)$.

Further, since Riemann surfaces are connected, we obtain:

Corollary 5. *An irreducible complex plane curve is connected.*

Corollary 6. *Let $C \subset \mathbb{CP}^2$ be a nonsingular algebraic curve. Then any meromorphic function on the Riemann surface of C is the pull-back of some rational function on $\mathbb{CP}^2$* (cf. Example 3 in Sect. 2.2).

§ 3. Topology of Riemann Surfaces

In this section, Riemann surfaces are considered from the topologist's point of view. We shall of course require some notions, methods, and results from algebraic topology. A more detailed treatment of the topology of surfaces can be found in Massey [1967,1977], and a treatment of algebraic topology as a whole in Dubrovin, Novikov & Fomenko [1984] and in Dold [1972].

3.1. Orientability. Orientability is a purely topological notion (cf. Dold [1972]). But, for simplicity, we shall restrict ourselves to its smooth variant. Let $(x_1, \dots, x_n)$ and $(y_1, \dots, y_n)$ be two real coordinate systems on a differentiable manifold M. We say that they have the same orientation if the Jacobian determinant, $\det\left(\frac{\partial y_i}{\partial x_j}\right)$, of the coordinate transformation is positive everywhere in the domain of definition. A differentiable or complex manifold M is said to be (smoothly) *orientable* if it has a differentiable atlas whose coordinate systems have identical orientations.

Proposition. *Any complex manifold M is orientable.*

Habitually one takes the underlying real atlas of some analytic atlas on M. This means that the complex coordinate systems $(z_1, \dots, z_n)$ of this atlas are replaced by the real coordinate systems $(x_1, y_1, \dots, x_n, y_n)$, where $z_i = x_i + \sqrt{-1}\, y_i$. The proof that these systems have identical orientations rests on the following fact from linear algebra. Let A be the complex $n \times n$-matrix of some $\mathbb{C}$-linear mapping $f\colon \mathbb{C}^n \to \mathbb{C}^n$. Then f corresponds to an $\mathbb{R}$-linear mapping $f_{\mathbb{R}}\colon \mathbb{R}^{2n} \to \mathbb{R}^{2n}$, whose (real) $2n \times 2n$-matrix $A_{\mathbb{R}}$ verifies:

$$\det A_{\mathbb{R}} = |\det A|^2$$

(see Kostrikin-Manin [1980]). The case $n = 1$ is obvious:

$$A = a + \sqrt{-1}\, b, \qquad A_{\mathbb{R}} = \begin{pmatrix} a & b \\ -b & a \end{pmatrix},$$

and

$$\det A_{\mathbb{R}} = a^2 + b^2 = \left|a + \sqrt{-1}\, b\right|^2.$$

Let S be a Riemann surface. Let $p \in S$ and consider an open neighbourhood U of p which is homeomorphic to the unit disk. Then $\pi(U - p) \simeq \mathbb{Z}$, and this fundamental group has a canonical generator, which is defined by a loop circling once around p in the positive direction. In fact, the local parameter z at p enables us to fix a small simple loop $z(t) = \varepsilon . e^{2\pi\sqrt{-1}\, t}$, $t \in [0, 1]$, which is described as positive. Any other simple loop on U around p is said to be positive if it admits a continuous deformation into our small loop. (In the Gaussian plane $\mathbb{C}$, these are counter-clockwise paths.)

The definition of positiveness does not depend on the choice of the local parameter. Indeed, all local parameters $z = x + \sqrt{-1}\, y$ can be deformed continuously into one another in a neighbourhood of p, since the Jacobian of the transformation on their real components, x and y, is positive. Intuitively, saying that S is orientable means this: if we pick a small disk and choose to travel in a certain direction along its circumference, we can move this disk continuously along any closed path on S and, when we are back to our starting point, we shall see that we are still travelling along the circumference in the same direction as before. The local coordinates on the Riemann surface allow the travelling direction to be controlled throughout the displacement of the disk.

Remark 1. The positiveness of a simple loop depends on the choice of a root $\sqrt{-1} \in \mathbb{C}$. Therefore $\sqrt{-1}$ is always assumed to be fixed.

Remark 2. There exist some non-orientable surfaces, such as the real projective plane $\mathbb{RP}^2$ or the Moebius strip. By the above proposition, these surfaces have no complex analytic structure.

3.2. Triangulability. A *triangle* on a Riemann surface S is a homeomorphic image T of an ordinary Euclidean triangle with the usual topology. The image of a vertex is called a vertex of T, and the image of a side is called an edge. A *triangulation* of S is a family $\{T_i\}$ of triangles on S such that

(a) $S = \bigcup T_i$;

(b) if two triangles meet then their intersection consists either of a common vertex or of a common edge;

(c) if $\{T_i\}$ is not a finite family, then we demand that it should be locally finite; this amounts to saying that only finitely many triangles have a common vertex and that their union defines a neighbourhood of that vertex (cf. Fig. 9).

Any triangulation of a compact Riemann surface is finite.

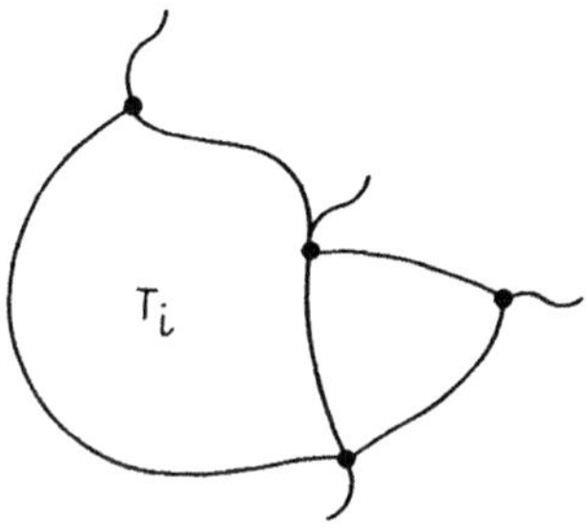

Fig. 9

Theorem (Rado; cf. Ahlfors-Sario [1960]). *Any Riemann surface is triangulable.*

In the smooth situation, for example for a Riemann or differentiable surface, triangulability is equivalent to the existence of a countable base for the topology, or to countability of the topology at infinity (cf. Rado [1925]). In particular, the theorem is obvious in the compact case.

3.3. Development; Topological Genus. In view of the finite triangulability property, a compact Riemann surface can be obtained by gluing together pairs of edges of some polygon M, which is called a *development* of S. This gluing together of edges is described by the symbol of the development, which is a sequence of letters designating the edges as we go around the boundary of M. The pairs of edges to be pasted together are denoted by the same letter. If two edges must be glued together in the same direction as we go around the boundary, then these edges are denoted by a letter with no exponent; otherwise, one of the letters is assigned the exponent -1. There are a number of standard operations on developments, through which a development with a reasonably simple symbol can be constructed (cf. Springer [1957]).

Theorem. *A compact Riemann surface S has a development with symbol*

(1) aa^{-1}, *or*

(2) $a_1b_1a_1^{-1}b_1^{-1}\dots a_gb_ga_g^{-1}b_g^{-1}$.

Corollary. *In case* (1), *the Riemann surface S is homeomorphic to a sphere; in case* (2), *to a sphere with g handles.*

Thus we see (but we have not proved) that the symbols in the Theorem are topological invariants of the Riemann surface.

Definition. The number g in (2) (and 0, in case (1)) is called the *(topological) genus* of the compact Riemann surface S. In other words, a Riemann surface S of genus g is homeomorphic to a sphere with g handles. The genus of S is denoted by $g(S)$, or simply g.

Example. An elliptic curve $\mathbb{C}/\Lambda$ has a development with symbol $aba^{-1}b^{-1}$, whence an elliptic curve is homeomorphic to a torus and $g(\mathbb{C}/\Lambda) = 1$.

3.4. Structure of the Fundamental Group

Theorem 1. *The fundamental group of a compact Riemann surface S of genus g is isomorphic to the quotient group of the free group on the generators $a_1, b_1, \ldots, a_g, b_g$ by the normal subgroup generated by the element $a_1b_1a_1^{-1}b_1^{-1} \ldots a_gb_ga_g^{-1}b_g^{-1}$.*

The case $g = 0$ is trivial. For $g \geq 1$, consider a development with symbol $a_1b_1a_1^{-1}b_1^{-1} \ldots a_gb_ga_g^{-1}b_g^{-1}$. The vertices of this development are all glued together into a single point $p \in S$. Every edge, a_i or b_i, therefore defines a loop on S, whose homotopy class defines an element of $\pi(S)$. Now the loop of the symbol $a_1b_1a_1^{-1}b_1^{-1} \ldots a_gb_ga_g^{-1}b_g^{-1}$ is clearly homotopic to the trivial one. Thus we have defined a map, which is the required isomorphism. The proof is based on the Seifert–van Kampen theorem (see Massey [1967,1977]).

Example. For an elliptic curve $\mathbb{C}/\Lambda$, the fundamental group $\pi(\mathbb{C}/\Lambda)$ is isomorphic to the group with generators a, b and commutation relation $aba^{-1}b^{-1} = 1$. Hence it is isomorphic to the free abelian group on two generators $\mathbb{Z} \oplus \mathbb{Z}$ (cf. Example 1 of Sect. 2.9).

To construct some finite mappings onto compact Riemann surfaces, it is useful to know the fundamental group of punctured surfaces (see Sect. 2.10).

Theorem 2. *Let S be a compact Riemann surface of genus g with a finite set of distinguished points, say $p_1, \ldots, p_n$. Then the fundamental group of the Riemann surface $S - \{p_i\}$ is isomorphic to the quotient group of the free group on the generators $a_1, b_1, \ldots, a_g, b_g, c_1, \ldots, c_n$ by the normal subgroup generated by the element $a_1b_1a_1^{-1}b_1^{-1} \ldots a_gb_ga_g^{-1}b_g^{-1}c_1 \ldots c_n$.*

The proof proceeds as in Theorem 1 (see Fig. 10).

Corollary 1. *For any compact Riemann surface S and any finite group G, there exists a finite, normal mapping of Riemann surfaces $f\colon S_1 \to S$ with automorphism group* $\operatorname{Aut} f \simeq G$.

Going over to extensions of meromorphic function fields (cf. Corollaries 9 and 11 in Sect. 4.14), we obtain:

Corollary 2. *Any field of transcendence degree* 1 *that is finitely generated over* $\mathbb{C}$ *has a finite normal extension with any preassigned finite Galois group.*

Remark. Corollary 2 solves the functional analogue of the inverse Galois problem. This, essentially topological, approach to the Galois problem has no complete analogue in the case of algebraic number fields.

3.5. The Euler Characteristic. Let S be a compact Riemann surface, with a triangulation $\{T_i\}$. We denote by v the number of vertices, by e the number of edges, and by t the number of triangles.

Lemma. *The number*

$$\chi(S) = v - e + t$$

is independent of the triangulation.

Moreover, the lemma remains true for any finite partition of S into polygons (i.e., homeomorphic images of convex polygons of the Euclidean plane); then, by t we mean the number of polygons in the partition.

Definition. $\chi(S)$ is known as the *Euler characteristic* of the surface S.

Corollary 1. *For a compact Riemann surface S of genus g we have*

$$\chi(S) = 2 - 2g.$$

In particular, the Euler characteristic is even and $\chi(S) \leq 2$.

For $g \geq 1$, the standard development of S (see Sect. 3.3) yields a partition with one polygon, one vertex, and $2g$ edges.

Corollary 2 (Euler). *The numbers V of vertices, E of edges, and F of faces of an arbitrary convex polyhedron are related by the formula $V - E + F = 2$.*

3.6. The Hurwitz Formulae. With the Euler characteristic we can control the behaviour of the genus under mappings of Riemann surfaces.

Hurwitz formula for the Euler characteristic. *Let $f\colon S_1 \to S_2$ be a non-constant mapping of compact Riemann surfaces. Then*

$$\chi(S_1) = \deg f \cdot \chi(S_2) - \deg R,$$

where R is the ramification divisor of f (see Example 2 of Sect. 2.6). *In particular, the number of ramification points – counting multiplicities – is always even, if by multiplicity we mean the ramification index.*

To prove this, we choose on S_2 a sufficiently fine triangulation $\{T_i\}$, whose vertices include the images of all the ramification points. On S_1 we take the triangulation $f^{-1}\{T_i\}$. Then the inverse image of any single triangle consists of $\deg f$ triangles. Counting the numbers of vertices, edges, and triangles, leads to the required relation (see Griffiths-Harris [1978]).

Corollary 1 of Sect. 3.5 allows us to rewrite these relations in terms of the genus.

Hurwitz's genus formula. $g(S_1) = \deg f \cdot g(S_2) + \frac{1}{2} \deg R - \deg f + 1.$

Corollary. $g(S_1) \geq g(S_2)$.

Example 1. Let S be a hyperelliptic Riemann surface, with projection map $\gamma\colon S \to \mathbb{CP}^1$, having n ramification points. Then $g(S) = \frac{1}{2}n - 1$. Hence on any orientable compact surface there is a Riemann surface structure, a hyperelliptic one, for instance.

Example 2. Let S be a compact Riemann surface of genus g, and $f\colon S \to \mathbb{CP}^1$ a nonconstant mapping of degree n. From Hurwitz's genus formula we get

$$\deg R = 2(n + g - 1) \geq 2n - 2,$$

where R is the ramification divisor of f. Conversely, for any even number $b \geq 2n - 2$, there is a finite mapping $f\colon S \to \mathbb{CP}^1$ with $\deg f = n$ and $\deg R = b$. Furthermore, it is possible to construct such an f with only simple (index 1) ramification above b distinct points $p_1, \ldots, p_b \in \mathbb{CP}^1$. We identify a generic fibre $f^{-1}(p)$ with the set $\{1, \ldots, n\}$. By moving the fibres of f along closed paths avoiding the p_i, we define a homomorphism

$$\mu\colon \pi(\mathbb{CP}^1 - \{p_i\}) \to \mathcal{S}_n,$$

called the *monodromy* of f, where $\mathcal{S}_n$ is the permutation group of $\{1, \ldots, n\}$. Thus to the generators c_i of the group $\pi(\mathbb{CP}^1 - \{p_i\})$ (see Theorem 2 of Sect. 3.4 and Fig. 10) correspond some permutations $\sigma_i \in \mathcal{S}_n$. The following conditions must hold:

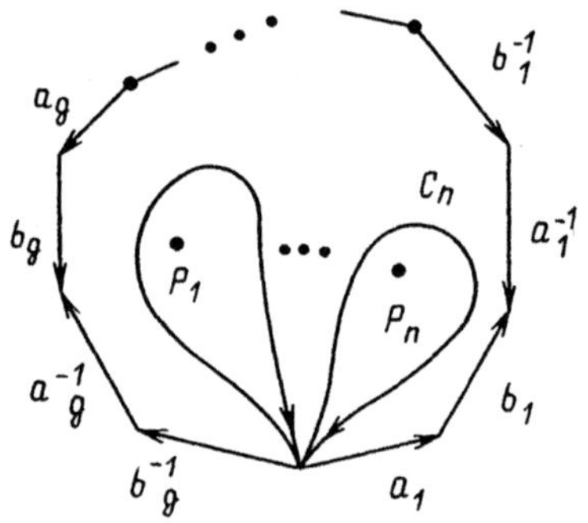

Fig. 10

(a) $\sigma_1 \ldots \sigma_b = id$ (by Theorem 2 of Sect. 3.4);
(b) the σ_i act transitively on $\{1, \ldots, n\}$ (by the connectedness of S);
(c) the σ_i are transpositions (since there is one simple ramification point over each p_i; cf. Example 6 of Sect. 1.6).

These conditions are fulfilled, for instance, by the transpositions

$$\sigma_1 = \sigma_2 = (1,2), \ \ldots, \ \sigma_{2n-3} = \sigma_{2n-2} = (1,n), \ \sigma_{2n-1} = \cdots = \sigma_b = (1,2).$$

It remains for us to establish the existence of the required mapping $f\colon S \to \mathbb{CP}^1$ with given monodromy μ, satisfying (b) and (c). To achieve this, one can apply Corollary 1 of Sect. 2.10 to the subgroup

$$\Gamma = \left\{\sigma \in \pi(\mathbb{CP}^1 - \{p_i\}) \mid \mu(\sigma)\, 1 = 1\right\},$$

which is of index n in $\pi(\mathbb{CP}^1 - \{p_i\})$. The points of the fibre $f^{-1}(p)$ can then be identified with the left cosets $\pi(\mathbb{CP}^1 - \{p_i\})/\Gamma$, and the monodromy μ with the natural homomorphism

$$\pi(\mathbb{CP}^1 - \{p_i\}) \to \operatorname{Aut}(\pi(\mathbb{CP}^1 - \{p_i\})/\Gamma)$$

measuring the action on these cosets.

The following method of constructing $f\colon S \to \mathbb{CP}^1$ from the monodromy μ may be more inspiring. We choose a finite number of nonintersecting arcs γ_j connecting the p_i, in such a way that discarding them leads to a simply connected surface $\mathbb{CP}^1 - \bigcup \gamma_j$. Further we assume that $p \in \mathbb{CP}^1 - \bigcup \gamma_j$. Every arc γ_j defines a permutation $\sigma_j \in \mathcal{S}_n$, provided we agree to call one of the sides of γ_j the upper edge, and the other the lower edge: σ_j is the monodromy $\mu(u)$ of a path u which crosses γ_j once only, going downwards. Now we take n disjoint copies $T_1, \ldots, T_n$ of the surface $\mathbb{CP}^1 - \bigcup \gamma_j$. For every j, we add to $\bigcup T_l$ a set $\{\gamma_j^l\}$ consisting of n copies of γ_j. Further we identify γ_j^l with the boundary of T_l along the upper edge of the cut of γ_j, and also with the boundary of $T_{\sigma_j(l)}$ along the lower edge of the cut of γ_j (see Fig. 11). As a result, we obtain the required surface S with a natural projection $f\colon S \to \mathbb{CP}^1$. This construction illustrates Riemann surface *surgery*, a method for constructing a new Riemann surface from given ones by 'cutting and gluing'.

Obviously, the number of mappings $f\colon S \to \mathbb{CP}^1$ with simple ramification above b given points p_i is finite. Hurwitz ([1891]) found a formula for the number of such mappings and established their topological equivalence.

Remark. The *monodromy group,* that is, the image of the monodromy μ, is the full permutation group $\mathcal{S}_n$, provided $f\colon S \to \mathbb{CP}^1$ is a sufficiently ramified mapping, for instance if $b \geq n(n-1)$. Knowing that the group $\mathcal{S}_n$ is not solvable if $n \geq 5$, we derive Abel's theorem to the effect that the general complex polynomial of degree ≥ 5 cannot be solved by radicals (see Alekseev [1976]).

It is assumed below that the reader is acquainted with such notions of combinatorial topology as homology, cohomology, and Betti numbers.

3.7. Homology and Cohomology; Betti Numbers. Let S be a compact Riemann surface of genus g.

Theorem. *There is a natural isomorphism*

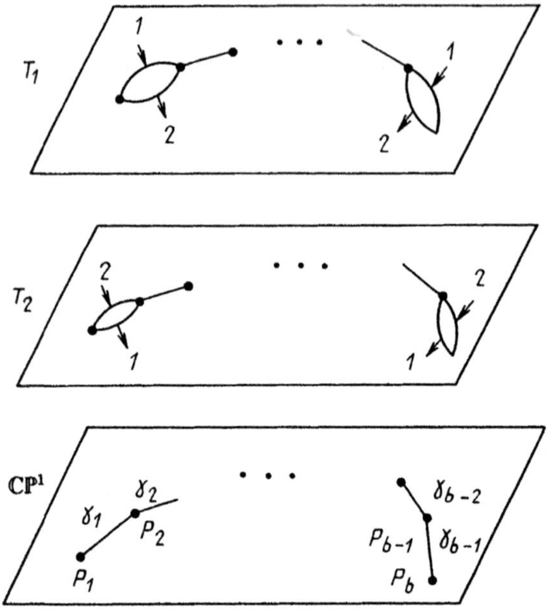

Fig. 11. A hyperelliptic projection

$$H_1(S,\mathbb{Z}) \xleftarrow{\sim} \pi(S)/\left[\pi(S),\pi(S)\right],$$

where $[\pi(S),\pi(S)]$ *is the commutator subgroup of the fundamental group* $\pi(S)$.

Every path is homotopically equivalent to a path along the edges of a triangulation and can therefore be represented by a simplicial 1-chain. Correspondingly, a loop can be represented by a 1-cycle. This is how the isomorphism is defined (cf. Springer [1957]). In particular, the edges a_i and b_i of a standard development (see Fig. 10 for $n = 0$) define one-dimensional homology classes, which are also denoted by a_i and b_i.

Example. For an elliptic curve $\mathbb{C}/\Lambda$, we have:

$$H_1(\mathbb{C}/\Lambda,\mathbb{Z}) \simeq \Lambda/\left[\Lambda,\Lambda\right] \simeq \Lambda.$$

Corollary. $H^1(S,\mathbb{Z}) \simeq H_1(S,\mathbb{Z}) = \mathbb{Z}a_1 \oplus \mathbb{Z}b_1 \oplus \ldots \oplus \mathbb{Z}a_g \oplus \mathbb{Z}b_g,$
$H^0(S,\mathbb{Z}) \simeq H_0(S,\mathbb{Z}) \simeq H^2(S,\mathbb{Z}) \simeq H_2(S,\mathbb{Z}) \simeq \mathbb{Z},$

whence

$$b^1 = b_1 = 2g, \quad b^0 = b_0 = b^2 = b_2 = 1 \quad \textit{and} \quad \chi(S) = 2 - 2g.$$

3.8. Intersection Product; Poincaré Duality. Two paths, $u(t)$ and $v(t)$, in $\mathbb{C}$ meet transversally at a point $p = u(t_0) = v(t_0)$ if the tangent vectors $\frac{du}{dt}(t_0)$

and $\frac{dv}{dt}(t_0)$ are linearly independent over the reals. Then the intersection product $(u.v)_p$ is taken to be $+1$ when $\frac{du}{dt}(t_0)$, $\frac{dv}{dt}(t_0)$ is a positive basis for $\mathbb{C}$, that is, when this coordinate system has the same orientation as the system (x, y), where $z = x + \sqrt{-1}\,y$. In the contrary case, $(u.v)_p = -1$. In a similar way, the local coordinates on a Riemann surface S enable us to define the transversality of an intersection, as well as the intersection product at a point of S.

Lemma. *Any two loops on S meet transversally, up to homotopy.*

The proof is based on the theorems of Arzelà and Sard (see Dubrovin, Novikov & Fomenko [1979]).

Definition. The *intersection product* of two loops u and v that meet transversally on a Riemann surface S, is defined by the formula

$$(u.v) \stackrel{\text{def}}{=} \sum (u.v)_p,$$

where p runs through the (finite) set of all intersection points of the loops.

The following result allows us to extend the notion of an intersection product to any pair of loops.

Proposition (on homotopic invariance). *The intersection product $(u.v)$ depends only on the homotopy class of the loops u and v.*

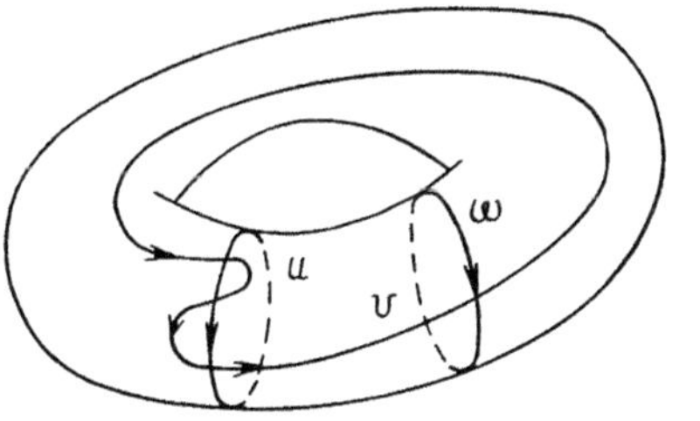

Fig. 12. Intersection products: $(u.v) = (w.v) = 1$

Note that it is essential to count not merely the intersection points, but how each of them contributes with a definite sign (see Fig. 12). The following properties of the intersection product are easy to check:

(1) $(uv.w) = (u.w) + (v.w)$ and $(u^{-1}.w) = -(u.w)$, where uv denotes the composition of the loops u and v, and u^{-1} is the inverse of u;

(2) $(u.v) = -(v.u)$ (skew-symmetry).

Thus, in the compact case, it follows from the Theorem in Sect. 3.7 that we have obtained a skew-symmetric bilinear form

$$(\ .\ ) \colon H_1(S, \mathbb{Z}) \times H_1(S, \mathbb{Z}) \to \mathbb{Z}. \tag{3}$$

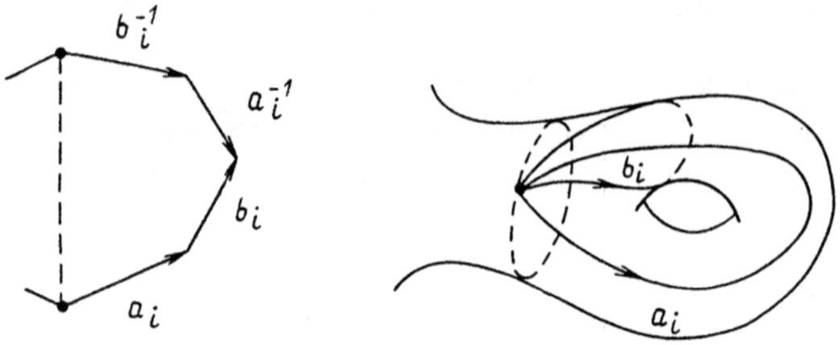

Fig. 13

Proposition. *The form* (3) *is unimodular.*

Hence the group $H_1(S, \mathbb{Z})$ has a basis $a_1, b_1, \ldots, a_g, b_g$ in which

$$(a_i . a_j) = (b_i . b_j) = 0 \quad \text{and} \quad (a_i . b_j) = \delta_{ij}.$$

Such, for example, is the basis given by the edges of a development with symbol $a_1 b_1 a_1^{-1} b_1^{-1} \ldots a_g b_g a_g^{-1} b_g^{-1}$, as we go around its boundary in the positive direction (see the Corollary of Sect. 3.7 and Fig. 13).

Corollary (cf. Dubrovin, Novikov & Fomenko [1984]). *The unimodular form* (3) *induces the isomorphism known as Poincaré duality:*

$$H_1(S, \mathbb{Z}) \simeq \mathrm{Hom}(H_1(S, \mathbb{Z}), \mathbb{Z}) = H^1(S, \mathbb{Z}).$$

§ 4. Calculus on Riemann Surfaces

This section covers three topics: calculus on smooth surfaces (Sections 4.1 to 4.7), meromorphic differentials on Riemann surfaces (4.8 to 4.10), and the main existence theorems (4.11 to 4.15). A more detailed treatment of calculus on differentiable manifolds can be found in Griffiths-Harris [1978], Narasimhan [1968], Spivak [1965], and Wells [1973]. We always assume infinite differentiability.

4.1. Tangent Vectors; Differentiations. Let S be a Riemann surface. By a differentiable complex-valued function on S, we mean a differentiable mapping of the form $S \to \mathbb{C}$ (see Sect. 1.7). Differentiable complex-valued functions on S form a $\mathbb{C}$-algebra $\mathcal{E}(S)$. A *differentiation* of $\mathcal{E}(S)$ at $p \in S$ is a $\mathbb{C}$-linear mapping $D \colon \mathcal{E}(S) \to \mathbb{C}$ that satisfies the Leibniz relation

$$D(f . g) = D(f) . g(p) + f(p) . D(g).$$

The differentiations at a point p form a complex vector space with the natural operations of addition and multiplication by constants. This vector space, denoted by $T_p(S)$, is called the *tangent space* to S at p.

Examples. Let $z = x + \sqrt{-1}\,y$ be a local coordinate at a point $p \in S$. Then the partial derivatives, $\dfrac{\partial f}{\partial x}(p)$ and $\dfrac{\partial f}{\partial y}(p)$, of the functions $f \in \mathcal{E}(S)$, written in the local coordinates x and y, determine differentiations $\left.\dfrac{\partial}{\partial x}\right|_p$ and $\left.\dfrac{\partial}{\partial y}\right|_p$ at p. Further examples of differentiations are provided by the operators of Wirtinger's calculus:

$$\left.\frac{\partial}{\partial z}\right|_p, \left.\frac{\partial}{\partial \bar{z}}\right|_p, \text{ with } \frac{\partial}{\partial z} \overset{\text{def}}{=} \tfrac{1}{2}\left(\frac{\partial}{\partial x} - \sqrt{-1}\,\frac{\partial}{\partial y}\right), \frac{\partial}{\partial \bar{z}} \overset{\text{def}}{=} \tfrac{1}{2}\left(\frac{\partial}{\partial x} + \sqrt{-1}\,\frac{\partial}{\partial y}\right).$$

We observe that a holomorphic function on an open set $U \subset \mathbb{C}$ is nothing but a differentiable function $f \in \mathcal{E}(U)$ that satisfies the Cauchy-Riemann equation $\dfrac{\partial}{\partial \bar{z}} f = 0$ (cf. Shabat [1969]).

Lemma. *The two vectors* $\left.\dfrac{\partial}{\partial x}\right|_p$ *and* $\left.\dfrac{\partial}{\partial y}\right|_p$ *(respectively,* $\left.\dfrac{\partial}{\partial z}\right|_p$ *and* $\left.\dfrac{\partial}{\partial \bar{z}}\right|_p$*) form a basis of the tangent space* $T_p(S)$*. Hence* $\dim_{\mathbb{C}} T_p(S) = 2$.

Remark. The algebra $\mathcal{E}(S)$ comprises a vast supply of functions. For example, any differentiable function defined in a neighbourhood of a point $p \in S$ can be continued to a differentiable function on the whole of S, with no modification in the vicinity of p. This and similar facts follow from the theorem on partitions of unity (see Spivak [1965]).

4.2. Differential Forms. Let S be a Riemann surface. A complex-valued function on the family of all tangent spaces $\bigcup T_p(S)$ which is linear on each space $T_p(S)$, is called a *differential* 1*-form,* or simply a *differential.*

Examples. Let $z = x + \sqrt{-1}\,y$ be a local coordinate on an open set $U \subset S$. Then there is a differential on U, which maps a tangent vector $a\dfrac{\partial}{\partial x} + b\dfrac{\partial}{\partial y}$ into its $\dfrac{\partial}{\partial x}$-coordinate a (respectively, its $\dfrac{\partial}{\partial y}$-coordinate b).This differential is denoted by dx (respectively, dy). In like manner we define the differentials dz and $d\bar{z}$. Obviously, $dz = dx + \sqrt{-1}\,dy$ and $d\bar{z} = dx - \sqrt{-1}\,dy$.

Lemma. *Any differential* 1*-form* ω *on* S *can be written locally as*

$$\omega|_U = f dz + g d\bar{z} = (f+g)dx + (f-g)\sqrt{-1}\,dy,$$

where f *and* g *are complex-valued functions on an open set* $U \subset S$*, and* $z = x + \sqrt{-1}\,y$ *is a local coordinate on* U.

A *differential* ω is said to be *differentiable* if, for every local representation $\omega = f dz + g d\bar{z}$, the functions f and g are differentiable. We denote by A^1 the (infinite-dimensional) complex vector space of differentiable 1-forms on S. A form ω is of type $(1,0)$ (respectively, $(0,1)$) if, locally, $\omega = f dz$ (respectively, $\omega = f d\bar{z}$). The forms of type $(1,0)$ (respectively, $(0,1)$) make up a complex vector space $A^{1,0}$ (respectively, $A^{0,1}$). Further, we have the decomposition $A^1 = A^{1,0} \oplus A^{0,1}$: every 1-form ω can be written uniquely as a sum $\omega = \omega_1 + \omega_2$, where ω_1 and ω_2 are forms of type $(1,0)$ and $(0,1)$, respectively.

If, in the definition of a differential, we replace each tangent space $T_p(S)$ by its product $T_p(S) \times \ldots \times T_p(S)$, and the linearity condition by multilinearity, we obtain the definition of a differential of degree ≥ 2. Further, if ω_1 and ω_2 are differential 1-forms, we define a form $\omega_1 \otimes \omega_2$, which maps each pair of tangent vectors $(t_1, t_2) \in T_p(S) \times T_p(S)$ into $\omega_1 \otimes \omega_2(t_1, t_2) = \omega_1(t_1) \cdot \omega_2(t_2)$. This form is called the tensor product of ω_1 and ω_2. The tensor product of any number of differentials is defined in a similar way. Locally, every differential ω can be written as a linear combination of tensor products of the differentials dz and $d\bar{z}$, with functions as coefficients. If each product has i factors then ω is said to be an i-form. Further, if all the coefficients are differentiable functions then we say that ω is differentiable. As with multilinear functions, one can impose conditions of symmetry, skew- or Hermitian-symmetry, and the like. For instance, if ω_1 and ω_2 are 1-forms then the 2-form $\omega_1 \wedge \omega_2 \stackrel{\text{def}}{=} \frac{1}{2}(\omega_1 \otimes \omega_2 - \omega_2 \otimes \omega_1)$ is skew-symmetric. Indeed, $\omega_2 \wedge \omega_1 = -\omega_1 \wedge \omega_2$. This form is called the exterior product of ω_1 and ω_2. Skew-symmetric differentials, also known as exterior differentials, are especially interesting for differential geometry. We shall denote by A^i the complex vector space of exterior differentiable i-forms on the Riemann surface S. By definition, $A^0 = \mathcal{E}(S)$. Of course, $A^i = 0$ for $i \geq 3$, since locally a Riemann surface has only two real coordinates. Every exterior 2-form ω on a Riemann surface is of type $(1,1)$ and has a local representation

$$\omega = f\, dz \wedge d\bar{z} = -2\sqrt{-1}\, dx \wedge dy.$$

For example, if locally $\omega_1 = f_1 dz + g_1 d\bar{z}$ and $\omega_2 = f_2 dz + g_2 d\bar{z}$, then

$$\omega_1 \wedge \omega_2 = \begin{vmatrix} f_1 & g_1 \\ f_2 & g_2 \end{vmatrix} dz \wedge d\bar{z}.$$

Further information on differentials and on tensors in general can be found in Dubrovin, Novikov & Fomenko [1979], Kostrikin-Manin [1980], and Spivak [1965].

4.3. Exterior Differentiations; de Rham Cohomology. There are three types of differentiations in the complex case:

$$d\colon A^0 \to A^1, \qquad \partial\colon A^0 \to A^{1,0} \qquad \text{and} \qquad \bar{\partial}\colon A^0 \to A^{0,1}.$$

Locally,

$$df \stackrel{\text{def}}{=} \frac{\partial f}{\partial z} dz + \frac{\partial f}{\partial \bar{z}} d\bar{z} = \frac{\partial f}{\partial x} dx + \frac{\partial f}{\partial y} dy,$$

$$\partial f \stackrel{\text{def}}{=} \frac{\partial f}{\partial z} dz \quad \text{and} \quad \bar{\partial} f \stackrel{\text{def}}{=} \overline{\partial \bar{f}} = \frac{\partial f}{\partial \bar{z}} d\bar{z}.$$

These mappings are $\mathbb{C}$-linear and have natural continuations to $\mathbb{C}$-linear maps $d, \partial, \bar{\partial}\colon A^1 \to A^2$. Locally,

$$d(f dz + g d\bar{z}) \stackrel{\text{def}}{=} df \wedge dz + dg \wedge d\bar{z} = \left(\frac{\partial g}{\partial z} - \frac{\partial f}{\partial \bar{z}} \right) dz \wedge d\bar{z},$$

$$\partial(f dz + g d\bar{z}) \stackrel{\text{def}}{=} \partial f \wedge dz + \partial g \wedge d\bar{z} = \frac{\partial g}{\partial z} dz \wedge d\bar{z},$$

and

$$\bar{\partial}(f dz + g d\bar{z}) \stackrel{\text{def}}{=} \bar{\partial} f \wedge dz + \bar{\partial} g \wedge d\bar{z} = -\frac{\partial f}{\partial \bar{z}} dz \wedge d\bar{z}.$$

The maps $d, \partial, \bar{\partial}\colon A^i \to A^{i+1}$, for $i \geq 2$, are defined in a similar way. However, for Riemann surfaces they are trivial, since $A^i = 0$ for $i \geq 3$. These maps are exterior differentiations, that is, $d(f\omega) = df \wedge \omega + f\, d\omega$ (and similarly for ∂ and $\bar{\partial}$). In addition the following relations hold: $dd = \partial\partial = \bar{\partial}\bar{\partial} = 0$, $d = \partial + \bar{\partial}$, and $\partial\bar{\partial} = -\bar{\partial}\partial$.

A *differential* $\omega \in A^i$ is said to be *closed* if $d\omega = 0$. Differentials of the form $d\omega$ (for some $\omega \in A^i$) are said to be *exact* or *cohomologically trivial.* (By definition, a function $f \in A^0$ is exact if $f \equiv 0$.) Every exact form is closed. Hence the exact i-forms constitute a complex vector subspace of the space of closed i-forms. The corresponding quotient space is denoted by $H^i_{DR}(S)$ and called the *de Rham cohomology group.*

Example 1. A function $f \in A^0$ is closed if and only if it is constant. Therefore $H^0_{DR}(S) \simeq \mathbb{C}$. On the other hand, f is $\bar{\partial}$-closed, that is, $\bar{\partial} f = 0$, if and only if it is holomorphic (see the Example in Sect. 4.1).

Example 2. Obviously, $H^i_{DR}(S) = 0$ for $i \geq 3$.

4.4. Kähler and Riemann Metrics. The symmetric product of two 1-forms is defined by the rule

$$\omega_1 \omega_2 \stackrel{\text{def}}{=} \tfrac{1}{2}(\omega_1 \otimes \omega_2 + \omega_2 \otimes \omega_1).$$

A Hermitian 2-form on a Riemann surface S is of type $(1,1)$ and can be written locally as $\omega = f\, dz\, d\bar{z} = f(dx^2 + dy^2)$, where f is some real-valued function. If f is differentiable and positive then the form ω is called a *Kähler metric* on S.

Lemma. *There exists a Kähler metric on any Riemann surface S.*

The existence of such a metric is obvious locally. To obtain it globally, one can apply the theorem on partitions of unity. Another approach is sketched in the Remark of Sect. 5.4.

Example 1. $dz\,d\bar{z} = dx^2 + dy^2$ is the *Euclidean metric* on $\mathbb{C}$. It is invariant under translations and therefore induces a Kähler metric on quotients of the form $\mathbb{C}/\Lambda$, where $\Lambda \subset \mathbb{C}$ is any discrete lattice.

Example 2. Let z be an affine coordinate on $\mathbb{CP}^1$. Then the 2-form $\dfrac{dz\,d\bar{z}}{(1+|z|^2)^2}$ can be continued to a Kähler metric on $\mathbb{CP}^1$. This is a special case of the *Fubini-Study metric* on $\mathbb{CP}^n$ (cf. Wells [1973]).

Example 3. The Poincaré model of Lobachevskian geometry on the unit disk has the metric $\dfrac{dz\,d\bar{z}}{(1-|z|^2)^2}$.

Consider now a differentiable surface S. Any real symmetric 2-form ω on S can be written locally as

$$\omega = f dx^2 + 2g\,dx\,dy + h dy^2.$$

Such a form is said to be a *Riemannian metric* on S if the functions f, g, and h are differentiable, and if ω itself is positive definite at every point $p \in S$. This is locally equivalent to the inequalities $f > 0$ and

$$\begin{vmatrix} f & g \\ g & h \end{vmatrix} > 0.$$

Theorem. *Let S be a differentiable, oriented surface with a Riemannian metric ω. Then there exists on S a unique Riemann surface structure such that ω is Kähler.*

This means that, for some local coordinate $z = x + \sqrt{-1}\,y$, we have $\omega = f(dx^2 + dy^2)$. The x- and y-*coordinates* such that ω is of this form are called *isothermal* or *conformal*. The orientation on S fixes a maximal atlas, whose coordinate systems have identical orientations. An isothermal coordinate system (x, y) for this atlas determines the required analytic coordinate $z = x + \sqrt{-1}\,y$. Now, the existence of isothermal coordinates reduces to solving the Beltrami differential equation (see Dubrovin, Novikov & Fomenko [1979]). The transition mappings between the corresponding local z-coordinates are conformal (angle- and orientation-preserving), and hence holomorphic (cf. Shabat [1969]).

Example 4. The ordinary Euclidean metric on a sphere of radius 1 leads to the Riemann sphere $\mathbb{CP}^1$ with the Fubini-Study metric multiplied by 4.

Remark. The connection between Riemannian and complex analytic geometries is substantially more complicated than appears in the theorem. A Riemannian metric on differentiable manifolds of dimension ≥ 3 has no good local representation in general.

4.5. Integration of Exterior Differentials; Green's Formula. The value of a function f at a point $p \in S$ is a very simple instance of integration:

$\int_p f \overset{\text{def}}{=} f(p)$. Next we have $\int_u \omega$, where $\omega \in A^1$ and $u\colon [0,1] \to S$ is a smooth path on a Riemann surface S. This integral is linear in ω and additive in u. By additiveness, the definition of the integral of a 1-form over any path reduces to the usual curvilinear integrals in the Gaussian plane $\mathbb{C}$ (cf. Springer [1957]). Using the following result one can define the integral of a closed 1-form over an arbitrary path.

Lemma. *Let u_1 and u_2 be smooth homotopic paths on a Riemann surface S, and let $\omega \in A^1$ be a closed form. Then*

$$\int_{u_1} \omega = \int_{u_2} \omega.$$

By considering the pull-back of ω with respect to a smooth homotopy, we reduce the proof to the corresponding fact for closed 1-forms on the square $[0,1] \times [0,1]$ (see also Springer [1957]).

Example 1. Any exact form $\omega = df$ can be integrated by means of the Newton-Leibniz formula

$$\int_u df = f(u(1)) - f(u(0)).$$

Example 2. Let $f(z) = \sum_{i=-n}^{+\infty} a_i z^i$ be the Laurent series of a meromorphic function defined in a neighbourhood of the origin. Let u be a simple positive loop around zero. Then (cf. Shabat [1969])

$$\frac{1}{2\pi\sqrt{-1}} \int_u f(z)dz = a_{-1}.$$

Corollary (criterion for exactness). *A closed form $\omega \in A^1$ on a Riemann surface S is exact if and only if $\int_u \omega = 0$ for any loop u on S.*

To establish the exactness of ω, one must find a function $f \in A^0$ with $df = \omega$. This is called the *antiderivative* (or *primitive*) of ω. Fix a point $p \in S$, and set $f(q) = \int_u \omega$, where u is any path from p to q. Then f is well-defined precisely when the conditions of the corollary hold for ω.

We now consider integrals of the form $\int_G \omega$, where $\omega \in A^2$ and G is a regular region of S. By a *regular region* on a Riemann surface S, we mean an open subset $G \subset S$, whose closure $\overline{G}$ in S is compact and whose boundary ∂G consists of a finite number of smooth paths. The orientation of the paths on the boundary is chosen so that an outward normal to G and a tangent vector to the path form a positive basis in a complex chart of S. In other words, when travelling along ∂G, the region G remains to the left of a tangent vector drawn in the direction of motion (see Fig. 14). The closure of a regular region G

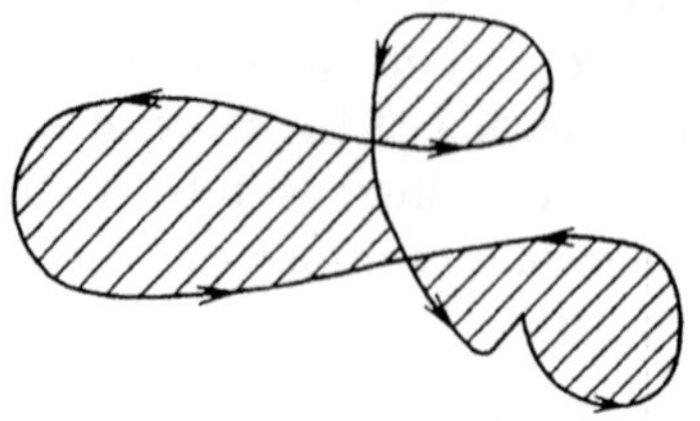

Fig. 14. A regular region

always has a (finite) triangulation $\{T_i\}$, all of whose edges are smooth paths. (The first rigorous proof of this fact is due to Kerns.) The triangulation can be chosen to be sufficiently fine, namely such that every triangle T_i lies in the domain of definition of some local coordinate z. Then one defines

$$\int_{T_i} \omega \stackrel{\text{def}}{=} \int_{z(T_i)} f\, dz \wedge d\bar{z}$$

to be the usual double integral over a curvilinear triangle $z(T_i) \subset \mathbb{C}$, and

$$\int_G \omega \stackrel{\text{def}}{=} \sum \int_{T_i} \omega.$$

This integral depends only on the region G and on the 2-form ω (cf. Springer [1957]). Further we remark that the integral $\int_u \omega$ over any path u (where $\omega \in A^1$) depends in fact only on the 1-form ω (even if it is not closed) and on the image $u([0,1]) \subset S$ and its orientation.

Green's formula. *For any regular region G on a Riemann surface S and any form $\omega \in A^1$, we have*

$$\int_G d\omega = \int_{\partial G} \omega.$$

This result follows immediately from the definition and from the Green's formula for regular regions of the Gaussian plane $\mathbb{C}$ (see Springer [1957]).

A form $\omega \in A^2$ is said to have compact support if it vanishes outside some compact subset of S. (The same definition applies to any i-forms.) Moreover, such a form vanishes outside some regular region G. Hence we can define the integral

$$\int_S \omega \stackrel{\text{def}}{=} \int_G \omega.$$

There is also another approach, which depends on the theorem on partitions of unity (cf. Springer [1957]).

Corollary. *If $\omega \in A^1$ is a differential form with compact support then we have: $\int_S d\omega = 0$. In particular, this holds for any form $\omega \in A^1$ on a compact Riemann surface S.*

4.6. Periods; de Rham Isomorphism. The *periods* of a closed 1-form ω on a Riemann surface S are the integrals $\int_u \omega$, where u runs through the loops on S. We assume that S is compact. In view of the homological (respectively, homotopic) invariance of these integrals, there is a *period homomorphism*

$$\Pi_\omega \colon H_1(S, \mathbb{Z}) \to \mathbb{C}$$

(respectively, $\Pi_\omega \colon \pi(S) \to \mathbb{C}$), defined by $c \mapsto \int_c \omega$. And this yields a $\mathbb{C}$-linear map

$$\begin{aligned} H^1_{DR}(S) &\to H^1(S, \mathbb{C}) = \operatorname{Hom}(H_1(S, \mathbb{Z}), \mathbb{C}), \\ \text{class of } \omega &\mapsto \Pi_\omega. \end{aligned} \tag{4}$$

De Rham's Theorem. *The mapping* (4) *is an isomorphism.*

In other words, every homomorphism $H_1(S, \mathbb{Z}) \to \mathbb{C}$ is the period homomorphism of some closed 1-form ω, which is uniquely determined up to an exact form. Uniqueness, which amounts to saying that $H^1_{DR}(S) \hookrightarrow H^1(S, \mathbb{C})$ is injective, follows from the exactness criterion. Existence is more difficult to prove. Fix some generators, a_i and b_i, of the homology group $H_1(S, \mathbb{Z})$, as in the Corollary of Sect. 3.7. The integrals over the loops a_i are called the *A-periods;* those over the b_i are the *B-periods.* It is thus required to find a closed 1-form ω with any preassigned A- and B-periods; or equivalently: to find $2g$ closed forms $\omega_1, \ldots, \omega_{2g} \in A^1$ such that $\int_{a_j} \omega_i = \int_{b_j} \omega_{g+i} = \delta_{ij}$ and $\int_{a_j} \omega_{g+i} = \int_{b_j} \omega_i = 0$, for $i, j = 1, \ldots, g$. For instance, we can take $\omega_i = df_i$, $\omega_{g+i} = dg_i$, where f_i and g_i are functions jumping by 1 over the loop a_i, respectively b_i, and differentiable otherwise (see Springer [1957]).

Corollary 1. *For a compact Riemann surface S of genus g we have*

$$\dim H^1_{DR}(S) = 2g.$$

Corollary 2. *If c is a 1-cycle (for example, a loop) such that $\int_c \omega = 0$ for all closed forms ω, then it is homologous to zero.*

Poincaré duality (see Sect. 3.8) defines an integral, unimodular bilinear form

$$\cup \colon H^1(S, \mathbb{Z}) \times H^1(S, \mathbb{Z}) \to \mathbb{Z},$$

which is called the *cup-product.* It extends by $\mathbb{C}$-linearity to the cohomology space $H^1(S, \mathbb{C})$. On the other hand, there is a natural bilinear form on de Rham cohomology, namely:

$$\begin{gathered} (\ ,\ )_{DR} \colon H^1_{DR}(S) \times H^1_{DR}(S) \to \mathbb{C} \\ (\omega_1, \omega_2)_{DR} \overset{\text{def}}{=} \int_S \omega_1 \wedge \omega_2. \end{gathered}$$

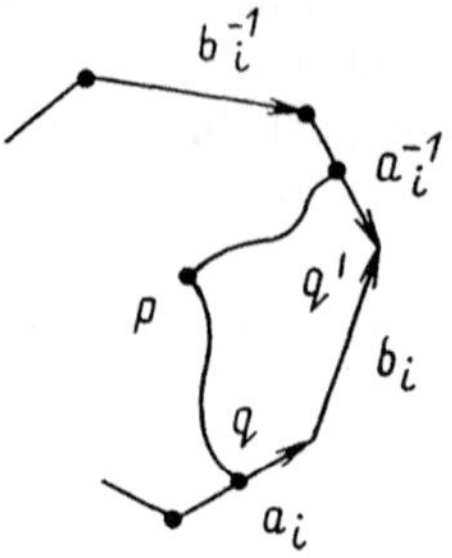

Fig. 15

Proposition. *The de Rham isomorphism* (4) *carries* $(\ ,\ )_{DR}$ *into the cup-product.*

This is readily seen on computing $(\omega_1, \omega_2)_{DR}$ via periods.

Lemma. *If* $\Pi_i^j \stackrel{\text{def}}{=} \int_{a_i} \omega_j$ *and* $\Pi_{g+i}^j \stackrel{\text{def}}{=} \int_{b_i} \omega_j$ *are, respectively, the A- and the B-periods of two closed forms* ω_j, $j = 1, 2$, *then*

$$(\omega_1, \omega_2)_{DR} = \sum_{i=1}^{g} (\Pi_i^1 \Pi_{g+i}^2 - \Pi_{g+i}^1 \Pi_i^2).$$

Let M be a development of the surface S, with symbol $a_1 b_1 a_1^{-1} b_1^{-1} \ldots a_g b_g a_g^{-1} b_g^{-1}$. Then ω_1 is exact on M, which is simply connected. We denote by $\pi(q)$ the primitive $\int_p^q \omega_1$ of ω_1, where p is the starting point, and q the end point, of an integration path on M. Then $\pi(q') - \pi(q) = \Pi_{g+i}^1$ (respectively, $\pi(q') - \pi(q) = -\Pi_i^1$), provided $q \in a_i$ and $q' \in a_i^{-1}$ (respectively, $q \in b_i$ and $q' \in b_i^{-1}$) glue into a single point of S (see Fig. 15). By the Green's formula, we have

$$\begin{aligned}(\omega_1, \omega_2)_{DR} &= \int_M \omega_1 \wedge \omega_2 = \int_{\partial M} \pi \omega_2 \\ &= \sum_{i=1}^{g} \left(\Pi_{g+i}^1 \int_{a_i^{-1}} \omega_2 - \Pi_i^1 \int_{b_i^{-1}} \omega_2 \right) = \sum_{i=1}^{g} (\Pi_i^1 \Pi_{g+i}^2 - \Pi_{g+i}^1 \Pi_i^2).\end{aligned}$$

For further details see Griffiths-Harris [1978].

4.7. Holomorphic Differentials; Geometric Genus; Riemann Bilinear Relations

Definition. A *differential* $\omega \in A^1$ on a Riemann surface S is said to be *holomorphic* if locally $\omega = g\,dz$, where g is a holomorphic function.

Holomorphic differentials form a complex space, denoted by Ω_S or simply Ω. Obviously, any holomorphic form is of type $(1,0)$ and is closed (even $\bar{\partial}$-closed: $\bar{\partial}\omega = 0$). The converse is also true (see Forster [1977]). On the other hand, the primitive of an exact holomorphic form (see Sect. 4.5) is holomorphic, and hence constant on a compact Riemann surface. Consequently, for a compact Riemann surface S, the natural map $\Omega \to H^1_{DR}(S)$ is injective and the space Ω of holomorphic differentials is finite-dimensional. Its (complex) dimension is known as the *(geometric) genus* of S.

Example 1. Let $f\colon S_1 \to S_2$ be a mapping of Riemann surfaces, and let ω be a holomorphic differential on S_2. The pull-back $f^*\omega$ is defined locally by the substitution $f^*\omega = g(f(z))\,df(z)$, where $w = f(z)$ is a local representation of f and $\omega = g\,dw$. The form $f^*\omega$ is well-defined and holomorphic. Thus the pull-back yields a $\mathbb{C}$-linear mapping $f^*\colon \Omega_{S_2} \to \Omega_{S_1}$, which is injective if f is nonconstant. This, in particular, gives us the Corollary of Sect. 3.6 for geometric genera.

Example 2. On an elliptic curve $\mathbb{C}/\Lambda$ there is a unique holomorphic form, up to multiplication by a constant. Indeed, the holomorphic differential dz on $\mathbb{C}$ is invariant under translations. So it induces a holomorphic differential ω on $\mathbb{C}/\Lambda$ with $\pi^*\omega = dz$, where $\pi\colon \mathbb{C} \to \mathbb{C}/\Lambda$ is the projection map. Any other holomorphic differential on $\mathbb{C}/\Lambda$ is of the form $f\omega$, where f is a holomorphic function on $\mathbb{C}/\Lambda$ and hence a constant. Thus the geometric genus of an elliptic curve is equal to 1.

The last example is remarkable in two respects. First, it shows that even on a compact Riemann surface there may exist some nonzero holomorphic differentials. Unfortunately, the obvious method of constructing such differentials as total differentials of holomorphic functions does not work, for lack of nonconstant holomorphic functions on a compact Riemann surface (cf. the argument following the Lemma in Sect. 4.8). Secondly, the example shows that the geometric and the topological genus of an elliptic curve coincide. As a matter of fact, this is true for any compact Riemann surface (cf. Example 1 of Sect. 4.8 and Corollary 1 in Sect. 4.13) and relates to the difficult and profound results on existence.

Let S be a compact Riemann surface of (topological) genus g, and let Π_i^* and Π_{g+i}^* be the A- and the B-periods of a holomorphic differential ω_*. From the Lemma of Sect. 4.6 we get:

Riemann's first bilinear relation.

$$\sum_{i=1}^{g} (\Pi_i^1 \Pi_{g+i}^2 - \Pi_{g+i}^1 \Pi_i^2) = 0.$$

Clearly, $\omega_1 \wedge \omega_2 = 0$ for $\omega_1, \omega_2 \in \Omega$.

Riemann's second bilinear relation.

$$\sqrt{-1}\sum_{i=1}^{g}(\Pi_i\overline{\Pi}_{g+i}-\Pi_{g+i}\overline{\Pi}_i)\geq 0.$$

If, locally, $\omega = f\,dz$ then $\sqrt{-1}\,\omega\wedge\overline{\omega} = 2\,|f|^2\,dx\wedge dy$. Furthermore, the expression can vanish only if $\omega = 0$.

Corollary. *If all the A-periods of a holomorphic form ω on S are zero, then $\omega = 0$.*

Riemann's bilinear relations play an important role in the theory of abelian varieties (see Example 2 in Chap. 3, Sect. 1.3).

4.8. Meromorphic Differentials; Canonical Divisors

Definitions. A *meromorphic differential* on a Riemann surface S is a holomorphic differential ω on some open subset $U \subset S$ whose complement $S - U$ is discrete in S, with the property that locally $\omega = f dz$, where f is a meromorphic function with poles in $S - U$. The points of $S - U$ are called the *poles* of ω. The *order* $\operatorname{ord}_p\omega$ of a nonzero meromorphic differential ω at a point $p \in S$ is defined locally by the formula $\operatorname{ord}_p\omega \stackrel{\text{def}}{=} \operatorname{ord}_p f$, where $\omega = f dz$. We say that $(\omega) \stackrel{\text{def}}{=} \sum \operatorname{ord}_p\omega . p$ is the *divisor* of the meromorphic differential ω. Divisors of the form (ω), where ω is a nonzero meromorphic differential on S, are called *canonical divisors* of the Riemann surface S and denoted by K_S, or simply K.

For example, dz is a meromorphic differential on the Riemann sphere $\mathbb{CP}^1$, and $(dz) = -2\,\infty$ is a canonical divisor.

Meromorphic differentials on a Riemann surface S form a complex space $\mathcal{M}^1(S)$. Multiplication by meromorphic functions makes $\mathcal{M}^1(S)$ into a vector space over the field of meromorphic functions $\mathcal{M}(S)$.

Lemma. *The dimension of $\mathcal{M}^1(S)$ over $\mathcal{M}(S)$ is at most equal to* 1. *More precisely, if ω is a nonzero meromorphic differential on S, then any other meromorphic differential on S is of the form $f\omega$, where $f \in \mathcal{M}(S)$. In this case, the dimension is equal to* 1.

In fact, equality holds in the lemma (see Corollary 6 in Sect. 4.14). But, to prove this, we must establish the existence of at least one nonzero meromorphic differential form on S. An obvious candidate is the total differential df of some nonconstant meromorphic function f on S, provided we can find one. However, we shall see later that it is more convenient to find meromorphic differentials than functions.

Corollary 1. *Any meromorphic differential on $\mathbb{CP}^1$ is rational, that is,* $\mathcal{M}^1(\mathbb{CP}^1) = \mathbb{C}(z)\,dz$ (cf. Example 1 in Sect. 2.2).

Corollary 2. *Let $f\colon S_1 \to S_2$ be a finite normal mapping of Riemann surfaces. Then $f^*\mathcal{M}^1(S_2) = \mathcal{M}^1(S_1)^{\operatorname{Aut} f}$, the invariant subspace of meromorphic differentials on S_1 under the action of* $\operatorname{Aut} f$. *Moreover, $f^*\Omega_{S_2} = \Omega_{S_1}^{\operatorname{Aut} f}$, that is, any holomorphic differential on S_1 which is invariant under* $\operatorname{Aut} f$, *is the pull-back of some holomorphic differential on S_2.*

Example 1. Let S be the hyperelliptic Riemann surface, of genus g (say), of the algebraic function $\sqrt{f}$, where f is a polynomial with pairwise distinct roots. Let $\gamma\colon S \to \mathbb{CP}^1$ be the hyperelliptic projection, and j the involution, as in the Example of Sect. 2.11. Consider the two-valued function $\sqrt{f}$ as a single-valued function on S. By construction, $\sqrt{f}$ is anti-invariant with respect to j: $j^*\sqrt{f} = -\sqrt{f}$. Let z be an affine coordinate on $\mathbb{CP}^1$. We assert that the differentials $\gamma^*(z^i dz)/\sqrt{f}$, where $i = 0, \ldots, g-1$, are holomorphic on S and form a basis for the space Ω. Indeed, any holomorphic differential on S is anti-invariant with respect to j. So it is of the form $\gamma^*(g(z)dz)/\sqrt{f}$, where $g(z)$ is a rational function. Now, a differential of this form is holomorphic precisely when $g(z)$ is a polynomial of degree $\leq g-1$. For particulars, see Griffiths-Harris [1978].

Example 2. Let $C \subset \mathbb{CP}^2$ be a nonsingular algebraic curve, given by an irreducible polynomial $f(x_0, x_1, x_2)$ of degree d. In affine coordinates $z_1 = x_0/x_2$, $z_2 = x_1/x_2$, the curve is given by the equation $F(z_1, z_2) = f(z_1, z_2, 1) = 0$. It follows from the above lemma and from Corollary 6 of Sect. 2.11 that all meromorphic differentials on the Riemann surface of C are rational. This means that any such differential is of the form $g(z_1, z_2)dz_1 + h(z_1, z_2)dz_2$, where g and h are rational functions of z_1 and z_2 (and z_1, z_2 are the restrictions of the coordinates z_1, z_2 to C). Moreover, every holomorphic differential can be written in the form

$$\omega = g(z_1, z_2)\,\frac{dz_2}{(\partial F/\partial z_1)(z_1, z_2)} = -g(z_1, z_2)\,\frac{dz_1}{(\partial F/\partial z_2)(z_1, z_2)},$$

where g is a polynomial of degree $\leq d-3$ (cf. Griffiths-Harris [1978]). Further, such a differential ω is evidently holomorphic in the domain where z_1 and z_2 are. When going over to other affine coordinates, ω preserves its shape. Hence it is holomorphic everywhere on C. Thus the genus of the Riemann surface of a nonsingular curve C is equal to $\frac{1}{2}(d-1)(d-2)$, which is the dimension of the space of polynomials in two variables of degree $\leq d-3$ (cf. Chap. 2, Sect. 3.12). In particular, a *cubic curve* ($d = 3$) is of genus 1.

Any two canonical divisors are in the same linear equivalence class (see Sect. 2.6). This is called the *canonical class.* Indeed, $(f\omega) = (f) + (\omega)$.

Hurwitz formula for canonical divisors. *If $f\colon S_1 \to S_2$ is a nonconstant mapping of Riemann surfaces, then $K_{S_1} \sim f^*K_{S_2} + R$, where R is the ramification divisor of f. More precisely, if $0 \neq \omega \in \mathcal{M}^1(S_2)$ then we have: $(f^*\omega) = f^*(\omega) + R$.*

The last formula reduces locally to $dz^n = nz^{n-1}dz$ (see the Lemma of Sect. 2.4). By virtue of Corollary 2 in Sect. 2.7, all canonical divisors on a compact Riemann surface S have the same degree. For example, a canonical divisor on the Riemann sphere $\mathbb{CP}^1$ is of degree -2, since $(dz) = -2\infty$. For a general compact Riemann surface, the degree of a canonical divisor is expressible in terms of the genus: $\deg K = 2g - 2$ (see Corollary 8 in Sect. 4.14). Hence it is a topological invariant of the surface. This is easy to obtain from the following numerical fact, ... up to the difficult statement about the existence of a nonconstant meromorphic function.

Hurwitz formula for the degree of a canonical divisor. *Let $f\colon S_1 \to S_2$ be a nonconstant mapping of compact Riemann surfaces. Then*

$$\deg K_{S_1} = \deg f \cdot \deg K_{S_2} + \deg R.$$

This is obvious from the preceding formula and the relation $\deg f^* K_{S_2} = \deg f \cdot \deg K_{S_2}$.

Remark 1. The existence of a canonical divisor on an arbitrary Riemann surface is non-obvious even in the compact case (cf. the above Lemma and Corollary 6 of Sect. 4.14).

Remark 2. Finding indefinite integrals $\int R(z, w(z))\,dz$, where R is a rational function and w an algebraic function, is one of the traditional problems in calculus. In terms of Riemann surface theory, this problem (to be precise, its complex version) amounts to discovering in general a multivalued primitive of the meromorphic differential $R(z, w)\,dz$ on the Riemann surface S of the algebraic function w. If S is isomorphic to $\mathbb{CP}^1$ then, by Corollary 1, this differential is rational and its primitive can be found by reduction to a sum of partial fractions. This principle underlies all known methods for making an integral rational by means of some algebraic irrationalities. For example, Euler's substitution $w = \sqrt{az^2 + bz + c}$ comes from an isomorphism of the closure of the conic $w^2 = az^2 + bz + c$ in $\mathbb{CP}^2$ with $\mathbb{CP}^1$ (see Example 4 in Sect. 1.6).

4.9. Meromorphic Differentials with Prescribed Behaviour at Poles; Residues. A *differential principal part* is a sum of the form $\omega_p = \sum_{i=-n}^{-1} a_i z^i\,dz$, where p is a point on a Riemann surface S and z is a local parameter at p. The principal part of a meromorphic differential at any point $p \in S$ is defined by means of Laurent series (cf. Sect. 2.3).

Mittag-Leffler's problem for meromorphic differentials. *Let $\{\omega_p\}$ be a set of differential principal parts given at a discrete set of points p on a Riemann surface S. It is required to find a meromorphic differential $\omega \in \mathcal{M}^1(S)$ with precisely this set of poles p and with the specified principal parts.*

Definition. Let $\sum_{i=-n}^{-1} a_i z^i \, dz$ be the principal part of a meromorphic differential ω at a point $p \in S$. The coefficient a_{-1} is called the *residue* of ω at p and denoted by $\operatorname{Res}_p \omega$.

$\operatorname{Res}_p \omega$ is independent of the choice of a local parameter z at p, a fact that can be proved by purely algebraic methods (see Forster [1977]). This can also be seen from the integral representation of the residue:

$$\operatorname{Res}_p \omega = \frac{1}{2\pi\sqrt{-1}} \int_u \omega,$$

where u is a small positive loop around the point $p \in S$.

Lemma (the Residue Theorem). *Let ω be a meromorphic differential on a compact Riemann surface S. Then $\sum \operatorname{Res}_p \omega = 0$.*

Example. Let f be a nonconstant meromorphic function on a compact Riemann surface S. Then $\operatorname{Res}_p(df/f) = \operatorname{ord}_p f$ and, by the lemma, $\deg(f) = \sum \operatorname{ord}_p f = 0$ (cf. Corollary 2 in Sect. 2.7).

The lemma can be proved directly from the Green's formula and the integral representation of the residue.

Necessary conditions for the solution of Mittag-Leffler's problems. *On a compact Riemann surface,*

(a) *Mittag-Leffler's problem for a set $\{\omega_p\}$ of differential principal parts can have a solution only if $\sum \operatorname{Res}_p \omega_p = 0$;*

(b) *Mittag-Leffler's problem for a set $\{f_p\}$ of functional principal parts can have a solution only if $\sum \operatorname{Res}_p(f_p \omega) = 0$ for all $\omega \in \Omega$.*

Remark 1. If $\omega_1, \ldots, \omega_g$ is a basis for Ω, then $\sum \operatorname{Res}_p(f_p \omega) = 0$ for all $\omega \in \Omega$ if and only if $\sum \operatorname{Res}_p(f_p \omega_i) = 0$ for $i = 1, \ldots, g$. Thus condition (b) on the system $\{f_p\}$ is equivalent to a set of g linear equations on the coefficients of the principal parts, where g is the genus of S.

Remark 2. The analogues of Mittag-Leffler's problems for higher-dimensional complex manifolds (Cousin's first and second problems) have played a vital role for the development of cohomological methods (see Gunning-Rossi [1965]).

4.10. Periods of Meromorphic Differentials

Definition. A meromorphic differential ω on a Riemann surface S is said to be

(1) *of the first kind* if it is holomorphic;

(2) *of the second kind* if all its residues are trivial;

(3) *of the third kind* if all its poles p are of order one, that is, $\operatorname{ord}_p \omega = -1$.

Meromorphic differentials are closed in any region where they are holomorphic. Besides, the integral over a loop of a differential ω of the (first and) second kind depends as much on the homotopy class of that loop on S as on $S - \{\text{poles of } \omega\}$. For example, $\int_u \omega = 2\pi\sqrt{-1}\,\mathrm{Res}_p\,\omega = 0$ for any small circuit u around a pole p. If a differential of the second kind is exact then its primitive is meromorphic. Conversely, the differential df of a meromorphic function f is of the second kind. It is natural to ask whether there exists a differential of the second kind with specified periods. The answer is simple: such a differential exists for any preassigned periods. Yet, why it does will become clear only in Sect. 6.3.

4.11. Harmonic Differentials. Since the time of Riemann, the solution of existence problems has been reduced to results from the theory of partial differential equations, particularly to those on elliptic type equations. Now, this is often obscured by a façade of cohomological constructions. In dealing with these problems we must consider real independent variables, which leads us to questions about the existence of harmonic functions and differentials with prescribed singularities or periods. To introduce some symmetry in the real x- and y-coordinates, where $z = x + \sqrt{-1}\,y$ is a local coordinate on a Riemann surface S, it is convenient to use the $\mathbb{C}$-linear conjugation operator $*\colon A^1 \to A^1$. Locally, $*(p\,dx + q\,dy) \stackrel{\text{def}}{=} -q\,dx + p\,dy$ or, equivalently, $*(f\,dz + g\,d\bar{z}) \stackrel{\text{def}}{=} \sqrt{-1}\,(-f\,dz + g\,d\bar{z})$. The operator $*$ is well-defined. Furthermore,

Lemma. *If* $\omega = \omega_1 + \omega_2 \in A^1$, *where* $\omega_1 \in A^{1,0}$ *and* $\omega_2 \in A^{0,1}$, *then*

$$*\omega = \sqrt{-1}\,(\omega_2 - \omega_1).$$

Definition. We say that $\omega \in A^1$ is a *harmonic differential* if

$$d\omega = d{*}\omega = 0.$$

The operator $d*$ is called a coclosure, and we say that ω is *coclosed* if $d{*}\omega = 0$.

Locally, $\omega = f\,dz + g\,d\bar{z}$ is closed if $\dfrac{\partial g}{\partial z} = \dfrac{\partial f}{\partial \bar{z}}$, and coclosed if $\dfrac{\partial g}{\partial z} = -\dfrac{\partial f}{\partial \bar{z}}$. Therefore, ω is harmonic if and only if $\dfrac{\partial f}{\partial \bar{z}} = \dfrac{\partial g}{\partial z} = 0$, that is, if and only if f is holomorphic and g antiholomorphic.

Proposition. $H^1 = \Omega \oplus \overline{\Omega}$, *where* H^1 *and* $\overline{\Omega} = \{\overline{\omega} \mid \omega \in \Omega\}$ *are, respectively, the spaces of harmonic and antiholomorphic differentials on the Riemann surface* S.

Remark. A differentiable function f on a Riemann surface is said to be harmonic if it is so locally: $\Delta f = 0$, where $\Delta = \dfrac{\partial^2}{\partial x^2} + \dfrac{\partial^2}{\partial y^2} = 4\dfrac{\partial}{\partial z}\cdot\dfrac{\partial}{\partial \bar{z}}$ is the (local) Laplace operator. It is easy to verify that the harmonicity of a function does not depend on the choice of a local coordinate $z = x + \sqrt{-1}\,y$. There is a close relationship between the harmonicity of a form and that of a function. In fact, a function f is harmonic if and only if its total differential, df, is. Moreover, a 1-form ω is harmonic if and only if, locally, $\omega = df$ for some harmonic function f. This significantly differs from a harmonic function viewed as a 0-form: any such 0-form is constant. This is just the reason why it is easier to look for meromorphic (read: harmonic) differentials rather than for functions.

4.12. Hilbert Space of Differentials; Harmonic Projection. Harmonic functions can be constructed by orthogonal projection to the harmonic component. The main result to be used below for solving existence problems is the orthogonal decomposition theorem.

Let ω_1 and ω_2 be two 1-forms with compact supports on a Riemann surface S. One defines a Hermitian inner product as follows:

$$(\omega_1, \omega_2) \stackrel{\text{def}}{=} \int_S \omega_1 \wedge *\bar{\omega}_2.$$

In fact, locally we have $\omega_1 \wedge *\bar{\omega}_2 = (p_1\bar{p}_2 + q_1\bar{q}_2)dx \wedge dy$, where $\omega_i = p_i dx + q_i dy$ for $i = 1, 2$. Hence the differentiable 1-forms with compact supports form a unitary space with norm $\|\omega\| = \sqrt{(\omega, \omega)}$. For a compact surface, this space contains all differentiable, and hence all harmonic forms. By contrast, in the noncompact case there are plenty of forms with compact support, but any harmonic and, in particular, any holomorphic form with compact support is trivial, by the uniqueness theorem. However, one has to consider some harmonic forms in the noncompact case as well, for example when constructing meromorphic differentials with prescribed singularities on a compact Riemann surface (see Sect. 4.14). These are bounded forms $\omega \in A^1$ such that $\|\omega\|^2 \stackrel{\text{def}}{=} \int_S \omega \wedge *\bar{\omega} < +\infty$. (This improper integral can be defined, for instance, as follows: $\int_S \stackrel{\text{def}}{=} \sup_U \int_U$, where U ranges over all open subsets of S with compact closure.) The bounded differentiable 1-forms make a unitary space B^1, which includes all 1-forms with compact support. On a compact Riemann surface, every differentiable 1-form is bounded and its support is compact. We denote by E the closure in B^1 of the space of exact differentials $d\varphi$ of all functions φ on S with compact support.

Theorem (on orthogonal decomposition). *Let $\omega \in B^1$ be a bounded differentiable 1-form on a Riemann surface S. There is a unique decomposition $\omega = \omega_h + df + *dg$, where ω_h is a bounded harmonic differential on S, and $f, g \in A^0$ and $df, dg \in E$.*

First of all, we observe that E and $*E$ are orthogonal to each other. Indeed, it is enough to check that, if φ and ψ are two differentiable functions with compact support then their differentials $d\varphi$ and $*d\bar{\psi}$ are orthogonal:

$$(d\varphi, *d\bar{\psi}) = \int_S d\varphi \wedge **d\psi = -\int_S d\varphi \wedge d\psi = -\int_S \psi\, dd\varphi + \int_S d(\psi\, d\varphi) = 0,$$

by the Corollary of Sect. 4.5. Now the essential point of the proof is that the space H of bounded harmonic differentials is orthogonal to both E and $*E$. More precisely, saying that a differential $\omega \in B^1$ is closed means that it is orthogonal to $*E$, and saying that it is coclosed means that it is orthogonal to E. For example, the latter statement follows from the fact that $d*$ is the conjugate of d:

$$(d*\omega, \varphi) \stackrel{\text{def}}{=} \int_S d*\omega \wedge \bar{\varphi} = \int_S d\bar{\varphi} \wedge *\omega = (d\bar{\varphi}, \overline{\omega}) = (\omega, d\varphi),$$

where $\omega \in B^1$ and φ is a differentiable function with compact support. Hence H is orthogonal to $E \oplus *E$. If we could show that $B^1 = H \oplus E \oplus *E$, we would immediately obtain the required orthogonal decomposition of $\omega \in B^1$. However, though at first sight it looks like a mere formality, a very fundamental difficulty arises at this point, namely the incompleteness of B^1. To overcome this problem, we proceed to completions with respect to $\| \ \|$: L^1 is the completion of B^1, $\check{E}$ of E, and $*\check{E}$ of $*E$ (since the $*$-operator is bounded). As before, the inner product on L^1 is denoted by $(\ , \ )$ and the norm by $\| \ \|$. Similarly, for every open subset $U \subset S$, we can define the space $L^1_U \supset B^1_U$, together with the inner product $(\ , \ )_U$ and the norm $\| \ \|_U$. The elements of L^1 may be interpreted as classes of measurable differentials ω on S such that the differential $\omega \wedge *\overline{\omega}$ is Lebesgue-integrable and $\int_S \omega \wedge *\overline{\omega} = \|\omega\|^2 < +\infty$ (cf. Springer [1957]). They may also be regarded as 1-currents, that is (cf. Griffiths-Harris [1978]), locally $L^1_U \ni l|_U = f\,dz + g\,d\bar{z}$, where z is a local coordinate, and f and g are complex-valued distributions (generalized functions) on U. Recall that, by a subspace of a Hilbert space, we mean a closed vector subspace. A finite sum of subspaces and the orthogonal complement of a subspace are again subspaces. Now the crux of the argument is the following

Theorem (on regularity). $H = (\check{E} \oplus *\check{E})^\perp$.

This truly remarkable fact indicates that completion does not make the space of harmonic differentials any larger. If the elements of L^1 are interpreted as classes of measurable 1-forms on S, then we have to establish that each element h of the orthogonal complement $(\check{E} \oplus *\check{E})^\perp$ is the class of a unique harmonic differential ω on S with $\|\omega\| < +\infty$. (The other inclusion, $H \subset (\check{E} \oplus *\check{E})^\perp$, is clear from the construction!) In terms of currents, the meaning is that h is the current of some form ω. This fact is local, so it may be assumed that $S = \mathbb{D}$ is the unit disk. In this case $h = f\,dz + g\,d\bar{z}$, where f and g are distributions on $\mathbb{D}$. By hypothesis, $(h, d\varphi)_\mathbb{D} = (h, *d\psi)_\mathbb{D} = 0$ for any differentiable functions φ and ψ with compact supports on $\mathbb{D}$. Using the differentiation of distributions (respectively, of currents), the first equation can be rewritten as $\dfrac{\partial g}{\partial z} = -\dfrac{\partial f}{\partial \bar{z}}$ (the property of being coclosed); the other one as $\dfrac{\partial g}{\partial z} = \dfrac{\partial f}{\partial \bar{z}}$ (the property of being closed). See the above reasoning about the orthogonality of harmonic forms to both E and $*E$. It follows that

$\dfrac{\partial f}{\partial \bar{z}} = \dfrac{\partial g}{\partial z} = 0$ in the sense of distribution theory. Of course, if f and g were differentiable functions, this would imply the harmonicity of $h = f\,dz + g\,d\bar{z}$. But, as compared with our preceding considerations, the question of differentiability of f and g is not purely formal and constitutes the essence of the regularity theorem. The solution of this problem involves the general regularity property of elliptic differential operators (see Griffiths-Harris [1978]). A first result in this direction, which will also suffice for our purposes is:

Weyl's lemma (cf. Forster [1977]). *Any distribution T satisfying the Laplace equation $\Delta T = 0$, is the distribution of some differentiable function.*

A proof can be found in standard books on partial differential equations (cf. Petrovskiĭ [1961]). In view of the relation $\dfrac{\partial f}{\partial \bar{z}} = \dfrac{\partial g}{\partial z} = 0$, the distributions f and g are solutions of the Laplace equation $\Delta f = \Delta g = 0$, and hence differentiable. This completes the proof of the regularity theorem.

Corollary. $L^1 = H \oplus \check{E} \oplus *\check{E}$.

In particular, every bounded differentiable form $\omega \in B^1$ has a unique orthogonal decomposition $\omega = \omega_h + \gamma + *\pi$, where $\omega_h \in H$ and $\gamma, \pi \in \check{E}$. The differential ω_h is bounded and is called the *harmonic projection* of ω. The orthogonal decomposition theorem asserts that the elements γ and π 'are' exact differentiable 1-forms. Differentiability is proved locally, just like the regularity theorem (cf. Springer [1957] for measurable differentials). (*Caution:* The orthogonal decomposition is not preserved under restriction, but the components of the decomposition differ by some harmonic forms.) To check that γ and π are exact differentials, one uses the following fact. For every smooth loop u on S there exists a (closed) differential $\eta_u \in A^1$ with compact support (whence $\eta_u \in B^1$) such that, for any form $\gamma \in A^1$, we have:

$$\int_u \gamma = \int_S \gamma \wedge *\bar{\eta}_u = (\gamma, \eta_u).$$

Therefore, by the known criterion, exact differentials remain exact in the limit in L^1. The differential η_u can be constructed as the exact differential of a function jumping by 1 along u (cf. Sect. 4.6). For a detailed treatment, see Springer [1957].

The main difficulties overcome, we may now proceed to the long-awaited applications.

4.13. Hodge Decomposition. Suppose S is a compact Riemann surface. Then $A^1 = B^1 \subset L^1$, and $H = H^1$ is the space of harmonic forms on S. According to the foregoing subsection, saying that a form is closed amounts to saying that it is orthogonal to $*E$. Hence, by the orthogonal decomposition theorem, a closed form $\omega \in A^1$ on S can be uniquely represented as $\omega = \omega_h + df$, where $\omega_h \in H^1$. Thus every one-dimensional de Rham cohomology class has a unique harmonic representative. This fact, together with the Proposition of Sect. 4.11, yields the following well-known result on the decomposition of cohomology classes. For convenience, we consider de Rham cohomology.

Theorem (Hodge). $H^1_{DR}(S) = \Omega \oplus \overline{\Omega}$.

From the combinatorial definition of cohomology and de Rham's theorem, we obtain:

Corollary 1. *For a compact Riemann surface S of topological genus g, we have* $\dim \Omega = g$. *Thus the geometric and the topological genus are always equal.*

Another, essentially similar, approach to this equality is related to the Riemann-Roch formula (see Remark 2 of Sect. 6.2). In what follows, g will denote the genus of S (all definitions being equivalent).

Corollary 2. *If c is an integral (or real) 1-cycle (for example, a loop) such that $\int_c \omega = 0$ for every $\omega \in \Omega$, then it is homologous to zero.*

Cf. Corollary 2 of Sect. 4.6.

Corollary 3. *There exists a holomorphic form ω on S with any preassigned A-periods $\Pi_i \in \mathbb{C}$.*

4.14. Existence of Meromorphic Differentials and Functions. With the theorem on partitions of unity or, more simply, by means of smoothing functions, we can construct a wealth of differentiable 1-forms with compact supports. The harmonic projections of these forms are harmonic, and their (1,0)-components are holomorphic (see Sections 4.2 and 4.11). If we want to obtain a nontrivial form, we must not start from a topologically trivial 1-form. In the compact case, the form is chosen in a nontrivial de Rham cohomology class (cf. the Hodge theorem). In the noncompact case, one prescribes the type of singularity (a principal part, say) of the differential at some point. Thus, let S be an arbitrary Riemann surface, and let z be a local parameter at a point $p \in S$.

Riemann's existence theorem (for harmonic differentials). *For every $n \geq 1$ on $S - p$, there exists an exact harmonic differential ω such that*
(a) *the differential* $\omega - d(1/z^n) = \omega + (n/z^{n+1})dz$ *is harmonic in some neighbourhood U of p; and*
(b) $\omega \in B^1_{S-\overline{U}}$, *that is,* $\|\omega\|_{S-\overline{U}} < +\infty$.

In accordance with the above considerations, let $\rho(z)$ be a differentiable function on S such that $\rho \equiv 0$ outside U and $\rho \equiv 1$ in some smaller neighbourhood of p. The form $\psi = d(\rho(z)/z^n)$ is meromorphic in a neighbourhood of p, with p as its only pole. Now the form $\psi - \sqrt{-1} * \psi$ is differentiable and has a compact support on S (for a suitable choice of U). By the orthogonal decomposition theorem, $\psi - \sqrt{-1} * \psi = \omega_h + df + *dg$, where ω_h is a harmonic form on S, while f and g are differentiable functions. It can be shown that the differential $\omega = \psi - df = \sqrt{-1} * \psi + \omega_h + *dg$ fulfils the conditions of the theorem (see Springer [1957]).

Corollary 1. *On a Riemann surface S, there exist differentials of the second kind with any preassigned finite set of poles p and any principal parts $\omega_p = \sum_{i=-n}^{-2} a_i z^i \, dz$, where $n \geq 2$.*

For instance, let ω be a harmonic differential as in the theorem. Then $\frac{1}{2}(\omega + \sqrt{-1} * \omega)$ is a differential of the second kind with principal part $(n/z^{n+1})dz$ at its single pole p.

On taking ratios of differentials of the second kind, one can establish the existence of various meromorphic functions. In particular,

Corollary 2. *On a Riemann surface S, there exists a meromorphic function that takes on any preassigned values at a finite set of points.*

Now we can strengthen some of the results of Sect. 2.11.

Corollary 3. *If $f: S_1 \to S_2$ is a finite mapping of Riemann surfaces, then the degree of the extension $f^*: \mathcal{M}(S_2) \hookrightarrow \mathcal{M}(S_1)$ is equal to* $\deg f$.

To show this, it is enough to find a function $g \in \mathcal{M}(S_1)$ that takes on pairwise distinct values $a_i = g(y_i)$, for $i = 1, \dots, \deg f$, at the points of a generic fibre $f^{-1}(p) = \{y_i\}$.

Since meromorphic functions separate points, we also get:

Corollary 4. *The nonconstant mappings of Riemann surfaces $f: S_1 \to S_2$ and the extensions of meromorphic function fields $f^*: \mathcal{M}(S_2) \hookrightarrow \mathcal{M}(S_1)$ are in one-to-one correspondence.*

Corollary 5. *Let S_2 be a Riemann surface and let $\varphi: \mathcal{M}(S_2) \hookrightarrow K$ be a finite extension. Then there is a unique finite mapping of Riemann surfaces, $f: S_1 \to S_2$, such that the inclusion map $f^*: \mathcal{M}(S_2) \hookrightarrow \mathcal{M}(S_1)$ is isomorphic to φ.* (cf. Theorem 2 of Sect. 2.11).

Example. Uniqueness allows one, in particular, to impart a geometrical meaning to hyperellipticity: a compact Riemann surface S with a mapping $\gamma: S \to \mathbb{CP}^1$ of degree 2 is hyperelliptic. Indeed, any quadratic extension of the field $\mathbb{C}(z)$ can be obtained by adjoining an algebraic function $\sqrt{f}$, where $f \in \mathbb{C}(z)$ is a polynomial with only simple roots. It follows also that a hyperelliptic Riemann surface is uniquely determined by the images of its ramification points.

Corollary 6. *For every Riemann surface S, we have* $\dim_{\mathcal{M}(S)} \mathcal{M}^1(S) = 1$. *So there exists a canonical divisor.*

To establish the following two important results, it suffices to show the existence of at least one nonconstant meromorphic function.

Corollary 7. *The field $\mathcal{M}(S)$ of meromorphic functions on a compact Riemann surface S is finitely generated over $\mathbb{C}$, with transcendence degree* 1.

Corollary 8 (Riemann-Hurwitz formula). *Let S be a compact Riemann surface of (topological) genus g. Then* $\deg K = -\chi(S) = 2g - 2$.

According to Sect. 4.8, this is true for the Riemann sphere. The general case can be deduced from Hurwitz's formulae for the degree of the canonical class (see Sect. 4.8) and for the Euler characteristic (see Sect. 3.6), as applied to a nonconstant mapping (meromorphic function) $f\colon S \to \mathbb{CP}^1$.

Remark. The relation $\deg K = -\chi(S)$ is dual to the Hopf theorem on the index of a vector field on a surface (see Dubrovin, Novikov & Fomenko [1979]). Another approach to the Riemann-Hurwitz formula, in which g is viewed as the geometric genus, is discussed in Remark 2 of Sect. 6.2 (see also Chap. 2, Sect. 2.9).

The proof of the remaining results does not require any new ideas.

Corollary 9. *Let $f\colon S_1 \to S_2$ be a finite mapping of Riemann surfaces. The functor $\sigma \mapsto \sigma^*$ is a contravariant*[3] *isomorphism of* $\operatorname{Aut} f$ *with the automorphism group of the associated field extension*

$$\operatorname{Aut}(f^*\colon \mathcal{M}(S_2) \hookrightarrow \mathcal{M}(S_1)) \stackrel{\text{def}}{=} \{\varphi \in \operatorname{Aut}(\mathcal{M}(S_1)) \mid \varphi \circ f^* = f^*\}.$$

It is convenient to first prove a slightly more general statement.

Corollary 10. *Let $f_1, f_2\colon S_1, S_2 \to S$ be finite mappings of Riemann surfaces, and let $\varphi\colon \mathcal{M}(S_2) \to \mathcal{M}(S_1)$ be a field isomorphism over $\mathcal{M}(S)$. That is, $\varphi \circ f_2^* = f_1^*$. Then there is a unique isomorphism $f\colon S_1 \to S_2$ over S (that is, $f_2 \circ f = f_1$) such that $f^* = \varphi$.*

Corollary 11. *A finite mapping of Riemann surfaces $f\colon S_1 \to S_2$ is normal if and only if the extension $f^*\colon \mathcal{M}(S_2) \to \mathcal{M}(S_1)$ is normal. In this case, $f^*\mathcal{M}(S_2) = \mathcal{M}(S_1)^{\operatorname{Aut} f}$ is the field of meromorphic functions on S_1 which are invariant under* $\operatorname{Aut} f$ (cf. Corollary 2 of Sect. 4.8).

Corollary 12. *Let $\varphi\colon \mathcal{M}(S_2) \xrightarrow{\sim} \mathcal{M}(S_1)$ be a $\mathbb{C}$-isomorphism of the meromorphic function fields of two Riemann surfaces S_1 and S_2. Then there is a unique isomorphism $f\colon S_1 \to S_2$ such that $f^* = \varphi$. In particular, a Riemann surface which is a model of a finitely generated $\mathbb{C}$-field of transcendence degree 1 is uniquely determined up to isomorphism.*

A compact Riemann surface whose field of meromorphic functions is isomorphic to the purely transcendental extension $\mathbb{C}(z)$ of $\mathbb{C}$, is said to be *rational*. Corollary 12 shows, in particular, that any rational Riemann surface is isomorphic to the Riemann sphere $\mathbb{CP}^1$.

[3] *Contravariant* means that $(\sigma \circ \delta)^* = \delta^* \circ \sigma^*$.

Corollary 13. *There is a natural contravariant isomorphism* $\sigma \mapsto \sigma^*$ *from the automorphism group* $\operatorname{Aut} S$ *of a compact Riemann surface* S *to the group* $\mathcal{M}(S)$ *of* $\mathbb{C}$*-automorphisms of its field of meromorphic functions.*

4.15. Dirichlet's Principle. It is easy to obtain uniqueness in Riemann's existence theorem for harmonic differentials. One further condition needs to be introduced, namely:

(c) $(\omega, dh) = 0$ *for any exact differential* $dh \in A^1$ *such that* $\|dh\| < +\infty$ *and* $dh \equiv 0$ *in some neighbourhood of* p.

For finding such a harmonic differential ω, the decomposition $H \oplus \check{E}$ is replaced by $\widetilde{H} \oplus \check{\widetilde{E}}$, where $\widetilde{E}$ is the space of bounded exact differentials $dh \in B^1$, and $\widetilde{H} = (\check{\widetilde{E}} \oplus *\check{E})^{\perp}$. Clearly, $E \subset \widetilde{E} \subset E + H$ and $\widetilde{H} \subset H$. Then the orthogonal decomposition becomes $\psi - \sqrt{-1} * \psi = \widetilde{\omega}_h + d\tilde{f} + *dg$, with $\widetilde{\omega}_h \in \widetilde{H}$ and $d\tilde{f} \in \widetilde{E}$. The differential $\omega = \psi - d\tilde{f}$ satisfies all the conditions (a), (b), and (c), and it is uniquely determined by them (cf. Springer [1957]). Condition (c) can be rephrased as follows. Suppose $dh \equiv 0$ in some neighbourhood N of p. Then

$$\begin{aligned}\|\omega + dh\|^2_{S-\overline{N}} &= (\omega, \omega)_{S-\overline{N}} + (dh, dh) + (\omega, dh) + \overline{(\omega, dh)} \\ &= \|\omega\|^2_{S-\overline{N}} + \|dh\|^2 \geq \|\omega\|^2_{S-\overline{N}}\end{aligned}$$

Thus the harmonic differential ω minimizes $\| \ \|_{S-\overline{N}}$ in the class of all differentials $\omega + dh$ such that $dh \equiv 0$ in N. This fact is called Dirichlet's principle. It obviously yields the uniqueness of ω. The crux of the matter is then the existence of ω, to which Riemann did not pay due regard. He was criticized for that by Weierstrass. In the compact case the first rigorous solutions of the existence problems are due to Schwarz and Neumann. Further development of the methods can be credited to Poincaré, Hilbert, Klein, and Koebe. The method of orthogonal decomposition was suggested by Weil in 1940, but its basic ideas go back to Riemann.

§ 5. Classification of Riemann Surfaces

The Corollary of Sect. 3.3 contains a topological classification of compact orientable surfaces: any compact Riemann (read: orientable) surface of genus g is homeomorphic to a sphere with g handles. A similar situation obtains when classifying differentiable surfaces (see Hirsch [1976]). But how many Riemann surfaces of genus g are there, up to isomorphism? Or, equivalently, how many analytic structures can be introduced on a sphere with g handles? This is a much more delicate question. That the Riemann sphere is the only Riemann surface of genus 0 is not typical of other genera. As a matter of fact, there are continuous families of Riemann surfaces of genus $g \geq 1$ whose members are non-isomorphic. This is the fundamental difference between the geometry of analytic manifolds, on the one hand, and differential geometry or topology, on the other. Besides discrete topological invariants (like the genus or the degree of a canonical divisor), Riemann surfaces must then possess some continuous invariants or, in coordinate form, some continuous parameters, called *moduli*. A universal example – the matrix of periods or, in invariant form, the Jacobian of a Riemann surface – is discussed at length in Chapter 3. The approach taken in this section, to the classification of Riemann surfaces, makes use of the universal cover. Its idea has been formulated in Sect. 2.9. The reader who wishes to go deeper into the classification problems, especially for noncompact Riemann surfaces, is referred to Sario-Nakai [1970].

5.1. Canonical Regions. There are two types of connected, simply connected surfaces up to homeomorphism and diffeomorphism: the real plane (noncompact case) and the two-dimensional sphere (compact case). Rather surprisingly, the classification of simply connected Riemann surfaces is not much more complicated.

Riemann mapping theorem. *A simply connected Riemann surface is isomorphic to one of the following regions of the Riemann sphere:* $\mathbb{CP}^1$, $\mathbb{C}$, *or* $\mathbb{H}$.

These are called *canonical regions,* though the choice of the last of them is not quite unique (cf. Springer [1957]). It is sometimes convenient to replace the upper half-plane $\mathbb{H}$ by the isomorphic disk $\mathbb{D}$. Thus, to establish that $\mathbb{C}$ and $\mathbb{H}$ are non-isomorphic, it is simpler to use $\mathbb{D}$ and Liouville's theorem.

Lemma. *The canonical regions* $\mathbb{CP}^1$, $\mathbb{C}$, *and* $\mathbb{H}$ *are pairwise non-isomorphic.*

The remaining cases are obvious from topological considerations. This yields a topological characterization of the rationality of a Riemann surface (cf. Sect. 4.14).

Corollary. *A Riemann surface is rational if and only if it is homeomorphic to a sphere or, equivalently, if it is compact and of genus* 0.

5.2. Uniformization. Given a multivalued analytic function $f(w)$ of one variable, we wish to find two single-valued analytic functions $\varphi(z)$ and $\psi(z)$, meromorphic in some region of the Riemann sphere, such that $\psi(z) = f(\varphi(z))$ and the image of $\varphi(z)$ is an everywhere dense subset of the domain of f.

Example 1. The function w^α for any complex α can be uniformized through the functions $\varphi(z) = e^z$ and $\psi(z) = e^{\alpha z}$.

Example 2. The function $\sqrt{1-z^2}$ is uniformized through $\varphi(z) = \sin z$ and $\psi(z) = \cos z$, or through $\varphi(z) = 2z/(1+z^2)$ and $\psi(z) = (1-z^2)/(1+z^2)$.

According to the general uniformization theorem – due to Klein, Poincaré, and Kœbe –, for any multivalued function $f(w)$ there exists a uniformization such that the domain of φ and ψ is equal to $\mathbb{CP}^1$, $\mathbb{C}$ or $\mathbb{H}$. Indeed, geometrically f is a subset $U = \{(w, f(w))\} \subset \mathbb{C} \times \mathbb{C}$, given locally by one analytic relation $F(w, f) = 0$ (and globally by one algebraic relation if f is an algebraic function). After removing the discrete set of its singular points ($\partial F/\partial w = \partial F/\partial f = 0$), one may think of U as a submanifold, and hence as a Riemann surface. (In the algebraic situation, one can make use of the desingularization of Sect. 2.11.) By the existence and simple connectedness of the universal cover, and by the Riemann mapping theorem, there is an (unramified) mapping of some simply connected canonical region onto $U \subset \mathbb{C} \times \mathbb{C}$. The first and second components of this mapping are the functions φ and ψ.

Remark. Uniformization has arisen in connection with the problem of integrating algebraic functions (cf. Remark 2 in Sect. 4.8). Uniformizing algebraic and analytic functions of an arbitrary number of variables is the content of Hilbert's twenty-second problem (cf. Hilbert's Problems [1969]). No significant progress, however, has been made so far in solving this problem, even for the case of two variables. Furthermore, due to the spreading of abstract structures (Riemann surfaces, complex manifolds, and algebraic varieties), the uniformization problem has lost some of the grandeur it had in its older formulation. Nevertheless, its substance has survived in the results and problems concerning the classification of complex manifolds and algebraic varieties, and in the theory of automorphic forms.

5.3. Types of Riemann Surfaces. Every universal covering is simply connected. So, by Riemann's theorem we have:

Theorem. *Every Riemann surface is isomorphic to a quotient surface S/Γ, where S is a simply connected canonical region, and Γ an automorphism group acting freely and discretely on S.*

The form of the canonical region is uniquely determined by the Riemann surface, for its universal covering is unique.

Definition. The canonical regions $\mathbb{CP}^1$, $\mathbb{C}$, and $\mathbb{H}$ are called *elliptic*, *parabolic*, and *hyperbolic*, respectively. More generally, a Riemann surface is said to be of *elliptic*, *parabolic* or *hyperbolic type* according to the type of its universal covering.

Caution: The word 'type' is sometimes omitted, and one simply says 'elliptic' or 'parabolic surface'. This may lead to some confusion, since an elliptic curve is then parabolic!

Corollary. *In the notation of the Theorem, the automorphism group of the Riemann surface S/Γ is isomorphic to N/Γ, where N is the normalizer of Γ in* $\operatorname{Aut} S$.

Remark. The subdivision of Riemann surfaces into types reflects the various metric geometries of the plane: Riemannian (no parallel lines), Euclidean (one parallel), Lobachevskian (at least two parallels).

5.4. Automorphisms of Canonical Regions

Theorem.

(a) Riemann sphere: *Every automorphism of* $\mathbb{CP}^1$ *is a linear fractional transformation:*

$$z \mapsto (az+b)/(cz+d), \qquad \begin{pmatrix} a & b \\ c & d \end{pmatrix} \in \mathrm{SL}(2,\mathbb{C}).$$

These transformations form a complex Lie group of dimension 3,

$$\text{Möb} \stackrel{\text{def}}{=} \operatorname{Aut} \mathbb{CP}^1 \simeq \mathrm{SL}(2,\mathbb{C})/\pm I,$$

known as the Moebius group.

(b) Gaussian plane: *Every automorphism of* $\mathbb{C}$ *is a (complex) affine transformation:*

$$z \mapsto az+b, \qquad a \in \mathbb{C}^\times, \quad b \in \mathbb{C}.$$

These transformations form a complex Lie group of dimension 2.

(c) Upper half-plane: *Every automorphism of* $\mathbb{H}$ *is a real linear fractional transformation:*

$$z \mapsto (az+b)/(cz+d), \qquad \begin{pmatrix} a & b \\ c & d \end{pmatrix} \in \mathrm{SL}(2,\mathbb{R}).$$

These transformations form a real Lie group $\mathrm{SL}(2,\mathbb{R})/\pm I$, *of dimension* 3.

(d) Unit disk: *Every automorphism of* $\mathbb{D}$ *is a linear fractional transformation of the form*

$$z \mapsto e^{\sqrt{-1}\vartheta}\frac{z-a}{1-\bar{a}z}, \qquad \vartheta \in \mathbb{R}, \quad a \in \mathbb{D}.$$

The automorphism group of the unit disk is isomorphic to that of the upper half-plane, since these regions are isomorphic (see Sect. 1.6). The proof of the theorem can be found in standard textbooks on analytic function theory (Shabat [1969], Hurwitz-Courant [1922, 1964]). That every automorphism of the Riemann sphere is linear fractional, is a combination of rationality (see Example 1 of Sect. 2.2) and injectivity. The case of the other canonical regions is more complicated and somewhat unexpected. In particular, any automorphism of these regions is linear fractional and can therefore be continued to the entire Riemann sphere.

The explicit description of the automorphisms enables one to verify the following remarkable fact, which explains the choice of metrics on different models of Lobachevskian geometry.

Corollary.

(a) Poincaré model $\mathbb{H}$: *Up to proportionality (choice of scale),*

$$dz\, d\bar{z}/(\operatorname{Im} z)^2$$

is the only Kähler metric on the upper half-plane which is invariant under all automorphisms.

(b) Poincaré model $\mathbb{D}$: *Up to proportionality,*

$$dz\, d\bar{z}/\left(1-|z|^2\right)^2$$

is the only Kähler metric on the unit disk which is invariant under all automorphisms.

Remark. In more abstract terms, on every simply connected Riemann surface of hyperbolic type there is – up to proportionality – a unique Kähler form which is invariant under automorphisms. It induces, in particular, a Kähler form on any Riemann surface of hyperbolic type. This metric is of constant Gaussian curvature $K < 0$, by which it is uniquely determined (see Dubrovin, Novikov & Fomenko [1979]). Further, in the parabolic case, the metric of 0-curvature is induced from the Euclidean one; and in the elliptic case, the metric of constant positive curvature is the Fubini-Study metric (see Dubrovin, Novikov & Fomenko [1979] and Sections 5.5 and 5.6 below). The terminology itself – elliptic, parabolic or hyperbolic type – originates from the subdivision of the metrics of constant Gaussian curvature after their signature: $K > 0, = 0, < 0$.

5.5. Riemann Surfaces of Elliptic Type

Proposition. *A Riemann surface of elliptic type is isomorphic to the Riemann sphere.*

By the Theorem of Sect. 5.3, such a surface is isomorphic to $\mathbb{CP}^1/\Gamma$. Now the group Γ acts freely. This is possible only for $\Gamma = \{id\}$, since a nontrivial linear fractional transformation always has a fixed point.

5.6. Riemann Surfaces of Parabolic Type. It is just as easy to prove the following

Proposition. *A Riemann surface of parabolic type is isomorphic either to* $\mathbb{C}$ *or to* $\mathbb{C}^\times$ *or to an elliptic curve* $\mathbb{C}/\Lambda$.

For a suitable choice of an affine coordinate on $\mathbb{C}$, we may assume that the lattice Λ is generated by 1 and $\tau \in \mathbb{H}$. The associated elliptic curve $\mathbb{C}/(\mathbb{Z}+\tau\mathbb{Z})$ is denoted by E_τ. (Formally, $\mathbb{C}^\times$ is isomorphic to E_0.) The parameter τ points to the presence of moduli for elliptic curves. However, τ is not uniquely determined by the isomorphism class of the surface E_τ, but only up to modular transformations of the form $\tau \mapsto (a\tau+b)/(c\tau+d)$, with

$\begin{pmatrix} a & b \\ c & d \end{pmatrix} \in \mathrm{SL}(2, \mathbb{Z})$. Indeed, the transformation $\tau' = (a\tau + b)/(c\tau + d)$ corresponds to an isomorphism $E_\tau \xrightarrow{\sim} E_{\tau'}$, which is induced by an affine mapping $z \mapsto z/(c\tau + d) + \text{const}$ of the universal covering $\mathbb{C}$. Using the properties of universal coverings and the description of the automorphisms of $\mathbb{C}$, it is easy to check that there are no other isomorphisms. Further, the automorphisms of an elliptic curve can be described in much the same way. There are some obvious automorphisms, namely the translations by elements $p \in E_\tau$. Thus $q \mapsto q + p$ is the addition of p in the sense of the group law on E_τ. In fact, the translation group is isomorphic to E_τ. Much more interesting is the quotient group $\mathrm{Aut}_0 E_\tau \overset{\text{def}}{=} \mathrm{Aut}\, E_\tau / E_\tau$, which can be identified with the automorphisms preserving $0 \in E_\tau$.

Theorem.
(a) $\mathrm{Aut}_0 E_\tau = \{\mathbb{C}/(\mathbb{Z} + \tau\mathbb{Z}) \xrightarrow{\pm z} \mathbb{C}/(\mathbb{Z} + \tau\mathbb{Z})\} = \mathbb{Z}/2$, *unless* E_τ *is isomorphic to either* $E_{\sqrt{-1}}$ *or* E_ρ, *where* $\rho = e^{\pi\sqrt{-1}/3} \in \mathbb{H}$.

(b) $\mathrm{Aut}_0 E_{\sqrt{-1}} = \{\mathbb{C}/(\mathbb{Z} + \sqrt{-1}\mathbb{Z}) \xrightarrow{\sqrt[4]{1}\, z} \mathbb{C}/(\mathbb{Z} + \sqrt{-1}\mathbb{Z})\} = \mathbb{Z}/4$.

(c) $\mathrm{Aut}_0 E_\rho = \{\mathbb{C}/(\mathbb{Z} + \rho\mathbb{Z}) \xrightarrow{\sqrt[6]{1}\, z} \mathbb{C}/(\mathbb{Z} + \rho\mathbb{Z})\} = \mathbb{Z}/6$.

The idea of the proof is to identify $\mathrm{Aut}_0(\mathbb{C}/\Lambda)$ with the group of rotations of the lattice Λ. Indeed, every automorphism of $\mathbb{C}/\Lambda$ that fixes the origin, is induced by a linear map $z \mapsto az$, $a \in \mathbb{C}^\times$, such that $a\Lambda = \Lambda$. For a generic lattice, this group consists of just two rotations, by the angles 0 and π, which proves (a). Moreover, there are only two exceptions, namely (b) and (c) (see Fig. 16). For a somewhat different approach, see Clemens [1980].

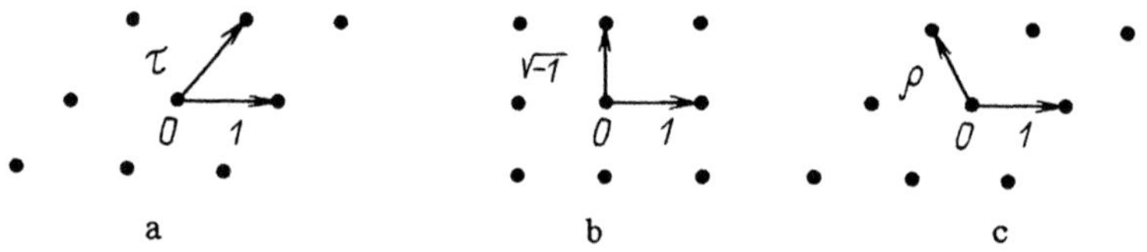

Fig. 16. Some plane lattices: (a) generic, (b) square, and (c) hexagonal

Remark 1. A mapping of an elliptic curve which preserves the origin, is a homomorphism. The group $\mathrm{Aut}_0(\mathbb{C}/\Lambda)$ can therefore be interpreted as the automorphism group of the Lie group $\mathbb{C}/\Lambda$.

Remark 2. A surjective endomorphism of an elliptic curve $\mathbb{C}/\Lambda$ is called an *isogeny*. The simplest example of an isogeny is multiplication by a positive integer:

$$\mathbf{N}\colon \mathbb{C}/\Lambda \xrightarrow{Nz} \mathbb{C}/\Lambda.$$

Remark 3. By the Proposition, every compact parabolic Riemann surface is of genus 1. The converse is stated in Corollary 2 of Sect. 5.7.

5.7. Riemann Surfaces of Hyperbolic Type. Every surface of this type is isomorphic to a quotient $\mathbb{H}/\Gamma$, where the subgroup $\Gamma \subset \mathrm{SL}(2,\mathbb{R})/\pm I$ acts freely and discretely on $\mathbb{H}$. This is equivalent to saying that Γ is discrete as a subgroup of the Lie group $\mathrm{SL}(2,\mathbb{R})/\pm I$ (cf. Shimura [1971]). The invariant metric $dz\,d\bar{z}/(\operatorname{Im} z)^2$ defines a distance $\rho(z_1, z_2)$ between any two points z_1 and z_2 in the upper half-plane $\mathbb{H}$, which is the distance in the Poincaré model of Lobachevskian geometry. A straight line in this model is a vertical ray or a semi-circle centred at a point on the real axis (see Fig. 17). From an intrinsic point of view, these are the geodesics for the invariant metric. Now we consider a more general situation. Suppose $\Gamma \subset \mathrm{SL}(2,\mathbb{R})/\pm I$ is a discrete subgroup. Fix a point $z_0 \in \mathbb{H}$ such that $g(z_0) \neq z_0$ for all $g \in \Gamma - \{id\}$. (If Γ acts freely, then z_0 is an arbitrary point of $\mathbb{H}$.) We set

$$D = \{z \in \mathbb{H} \mid \rho(z, z_0) \leq \rho(gz, z_0),\ g \in \Gamma\}.$$

Theorem. *The set D is a convex polygon in Lobachevskian geometry* (see Fig. 17), *possibly with an infinite number of sides, and its interior D° is a fundamental domain for the group Γ.*

D is called the *normal polygon* of Γ with centre $z_0 \in \mathbb{H}$. A proof of the theorem for the Poincaré model of $\mathbb{D}$ can be found in Springer [1957].

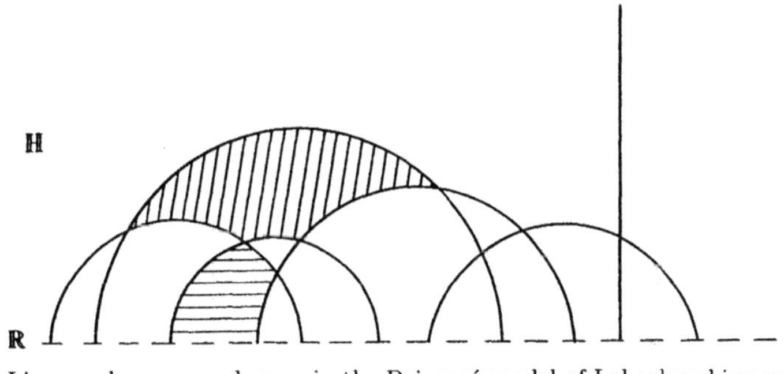

Fig. 17. Lines and convex polygons in the Poincaré model of Lobachevskian geometry

Example. The normal polygon of the full modular group $\Gamma_1 = \mathrm{SL}(2,\mathbb{Z})/\pm I$, with centre $z_0 = a\sqrt{-1}$, $a \in \mathbb{R}$, $a > 1$, is the familiar *modular figure:*

$$\left\{z \in \mathbb{H} \;\middle|\; |z| \geq 1,\ -1/2 \leq \operatorname{Re} z \leq 1/2\right\}$$

(see Fig. 18).

Corollary 1. *The quotient space $\mathbb{H}/\Gamma$ is compact if and only if the normal polygon D is bounded.*

Furthermore, in the case of a compact quotient, the polygons gD, with $g \in \Gamma$, are closed, have a finite number of sides, and form a tesselation of the Lobachevskian plane. As D is fundamental, each side s of D defines

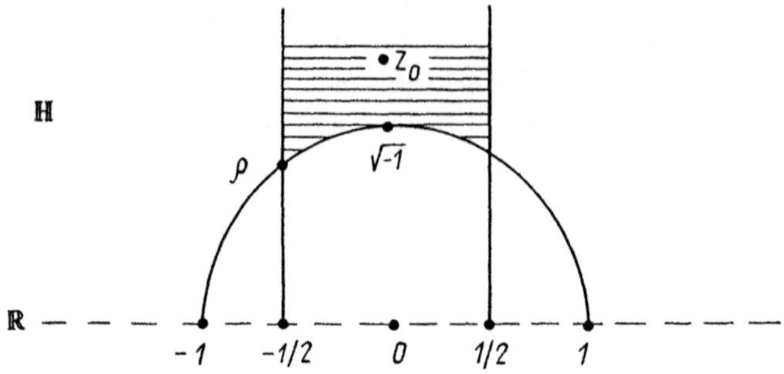

Fig. 18. The modular figure

exactly one non-identical transformation $t_s \in \Gamma$, which carries that side into the Γ-equivalent side $t_s(s)$ of the same polygon D. The group Γ is generated by these elements t_s (see Springer [1957]). Returning to the original situation, suppose in addition that Γ acts freely. The polygon D may be regarded as a development of the compact Riemann surface $\mathbb{H}/\Gamma$. In this case, gluing together sides and vertices is the same as Γ-equivalence. This development has the following properties:

(a) the sides to be glued together are equal, with opposite orientations;

(b) the sum of the angles at all vertices merging into a single point is equal to 2π.

Conversely, the interior of a development D with properties (a) and (b) is a fundamental domain for some discrete group Γ. Moreover, Γ acts freely on $\mathbb{H}$, and the quotient $\mathbb{H}/\Gamma$ is a compact Riemann surface with development D.

Caution: The original polygon in the last construction need not be normal. This is what usually happens in the case of the Poincaré developments with standard symbol $a_1 b_1 a_1^{-1} b_1^{-1} \dots a_g b_g a_g^{-1} b_g^{-1}$. Let $2e$ be the number of edges in the development D and let v be the maximum number of Γ-nonequivalent vertices of D. Then the Euler characteristic $\chi(\mathbb{H}/\Gamma)$ is equal to $v - e + 1$ (see Sect. 3.5). On the other hand,

$$\begin{aligned}\text{area of } D &= \textit{defect} \\ &\overset{\text{def}}{=} \text{sum of angles of a } 2e\text{-gon in Euclidean geometry} \\ &\qquad - \text{sum of angles of } D \text{ in Lobachevskian geometry} \\ &= (2e-2)\pi - 2\pi v = -2\pi\,\chi(\mathbb{H}/\Gamma).\end{aligned}$$

Further we observe that areas, in our model, are computed by the form $\sqrt{-1}\,dz \wedge d\bar{z}/2(\operatorname{Im} z)^2 = \dfrac{dx \wedge dy}{y^2}$, where $z = x + \sqrt{-1}\,y$. Hence we obtain the following

Proposition. *Let $\Gamma \subset \mathrm{SL}(2,\mathbb{R})/\pm I$ be a discrete subgroup, acting freely, with compact quotient $\mathbb{H}/\Gamma$. Let D be a fundamental domain for Γ, for instance a normal polygon. Then*

$$\chi(\mathbb{H}/\Gamma) = \frac{1}{4\pi\sqrt{-1}} \int_D \frac{dz \wedge d\bar{z}}{(\operatorname{Im} z)^2}.$$

This is a special case of the Gauss-Bonnet formula (see Eisenhart [1947], Griffiths-Harris [1978], and Shimura [1971]). At this point the reader will not be surprised by the topological invariance of the type in the compact case.

Corollary 2. *Let S be a compact Riemann surface of genus g. It is*
(a) *of elliptic type if $g = 0$;*
(b) *of parabolic type if $g = 1$;*
(c) *of hyperbolic type if $g \geq 2$.*

The proof uses the results of the preceding subsections and the above Proposition, which implies that $\chi(S) = 2 - 2g < 0$ for a hyperbolic surface.

Remark 1. It follows from Corollary 2 that a Riemann surface S of genus 1 is isomorphic to an elliptic curve. In particular, this is true for a plane cubic curve. An explicit description of the isomorphism is given by Abel's theorem (see Chap. 3, Sect. 2.6). How to find τ is clear *a posteriori*: $S \simeq E_\tau$ for $\tau = \int_{b_1} \omega / \int_{a_1} \omega$, where $\omega \neq 0$ is a holomorphic differential on S, and a_1, b_1 is a standard basis for the homology group $H_1(S, \mathbb{Z})$ (verifying $(a_1, b_1) = +1$). Transition to another such basis is given by a matrix $\begin{pmatrix} a & b \\ c & d \end{pmatrix} \in \mathrm{SL}(2, \mathbb{Z})$, which corresponds to a linear fractional transformation on the period ratios: $\tau \mapsto (d\tau + c)/(b\tau + a)$ (cf. Sect. 5.6).

Remark 2. The classification of Riemann surfaces of elliptic or parabolic type up to isomorphism leads to a finite number of families (four, to be precise): the Riemann sphere, the Gaussian plane, $\mathbb{C}^\times$, and the family of elliptic curves E_τ. Now, the hyperbolic type is composed of an infinite series of families, even in the compact case, corresponding to Riemann surfaces of genus $g \geq 2$ (see Sect. 5.10). Therefore we sometimes say that this is the *general* type.

We lack a complete description for the automorphism groups of Riemann surfaces of hyperbolic type, as for the surfaces themselves. One has to be content with qualitative results. Anyway, this is typical of any general situation. It is customary to label such results as *fundamental.*

Theorem (Schwarz). *A compact Riemann surface of genus $g \geq 2$ or, equivalently, of hyperbolic type has a finite automorphism group.*

The proof depends on the description of hyperbolic Riemann surfaces with a nondiscrete automorphism group. These are $\mathbb{D}$ and the annuli $\{r < |z| < 1\}$, for $0 \leq r < 1$ (cf. Springer [1957]). In contrast, the automorphism group of Riemann surfaces of elliptic or parabolic type always has positive dimension (as a complex Lie group). Thus the general type is characterized as having fewer symmetries.

Remark 3. As a matter of fact, the order of the automorphism group of a Riemann surface of genus $g \geq 2$ does not exceed $84(g-1)$ (cf. Fricke-Klein [1897, 1912]). This is a consequence of the following remarkable fact: the area of a fundamental domain – for example of the normal polygon – of a discrete subgroup $\Gamma \subset \mathrm{SL}(2,\mathbb{R})/\pm I$ cannot be less than the absolute constant $\pi/21$ (see Shimura [1971]). Another, more algebraic, approach involves Hurwitz's formula (see Arbarello et al. [1985] and Sect. 2.9 of Chap. 2). We note, further, that a sufficiently general Riemann surface of genus $g \geq 3$ has no automorphisms at all (see the Corollary of Sect. 5.10).

5.8. Automorphic Forms; Poincaré Series. Historically, the earliest approach to the existence theorems for meromorphic functions and differentials on compact Riemann surfaces made use of the universal covering (see the Theorem of Sect. 5.3). The main difficulty was to prove the Riemann mapping theorem. In fact, the Riemann sphere is the only surface of elliptic type, and its meromorphic functions and differentials are rational (see Example 1 of Sect. 2.2 and Corollary 1 of Sect. 4.8). The construction of functions and differentials on elliptic curves has already been dealt with in Sections 2.2 and 4.7 (see also Sect. 1.7 of Chap. 3). We look now at the compact hyperbolic case $\mathbb{D}/\Gamma$, and turn to the more general question of finding meromorphic (symmetric) differentials of degree m, where m is a nonnegative integer. This is equivalent to finding such differentials on the covering $\mathbb{D}$, which are invariant under Γ. Further, any differential of degree m on $\mathbb{D}$ can be written as $f\,dz^m$. It is meromorphic if f is; and the invariance means that

$$f(g(z)) = f(z)\left(\frac{dg(z)}{dz}\right)^{-m} \qquad \text{for all } g \in \Gamma. \tag{5}$$

Definition. A meromorphic function f on $\mathbb{D}$ with property (5) is called an *automorphic form* of weight $2m$ with respect to Γ. Automorphic forms of weight 0 are called *automorphic functions.*

Example 1. Let $m \geq 2$ and let h be a holomorphic, bounded function on $\mathbb{D}$. Then the *Poincaré series*

$$f(z) = \sum_{g \in \Gamma} h(g(z))\left(\frac{dg(z)}{dz}\right)^{m}$$

is absolutely and uniformly convergent on every compact subset of $\mathbb{D}$. It therefore defines a holomorphic function f on $\mathbb{D}$, and it is easy to see that (5) holds. Hence it is an automorphic form of weight $2m$ with respect to Γ. To prove convergence, it suffices to know that Γ is discrete and that $\mathbb{D}$ is bounded (see Shafarevich [1972]). This explains the choice of $\mathbb{D}$ rather than $\mathbb{H}$ in this subsection.

Example 2. Some forms can be constructed from others by arithmetic operations. Thus the sum of automorphic forms of the same weight is an automorphic form of that weight. The product of two automorphic forms is an automorphic form, whose weight is the sum of the weights of the components. Finally, their ratio is a form whose weight is the difference of the weights. In particular, the ratio of two Poincaré series of the same weight is an automorphic function.

Let $f_0, \ldots, f_n$ be nonzero automorphic forms of the same weight with respect to Γ. Then there is a holomorphic mapping

$$\begin{aligned} f\colon \mathbb{D}/\Gamma &\to \mathbb{CP}^n \\ \Gamma\text{-orbit of } z &\mapsto (f_0(z) : \ldots : f_n(z)). \end{aligned} \tag{6}$$

This is clearly well-defined, in view of the automorphic property. Further, holomorphy is evident at a generic point z_0, where all the f_i are holomorphic in some neighbourhood and at least one of them does not vanish at z_0. At the remaining points the indetermination can be removed by dividing all f_i by $(z - z_0)^{\min \operatorname{ord}_{z_0} f_i}$.

Theorem. *Let Γ be an automorphism group acting freely and discretely on $\mathbb{D}$, with compact quotient $\mathbb{D}/\Gamma$. Then there exist finitely many automorphic forms of the same weight such that* (6) *is an embedding of $\mathbb{D}/\Gamma$ into $\mathbb{CP}^n$.*

An embedding means an isomorphism onto the image (cf. the Proposition of Sect. 6.1). For proving, and hence also for enunciating, the theorem one can work with Poincaré series (see Shafarevich [1972]). One can conclude from this that every automorphic form is a rational function of Poincaré series (see the rationality theorem in Sect. 6.5). Hence the constructions in the examples are sufficiently general. A more algebraic, and more precise, version of the theorem is given in Sect. 6.4 below.

Remark. A detailed treatment of this approach to the existence theorems can be found in Kra [1972].

5.9. Quotient Riemann Surfaces; the Absolute Invariant. We begin with a general construction. Let $\Gamma \subset \mathrm{SL}(2, \mathbb{R})/\pm I$ be a discrete transformation group of the upper half-plane $\mathbb{H}$. Such transformation groups of $\mathbb{H}$ are called *Fuchsian.* As in the case of a free action, there is on $\mathbb{H}/\Gamma$ a unique Riemann surface structure such that the quotient map $\mathbb{H} \to \mathbb{H}/\Gamma$ is holomorphic. Furthermore, $\mathbb{H}$ can be replaced by an arbitrary Riemann surface S, and Γ by a group acting on S discretely. The difficulty in introducing an analytic structure on S/Γ lies in constructing quotients of neighbourhoods of the points $p \in S$ having a nontrivial stabilizer $\Gamma_p \overset{\text{def}}{=} \{g \in \Gamma \mid g(p) = p\}$. By discreteness, Γ_p is a finite rotation group of p in a constant curvature metric (see the Remark of Sect. 5.4). Hence, for a suitable choice of a local parameter z in a neighbourhood of p, we have $\Gamma_p = \{z \mapsto \sqrt[n]{1}\, z\}$, where n is the order of Γ_p. The function z^n can be taken as a local parameter on S/Γ in some neighbourhood of the orbit Γ_p. Now, unless it is the identity, a transformation $g(z) = (az+b)/(cz+d)$, with $\begin{pmatrix} a & b \\ c & d \end{pmatrix} \in \mathrm{SL}(2, \mathbb{R})$, has at most one fixed point in $\mathbb{H}$ (at most two on $\mathbb{CP}^1$). A *transformation* $g(z)$ having a fixed point $z \in \mathbb{H}$ (two points $z, \bar{z} \in \mathbb{CP}^1$) is said to be *elliptic* (cf. Shimura [1971]). Correspondingly, the points $z \in \mathbb{H}$ whose stabilizer Γ_z is nontrivial (and their images on $\mathbb{H}/\Gamma$) are called *elliptic points* of the group Γ. Their *order* is the order of the group Γ_z.

Example. The most important example of a Fuchsian group is the full modular group Γ_1 (see the Example of Sect. 5.7). The quotient surface $\mathbb{H}/\Gamma_1$ has two elliptic points: the orbits $\Gamma_1\sqrt{-1}$ and $\Gamma_1\,\rho$, of order 2, respectively 3. Considering the modular figure (see Fig. 18) as a development of $\mathbb{H}/\Gamma_1$, it is easy to see that $\mathbb{H}/\Gamma_1$ is homeomorphic to a plane. What is more, the surface $\mathbb{H}/\Gamma_1$ is isomorphic to the Gaussian plane $\mathbb{C}$. By the Riemann mapping theorem, this is equivalent to saying that $\mathbb{H}/\Gamma_1$ has a one-point compactification. Again this is easy to see from the modular figure. The local parameter at the compactifying point $\Gamma_1\infty$ is given by the Γ_1-invariant function $e^{2\pi\sqrt{-1}\,z}$, provided $\operatorname{Im} z \gg 0$.

This example is remarkable in at least two respects. First, it implies the existence of an unramified covering of a bounded region of the Gaussian plane $\mathbb{C}$ onto $\mathbb{C} - \{\text{two points}\}$. Besides, this also follows from the fact that the latter region is hyperbolic. This circumstance is of key importance for the proof of Picard's first theorem, to the effect that a nonconstant holomorphic function on $\mathbb{C}$ assumes all complex values, with at most one exception (Shabat [1969]). Secondly, there is a unique global coordinate z on $\mathbb{H}/\Gamma_1 \simeq \mathbb{C}$ such that $z(\Gamma_1\sqrt{-1}) = 1$ and $z(\Gamma_1\,\rho) = 0$. Equivalently, there is on $\mathbb{H}$ a unique holomorphic function $j(\tau)$ which is Γ_1-invariant, (assumes every complex value and) takes on the same value only at points belonging to the same orbit, and is such that $j(\sqrt{-1}) = 1$ and $j(\rho) = 0$. This function is called the *absolute invariant* because, contrary to $\tau \in \mathbb{H}$, the value $j(\tau)$ is a true invariant of the elliptic curve E_τ, and any other invariant can be expressed in terms of it. Indeed, $j(\tau)$ determines E_τ up to isomorphism.

5.10. Moduli of Riemann Surfaces. The idea of moduli as numerical parameters of Riemann surfaces of the same topological type stems from the following example.

Example 1. Any Riemann surface S of genus 1 can be assigned a number, its *absolute invariant* $j(S) = j(\tau)$, where $S \simeq E_\tau$, $\tau \in \mathbb{H}$. Obviously,

(a) two Riemann surfaces, S_1 and S_2, of genus 1 are isomorphic if and only if their absolute invariants are equal: $j(S_1) = j(S_2)$;

(b) there exists a Riemann surface of genus 1 with any preassigned complex value of the absolute invariant.

This example, however, not so much settles the question of describing Riemann surfaces of genus 1 up to isomorphism as brings up other questions. In what sense is the above parametrization natural? Are there essentially different parametrizations, say, with a larger number of independent complex parameters? Matters are made even worse by the negative solution of the moduli problem: Riemann surfaces of sufficiently large genus (≥ 40) have no natural parametrization (in what sense is not yet clear) with independent parameters (see Harris-Mumford [1982] and Harris [1984]). Thus, in the general situation, a Riemann surface of genus g is described by a collection of numerical parameters with some relations and possible identifications under a group action or modulo some equivalence relation (cf. Chap. 2, Sect. 2.10). Even in cases like the moduli of hyperelliptic Riemann surfaces (see Katsylo [1984]) – where independent parameters are known to exist –, the choice of parameters is itself extremely inefficient. All this leads to the idea that what is important and accessible is most probably not the concrete numerical realization of parameters, but their geometry: the number of parameters up to relations and identifications, 'how close' are two Riemann surfaces with close parameters, does there exist a parametrization with independent parameters, etc. In other words, one should provide the *a priori* discrete set $\mathcal{M}_g$ of isomorphism classes of Riemann surfaces of genus g with a topology and a complex analytic or some other structure. The coordinate systems on $\mathcal{M}_g$ correspond to some natural parameters, the moduli. The existence of global coordinates is equivalent to the existence of independent parameters. This leads to the idea of a moduli space $\mathcal{M}_g$. The complex analytic structure of this is defined and constructed using (analytic) families of Riemann surfaces.

Definition 1. A mapping of complex manifolds $f\colon M \to B$ is said to be a *family of Riemann surfaces* of genus g with base B if the fibre $f^{-1}(b)$ above each point $b \in B$ is a Riemann surface of genus g.

Example 2. Let $\Gamma \subset \mathrm{SL}(2,\mathbb{Z})$. There is an extension $\Gamma \times \mathbb{Z} \times \mathbb{Z}$ of Γ (with normal subgroup $\mathbb{Z} \times \mathbb{Z}$), which acts freely and discretely on the product $\mathbb{H}' \times \mathbb{C}$ according to the rule

$$(\tau, z) \xmapsto{(g,n,m)} \left(\frac{a\tau + b}{c\tau + d}, \frac{z + \tau n + m}{c\tau + d}\right),$$

where $g = \begin{pmatrix} a & b \\ c & d \end{pmatrix} \in \Gamma$, $n, m \in \mathbb{Z}$, and

$$\mathbb{H}' = \mathbb{H} - \{\text{elliptic points of the group } \Gamma/\pm I\}.$$

It is easy to check that

$$f_\Gamma\colon \mathbb{H}' \times \mathbb{C}/\Gamma \times \mathbb{Z} \times \mathbb{Z} \to \mathbb{H}'/\Gamma$$

is a family of Riemann surfaces of genus 1.

A family $f\colon M \to B$ of Riemann surfaces of genus g induces a mapping of the base $B \to \mathcal{M}_g$, which carries a point $b \in B$ into the isomorphism class of the fibre $f^{-1}(b)$.

Definition 2. The set $\mathcal{M}_g$, equipped with a complex analytic structure such that any induced mapping $B \to \mathcal{M}_g$ is holomorphic, is said to be a *coarse moduli space* for Riemann surfaces of genus g.

Theorem. *The moduli space* $\mathcal{M}_1$ *exists and is canonically isomorphic to* $\mathbb{H}/\Gamma_1 \simeq \mathbb{C}$.

The family $f_{\mathrm{SL}(2,\mathbb{Z})}$ with base $\mathbb{H}/\Gamma_1 - \{\text{two elliptic points}\}$ induces an embedding, which can be continued to an isomorphism of $\mathbb{H}/\Gamma_1$ with $\mathcal{M}_1$. In coordinate form the isomorphism $\mathcal{M}_1 \simeq \mathbb{C}$ is given by the absolute invariant.

For Riemann surfaces of genus $g \geq 2$, the coarse moduli space exists as a complex analytic space and not as a complex manifold. What makes a complex space differ from a complex manifold is the presence of singularities. For a precise definition, see Griffiths-Harris [1978], Gunning-Rossi [1965], or Mumford [1976]. Under some natural restrictions on singularities the coarse moduli space is unique. It is denoted by $\mathcal{M}_g$. Its existence and uniqueness are proved in a rather technical way. The most familiar approach uses Teichmüller spaces (cf. Bers [1960]). However, it is not difficult to determine the number of parameters. Every Riemann surface of genus $g \geq 2$ is hyperbolic. Hence, up to isomorphism, it is a quotient $\mathbb{H}/\Gamma$, where Γ is a Fuchsian group acting freely on $\mathbb{H}$. We note, further, that Γ is isomorphic to the fundamental group of a Riemann surface of genus g (see Sect. 3.4). Therefore Γ has $2g$ generators $A_1, B_1, \ldots, A_g, B_g \in \mathrm{SL}(2,\mathbb{R})/\pm I$, connected by the relation:

$$A_1 B_1 A_1^{-1} B_1^{-1} \ldots A_g B_g A_g^{-1} B_g^{-1} = \pm I. \tag{7}$$

It is easy to derive the following assertions.

Lemma (Ahlfors [1966]). *The sequences* $A_1, B_1, \ldots, A_g, B_g \in \mathrm{SL}(2,\mathbb{R})/\pm I$ *that satisfy* (7) *form a real analytic manifold of dimension* $6g-3$. *Those corresponding to Riemann surfaces of genus* g *make up an open subset.*

Proposition (Ahlfors [1966]). $\dim_{\mathbb{R}} \mathcal{M}_g = 6g - 6$, *provided* $g \geq 2$.

The proof makes essential use of Schwarz's theorem, which implies that close generator sequences yield isomorphic Riemann surfaces if and only if they can be obtained from one another by an inner automorphism:

$$A_1, B_1, \ldots, A_g, B_g \mapsto aA_1a^{-1}, aB_1a^{-1}, \ldots, aA_ga^{-1}, aB_ga^{-1},$$

$a \in \mathrm{SL}(2,\mathbb{R})/\pm I$. Subtracting $\dim_{\mathbb{R}} \mathrm{SL}(2,\mathbb{R})/\pm I = 3$ from $6g-3$, we get what was required.

The existence of a moduli space gives a meaning to statements to the effect that some property P is satisfied for a generic Riemann surface of genus g. This means that all Riemann surfaces of genus g for which P fails to hold are included in a complex analytic subspace of smaller dimension (for instance, in the topological sense) than the dimension of the moduli space ($= 6g-6$ if $g \geq 2$; cf. Sect. 1.4 of Chap. 2). In other words, a sufficiently small, arbitrary variation of a Riemann surface of genus g yields a Riemann surface of genus g having property P. One way of proving a generic property is to count parameters. In this way, for example, we prove the following corollary.

Corollary. *The automorphism group of a generic Riemann surface of genus $g \geq 3$ is trivial. In particular, the generic Riemann surface of genus $g \geq 3$ is non-hyperelliptic.*

More precisely, using Hurwitz's genus formula, one checks that the Riemann surfaces of genus $g \geq 1$ with an automorphism of order ≥ 2 form a subspace of real dimension $\leq 4g - 2$, which is less than $6g - 6$ when $g \geq 3$.

Remark. The existence of an analytic structure on $\mathcal{M}_g$ is essentially due to the fact that we classify compact Riemann surfaces. In the noncompact case, the best we can hope for is real analyticity. For instance, any Riemann surface which is homeomorphic to an annulus is isomorphic to some annulus $\{r < |z| < 1\}$. The real number $0 \leq r < 1$ is its absolute invariant. Hence the semi-open interval $[0, 1)$ is a moduli space for these surfaces.

§ 6. Algebraic Nature of Compact Riemann Surfaces

An isomorphism of a Riemann surface onto a one-dimensional submanifold of a complex manifold is called an embedding. The main theme of this section is how to construct embeddings of a compact Riemann surface in some projective spaces. We begin by introducing the necessary tools: the mapping associated with a divisor and the Riemann-Roch formula. Then, before embarking on the discussion of embeddings, we solve some of the existence problems set earlier. The final subsection, 6.6, plays an illustrative role.

Throughout this section, S is a compact Riemann surface of genus g, and K denotes a canonical divisor (see Corollary 6 of Sect. 4.14).

6.1. Function Spaces and Mappings Associated with Divisors. A nonzero meromorphic function on S maps S into $\mathbb{CP}^1$. An immediate generalization is any mapping of the form

$$\begin{aligned} S &\to \mathbb{CP}^n \\ p &\mapsto (f_0(p) : \ldots : f_n(p)), \end{aligned} \tag{8}$$

where $f_0, \ldots, f_n$ are nonzero meromorphic functions on S. This map is defined and holomorphic everywhere on S, by similar considerations as for the mapping (6) in Sect. 5.8. The most convenient case is when the functions f_i form a basis of the complex vector space

$$\begin{aligned} L(D) &\stackrel{\text{def}}{=} \{f \in \mathcal{M}(S) \mid f \equiv 0 \text{ or } (f) + D \geq 0\} \\ &= \{f \equiv 0 \text{ or } \operatorname{ord}_{p_i} f \geq -a_i \text{ for all } p_i\}, \end{aligned}$$

where $D = \sum a_i p_i$ is some divisor on S.

Definition 1. $L(D)$ is called the *space of meromorphic functions* on S which is *associated with* the divisor D.

It is clear that $L(D)$ is a vector space. It is slightly more difficult to prove the following

Theorem. *$L(D)$ is a finite dimensional vector space. Moreover,* $\dim L(D) \leq \deg D + 1$, *provided* $\deg D \geq -1$.

A very interesting and illustrative case is that of an effective divisor $D = \sum a_l p_l$. Then the space $L(D)$ consists of meromorphic functions having poles only at the p_l and with principal parts $\sum_{i=-a_l}^{-1} c_{li} z_l^i$, where z_l is a local parameter at p_l (cf. Sect. 2.3). The linear map that associates with such a function the sequence of coefficients (c_{li}) has in its kernel only holomorphic, and hence constant, functions on S. Therefore $\dim L(D) \leq 1 + \sum a_l = \deg D + 1$. The general case can be obtained from the following properties of $L(D)$.

Lemma. (a) *Suppose D and D' are linearly equivalent, that is, $D = D' + (g)$, where $g \in \mathcal{M}(S)$. Then the map $L(D) \to L(D')$, defined by $f \mapsto g \cdot f$, is a $\mathbb{C}$-linear isomorphism.*

(b) $\dim L(D) > 0$ *if and only if D is linearly equivalent to an effective divisor. This implies in particular that* $\deg D \geq 0$.

The dimension of $L(D)$ is denoted by $l(D)$.

Definition 2. Let $f_0, \ldots, f_n$ be a basis for $L(D)$. Then the *mapping* (8) is said to be *associated with* the divisor D. It is denoted by φ_D. Note that $n = l(D) - 1$ and that φ_D is defined only for $l(D) \geq 1$.

Up to isomorphism, the mapping φ_D remains unchanged if we choose a different basis for $L(D)$ or if we replace D by a linearly equivalent divisor. This follows from the lemma.

Remark. If, instead of a basis f_i, we simply take a system of linearly independent functions then the corresponding mapping (8) will be the composition of φ_D followed by projection with centre in a suitable subspace of $\mathbb{CP}^n$. This exhausts essentially all mappings of a Riemann surface S into projective spaces (cf. Chap. 2, Sect. 3.2).

Example 1. Let $S = \mathbb{CP}^1$. Then $L(d\,\infty)$ is the space of polynomials in z with degree $\leq d$, whence $l(d\,\infty) = d + 1$ for $d \geq 0$, and 0 otherwise. As a basis for $L(d\,\infty)$ if $d \geq 0$, one can take the monomials $1, \ldots, z^d$. The associated map $\varphi_{d\infty}\colon z \mapsto (1 : z : \ldots : z^d)$, which is a special case of the Veronese mapping, is denoted by v_d. If $d \geq 1$, it is an embedding. Any divisor D on $\mathbb{CP}^1$ is linearly equivalent to the divisor $d\,\infty$, where $d = \deg D$ (see Example 4 of Sect. 2.6). Therefore $l(D) = \deg D + 1$ for $\deg D \geq 0$, and 0 otherwise. Conversely, if a Riemann surface S contains a divisor D such that $\deg D \geq 1$ and $l(D) = \deg D + 1$, then S is isomorphic to $\mathbb{CP}^1$. By the lemma, we can assume that

D is effective. Then the equality $l(D) = \deg D + 1$ amounts to saying that Mittag-Leffler's problem is solvable for any (c_{li}). In particular, there is a meromorphic function f with a unique pole of order one in $\operatorname{Supp} D$. Now, this function gives the required isomorphism of S with $\mathbb{CP}^1$. Thus, on a Riemann surface S of genus ≥ 1, we have $l(D) \leq \deg D$ for any divisor D of degree ≥ 1. This result reflects the nontriviality of Mittag-Leffler's problem for $g \geq 1$, which is not surprising given the necessary condition for its solvability (see Sect. 4.9).

Example 2. Let $\omega_0, \ldots, \omega_n$ be a basis for the space Ω^d of holomorphic (symmetric) differentials of degree $d \geq 1$. Then we have the *pluricanonical mapping*

$$\begin{aligned} \varkappa_d\colon S &\to \mathbb{CP}^n \\ p &\mapsto (\omega_0(p) : \ldots : \omega_n(p)) \end{aligned}$$

(cf. (6) of Sect. 5.8). This is defined for $g \geq 1$. The special case of the *canonical mapping*, $\varkappa \stackrel{\text{def}}{=} \varkappa_1\colon S \to \mathbb{CP}^{g-1}$, is particularly important. (Recall that $\dim \Omega = g$.) Clearly, $\omega_0/\omega^d, \ldots, \omega_n/\omega^d$ is a basis for $L(dK)$, where $K = (\omega)$ is the canonical divisor of a holomorphic or meromorphic differential ω of degree 1 (see Sect. 4.8) and $\varkappa_d = \varphi_{dK}$. However, the above direct description in terms of differentials is more canonical. In fact, the properties of the pluricanonical mapping $\varkappa_d$ reflect the intrinsic properties of the Riemann surface S, rather than those of a particular divisor dK. Nevertheless, φ_{dK} is canonical, in that it can be defined up to isomorphism by the linear equivalence class of dK (cf. Example 5 of Chap. 2, Sect. 2.9).

Example 3. Suppose S is a hyperelliptic Riemann surface, with hyperelliptic projection γ. Using the basis for Ω given in Example 1 of Sect. 4.8, it is easy to check in this case that the canonical mapping factorizes as

$$\begin{array}{ccc} S & \xrightarrow{\varkappa} & \mathbb{CP}^{g-1} \\ & {\scriptstyle\gamma}\searrow \quad \nearrow{\scriptstyle v_{g-1}} & \\ & \mathbb{CP}^1 & \end{array}$$

This implies the uniqueness of a hyperelliptic structure – projection or involution – for $g \geq 2$. Moreover, the presence of a hyperelliptic structure for $g \geq 1$ is characterized by the existence on S of a divisor D of degree 2 such that $l(D) = 2$: $D = \gamma^* z$, and $\gamma = \varphi_D$ for such a divisor D. In particular, a Riemann surface of genus 2 is always hyperelliptic: $\gamma = \varkappa$ (cf. the Corollary in Sect. 5.10).

The ease of working with the maps φ_D accounts for the

Proposition. *φ_D is an embedding if*

$$l(D - p - q) = l(D) - 2 \quad \textit{for any } p, q \in S. \tag{9}$$

For $p \neq q$, it follows from (9) that $\varphi_D(p) \neq \varphi_D(q)$. For $p = q$, (9) implies that φ_D is regular[4] at p. Thus φ_D is a one-to-one mapping onto a one-dimensional submanifold $\varphi_D(S) \subset \mathbb{CP}^n$ (see Griffiths-Harris [1978]). Hence it is an isomorphism onto its image (cf. Proposition 2 of Sect. 2.5).

So, in order to produce some embeddings, one should know how to compute the dimension $l(D)$ as a function of D. The key to this problem is the well-known Riemann-Roch formula.

6.2. Riemann-Roch Formula; Reciprocity Law for Differentials of the First and Second Kind

Theorem (Riemann, Roch). $l(D) - l(K - D) = \deg D - g + 1$.

This is one version of the Riemann-Roch formula. It seems useless on the face of it, since it reduces the calculation of $l(D)$ to that of $l(K - D)$. But $l(K - D) = 0$ for divisors D of degree $> \deg K = 2g - 2$ (see (*b*) of the Lemma in Sect. 6.1).

Corollary 1. $l(D) = \deg D - g + 1$ *if* $\deg D \geq \deg K + 1 = 2g - 1$.

In the general case we have the following result.

Corollary 2 (Riemann's inequality). $l(D) \geq \deg D - g + 1$.

Among the many perfections of the Riemann-Roch formula are its *symmetry:* the Riemann-Roch formula for D is equivalent to that for $K - D$, and the *topological invariance* of its right-hand side: it depends only on $\deg D$ and on g, and it does so linearly.

It is convenient to prove the Riemann-Roch formula in the following form:

$$\dim L(D) - \dim \Omega(-D) = \deg D - g + 1, \tag{10}$$

where

$$\Omega(-D) \stackrel{\text{def}}{=} \{\omega \in \mathcal{M}^1(S) \mid \omega \equiv 0 \text{ or } \omega - D \geq 0\}.$$

To see that this is equivalent to the preceding formulation, one uses the isomorphism $\Omega(-D) \to L(K - D)$, $\eta \mapsto \eta/\omega$, where $(\omega) = K$. The challenging and most interesting part of the proof of the Riemann-Roch formula is proving formula (10) for $D \geq 0$. A sketch of proof is given below. The analysis of the remaining cases is simple and is based on the symmetry of the Riemann-Roch formula (see Griffiths-Harris [1978] and Springer [1957]). Here, too, however, there is a hidden layer connected with existence theorems. An important role is played, for example, by the Riemann-Hurwitz formula: $\deg K = 2g - 2$.

To check formula (10) for an effective divisor $D = \sum d_l p_l$, we first consider the $\mathbb{C}$-linear mapping $L(D) \to \mathcal{M}^1(S)$, $f \mapsto df$. The kernel of this map is the set of constants, and hence it is one-dimensional. The image V consists of differentials of the second kind with zero-periods, and poles only at the

[4] In the present section, this means that the tangent mapping is of maximal rank.

p_l, of order $\leq d_l + 1$. We fix a standard basis $a_1, b_1, \ldots, a_g, b_g$ for the group $H_1(S, \mathbb{Z})$. The key point of the proof is the assertion that, for any point p_l with local parameter z_l, there exists a unique differential of the second kind ω_{lj} ($j \geq 2$) with principal part $z_l^{-j} dz_l$ and vanishing A-periods. Existence follows from Corollary 1 of Sect. 4.14 in conjunction with Corollary 3 of Sect. 4.13; and uniqueness from the corollary of Sect. 4.7. Obviously, V is the kernel of the $\mathbb{C}$-linear mapping

$$\psi\colon \oplus_l (\mathbb{C}\omega_{l2} \oplus \ldots \oplus \mathbb{C}\omega_{l\, d_l+1}) \to \mathbb{C}^g$$

$$\eta \mapsto \Big(\int_{b_1} \eta, \ldots, \int_{b_g} \eta\Big).$$

This is enough to yield the Riemann inequality. A more accurate estimate is provided by the following statement.

Reciprocity law for differentials of the first and second kind. *Let ω, respectively η, be a differential of the first, respectively second, kind. Then*

$$\sum_{i=1}^{g} (\Pi_i N_{g+i} - \Pi_{g+i} N_i) = 2\pi\sqrt{-1} \sum_{l,j} \frac{a_{lj} c_{l\,j+2}}{j+1},$$

where Π_i, Π_{g+i}, N_i, and N_{g+i} are A- and B- periods of the forms ω and η, respectively; the c_{lj} are the coefficients of the principal parts of η at p_l:

$$\eta(z_l) = (c_{l2} z_l^{-2} + \ldots) dz_l,$$

and the a_{lj} are the coefficients of the Taylor expansion for ω at p_l:

$$\omega(z_l) = (a_{l0} + a_{l1} z_l + \ldots) dz_l.$$

This result is a mere generalization of the Lemma in Sect. 4.6, taking into account the contribution of residues (the sum on the right) (cf. Griffiths-Harris [1978]). In the space Ω of differentials of the first kind, one can choose a basis $\omega_1, \ldots, \omega_g$ such that $\Pi_j^i \overset{\text{def}}{=} \int_{a_j} \omega_i = \delta_{ij}$ for $1 \leq i, j \leq g$. We say that this basis is *normalized* with respect to the base $a_1, b_1, \ldots, a_g, b_g$. The reciprocity law for ω_i and $\omega_{l\,j+2}$ yields the entries of the matrix of ψ:

$$\int_{b_i} \omega_{l\,j+2} = 2\pi\sqrt{-1}\,\frac{a_{lji}}{j+1}, \quad 0 \leq j \leq d_l - 1,$$

where the a_{lji} are the Taylor coefficients of the form ω_i at p_l. Every relation between the rows of the matrix has a natural interpretation as a differential of the first kind $\omega = \sum \lambda_i \omega_i$, whose Taylor coefficients a_{lj} vanish at the points p_l for $j \leq d_{l-1}$, that is, $\omega \in \Omega(-D)$. To sum up,

$$\dim L(D) = \dim V + 1 = \dim(\ker\psi) + 1 = (\sum d_l) - \operatorname{rank}\psi + 1$$
$$= \deg D - (g - \dim\Omega(-D)) + 1 = \dim\Omega(-D) + \deg D - g + 1.$$

Remark 1. The dimension $i(D) \stackrel{\text{def}}{=} l(K-D)$ is commonly called the *irregularity* of the divisor D.

Remark 2. The more recent proofs of the Riemann-Roch formula are based on cohomology techniques (see Forster [1977] and Gunning [1972]). The Riemann-Roch formula is written in the form

$$\dim H^0(D) - \dim H^1(D) = \deg D - g + 1, \tag{11}$$

where $H^i(D)$ is the i-th cohomology group (using, for instance, Čech cohomology) with coefficients in the sheaf associated with the divisor D. Further, $g \stackrel{\text{def}}{=} \dim H^1(0)$ is the *arithmetic genus* of S. The hard point in the proof of (11) is that $H^1(0)$ is finite dimensional (finiteness of the arithmetic genus). It is easy to derive from this that $H^1(D)$ is finite dimensional for any divisor D. The following two facts are used when going from (11) to (10). One of them is quite simple: the spaces $H^0(D)$ and $L(D)$ are isomorphic. The other fact is deep and delicate, namely the *Serre duality* between $H^1(D)$ and $\Omega(-D)$; in particular, between $H^1(0)$ and Ω. On the other hand, from (10) it is easy to derive the Riemann-Hurwitz formula, $\deg K = 2g - 2$, and the equality of genera in all senses: $\dim H^1(0) = \dim\Omega = \frac{1}{2}\operatorname{rank} H_1(S, \mathbb{Z})$. And from this, one can obtain all known results about existence, together with their corollaries (see the end of § 4 and the next subsection), at least in the compact case.

6.3. Applications of the Riemann-Roch Formula to Problems of Existence of Meromorphic Functions and Differentials

Theorem 1. *In the compact case, the Mittag-Leffler problem is solvable if and only if the conditions of Sect. 4.9 are met.*

That the conditions are necessary was established in Sect. 4.9. That they are also sufficient, in the case of differentials, means that for any effective divisor $D = \sum d_l p_l \neq 0$ the mapping

$$\Omega(D) \to \bigoplus_l (\mathbb{C} z_l^{-1} dz_l \oplus \ldots \oplus \mathbb{C} z_l^{-d_l} dz_l)$$
$$\omega \mapsto \sum_l (c_{l1} z_l^{-1} + \ldots + c_{l\,d_l} z_l^{-d_l}) dz_l,$$

which sends a differential to its principal parts at the p_l, has for image a subspace of codimension 1. This is easy to check by means of the Riemann-Roch formula. Moreover, the image is given by the relation $\sum \operatorname{Res}_{p_l} \omega = \sum c_{l1} = 0$, which is just the necessary condition. The case of functions can be analysed in a similar way.

Theorem 2. *There exists a differential of the second kind with any preassigned periods.*

Any A-periods can be obtained by adding holomorphic differentials. To find differentials of the second kind with zero A-periods and any B-periods, it is enough to check that the map ψ of Sect. 6.2 is surjective for some suitable effective divisor D. Now, this is equivalent to saying that $\dim \Omega(-D) = l(K - D) = 0$, and this relation holds for $\deg D \geq 2g - 1$. For a less wasteful choice of D, see Springer [1957].

Corollary. *There is a natural isomorphism* $H^1_{DR}(S) \simeq \Omega_2/d\mathcal{M}(S)$, *where* Ω_2 *is the space of differentials of the second kind on* S.

Modulo an exact differential this isomorphism maps a closed differentiable 1-form to a differential of the second kind with the same periods, defined up to the total differential of some meromorphic function.

Remark. The isomorphism of the corollary brings us to the algebraic theory of de Rham cohomology (cf. Grothendieck [1966]).

6.4. Compact Riemann Surfaces are Projective. It is obvious from the Proposition of Sect. 6.1 and Corollary 1 of Sect. 6.2 that any compact Riemann surface can be embedded in some projective space. More precisely,

Theorem. φ_D *is an embedding if* $\deg D \geq 2g + 1$.

Thus every compact Riemann surface S is isomorphic to a closed one-dimensional submanifold of some projective space. This is called a *projective model* of S. To describe that there exists a projective model, one says that S is *projective*.

Remark 1. A complex manifold of dimension ≥ 2 need not be projective, even if it is compact (see Chap. 3, Sect. 1.3).

Example 1. It is easy to see from the theorem that the pluricanonical map $\varkappa_d$ is an embedding if $d \geq 3$ and $g \geq 2$, or if $d \geq 2$ and $g \geq 3$. The corresponding projective model is called *pluricanonical.*

Example 2. By means of the Riemann-Roch formula one can also show that the canonical map $\varkappa$ is an embedding when S is non-hyperelliptic. This holds in particular for a generic Riemann surface of genus ≥ 3. Indeed, $l(K - p - q) = g - 2$ if and only if $l(p + q) = 1$ (cf. Example 3 in Sect. 6.1). The corresponding model is called *canonical.*

On projecting S from generic points, the theorem can be strengthened.

Corollary. *Every compact Riemann surface* S *can be embedded in* $\mathbb{CP}^3$.

By the theorem we may assume that $S \subset \mathbb{CP}^n$. Now, projecting S from a point $p \in \mathbb{CP}^n$ into a hyperplane $\mathbb{CP}^{n-1} \subset \mathbb{CP}^n$ (cf. Example 5 of Chap. 2, Sect. 1.2) yields an embedding, provided p does not lie on any chord (complex straight line through a pair of points of S) or tangent to S (cf. Example 5 of

Chap. 2, Sect. 1.7). A simple count of parameters shows that such points p always exist for $n \geq 4$.

Remark 2. A generic Riemann surface cannot be embedded in $\mathbb{CP}^2$. A first obstruction is the genus formula for a plane curve (see Example 2 of Sect. 4.8). Nonetheless there is always an immersion, that is, a regular mapping $S \to \mathbb{CP}^2$ which is one-to-one almost everywhere (see the Example in Sect. 3.11 of Chap. 2).

6.5. Algebraic Nature of Projective Models; Arithmetic Riemann Surfaces. A subset of projective space is said to be algebraic if it is the zero set of some family of homogeneous polynomials in the homogeneous coordinates of the space. Of course the values of these homogeneous polynomials are not well defined on projective space, but their zeros are. Since polynomial rings over fields are noetherian, we may assume without loss of generality that the collection of polynomials defining our algebraic set is finite.

Theorem (Chow). *An embedded Riemann surface $S \subset \mathbb{CP}^n$ is algebraic, that is, its points make up an algebraic subset.*

This is a special case of a theorem of Chow (cf. Sect. 1.4 of Chap. 3), at least so far as its formulation is concerned. In fact, the method of proof is different. We prove, more generally, that the image of a holomorphic mapping from a compact Riemann surface to projective space is algebraic. By successive projections, this statement reduces to the case of mappings into $\mathbb{CP}^2$ (see Corollary 3 of Sect. 2.11). And this case is a consequence of Theorem 1 of Sect. 2.11 on the algebraic nature of finite mappings.

Thus, embedded Riemann surfaces $S \subset \mathbb{CP}^n$ can be described algebraically. It turns out that meromorphic functions on S can also be described in purely algebraic terms. Though a nonzero homogeneous polynomial does not correctly define a function on projective space, the ratio of two such polynomials of the same degree is a rational function, defined outside the zeros of the denominator. If S is not contained entirely in the set of indeterminacy of a rational function on $\mathbb{CP}^n$, it is easily seen that the restriction of that function to S is meromorphic. We also say that it is a *rational function* on S. Locally, by meromorphic function we mean a ratio of holomorphic functions. In this sense, every rational function on $\mathbb{CP}^n$ is meromorphic.

Theorem (on rationality). *A meromorphic function on an embedded Riemann surface $S \subset \mathbb{CP}^n$ is rational.*

Some suitable projections reduce the theorem to the already familiar case where $S = \mathbb{CP}^1$ (see Example 1 of Sect. 2.2 and cf. Corollary 6 of Sect. 2.11).

Remark. The topology of a Riemann surface is nevertheless transcendental in nature. This prevents the notion of a Riemann surface from being made completely algebraic, as well as an immediate algebraic translation of numerous methods for studying Riemann surfaces (cf. Chap. 2, Sect. 1.9).

From an arithmetic viewpoint, the most interesting objects are embedded Riemann surfaces $S \subset \mathbb{CP}^n$ given as zero sets of polynomials with rational coefficients or, more generally, with coefficients in an algebraic number field. We say that these Riemann surfaces are *arithmetic*. Obviously they form a countable set, up to isomorphism. So, the property of being arithmetic is quite a rare phenomenon. Moreover, checking that a Riemann surface is arithmetic is a tedious matter: first of all, a very special embedding in projective space must be selected and, secondly, one has to choose an equally special system of homogeneous coordinates in that space. The fact that arithmetic Riemann surfaces can be characterized topologically seems all the more unexpected.

Theorem (G. Belyĭ [1979]). *A Riemann surface S is arithmetic if and only if there is a mapping $S \to \mathbb{CP}^1$ which is ramified over three points.*

Another striking example of the dependence of arithmetic on topology is the well-known Mordell-Faltings theorem (see Zarkhin-Parshin [1986]).

6.6. Models of Riemann Surfaces of Genus 1. We have met the following special Riemann surfaces of genus 1:

(a) elliptic curves (see Example 2 of Sect. 1.6, Sect. 5.6, and Remark 1 in Sect. 5.7);

(b) hyperelliptic Riemann surfaces with four branch points (see the Example in Sect. 2.11, Example 1 of Sect. 3.6, and Example 1 of Sect. 4.8);

(c) plane cubics (see Example 2 of Sect. 4.8).

A Riemann surface S of genus 1 has a model of each of these types. By Corollary 2 of Sect. 5.7, $S \simeq E_\tau$. Therefore, up to a translation, the hyperelliptic projection $\varphi_{2p}\colon S \to \mathbb{CP}^1$ does not depend on the choice of $p \in S$. On the other hand, it can be reconstructed in a unique way from the images $z_1, z_2, z_3, z_4 \in \mathbb{CP}^1$ of the ramification points (see the Example in Sect. 4.14). Hence isomorphism classes of Riemann surfaces of genus 1 are in one-to-one correspondence with quadruplets of points $\{z_i\}$, modulo linear fractional transformations. Now, the ordered quadruplets (z_1, z_2, z_3, z_4) of points in $\mathbb{CP}^1$ are classified by the cross-ratio

$$\lambda = \frac{z_3 - z_1}{z_3 - z_2} : \frac{z_4 - z_1}{z_4 - z_2}.$$

If z is a coordinate of $\mathbb{CP}^1$ such that $z_1 = \infty$, $z_2 = 0$, and $z_3 = 1$ then $z_4 = \lambda$. Thus S is isomorphic to the Riemann surface of the algebraic function $\sqrt{z(z-1)(z-\lambda)}$ or, equivalently, to the cubic curve $y^2 = z(z-1)(z-\lambda)$ in affine coordinates. Further, any permutation of the z_i corresponds to one of the transformations $\lambda \mapsto \lambda,\ 1/\lambda,\ 1-\lambda,\ 1/(1-\lambda),\ \lambda/(\lambda-1),\ (\lambda-1)/\lambda$. (A permutation of type $(2,2)$ corresponds to the identity transformation $\lambda \mapsto \lambda$.) It is not hard to check that

$$j(S) = \frac{4}{27} \cdot \frac{(\lambda^2 - \lambda + 1)^3}{\lambda^2(1-\lambda)^2}.$$

Example 1. The elliptic curve $E_{\sqrt{-1}}$ corresponds to $\lambda = -1$, $j = 1$. It is isomorphic to the plane cubic $y^2 = z(z^2 - 1)$, which has an automorphism of order 4, namely: $(z, y) \mapsto (-z, \sqrt{-1}\, y)$ (cf. the Theorem of Sect. 5.6).

The embedding $\varphi_{3p}\colon S \hookrightarrow \mathbb{CP}^2$ (see the Theorem of Sect. 6.4) has for image a cubic curve. For $S = E_\tau$ and $p = 0$, one uses the explicit description by means of the Weierstrass function $\wp(z) = \wp(z, \mathbb{Z} + \tau\mathbb{Z})$. The functions $\wp'_z$, $\wp$, and 1 make up a basis for $L(3p)$. In addition, the functions $(\wp'_z)^2$, $\wp^3$, $\wp'_z \cdot \wp$, $\wp^2$, $\wp'_z$, $\wp$, and $1 \in L(6p)$ are linearly dependent since $l(6p) = 6$. This is how we get a relation of the third degree. Moreover, the Laurent expansion $\wp(z) = 1/z^2 + \{\text{terms of even order} \geq 2\}$ shows that

$$(\wp'_z)^2 = 4\wp^3 - g_2\, \wp - g_3, \qquad g_2, g_3 \in \mathbb{C},$$

which is a cubic in *Weierstrass normal form.* The absolute invariant of this cubic is furnished by the formula:

$$j = \frac{g_2^3}{\Delta} = 1 + \frac{27 g_3^2}{\Delta},$$

where $\Delta = g_2^3 - 27g_3^2 = \frac{1}{4}\,(\text{discriminant of } 4z^3 - g_2\, z - g_3)$.

Example 2. The elliptic curve E_ρ with $j = 0$ is isomorphic to a cubic of the form $y^2 = 4z^3 - g_3$, $g_3 \neq 0$. The automorphism $(z, y) \mapsto (\rho^2 z, -y)$ corresponds to a generator of the group $\mathrm{Aut}_0\, E_\rho$ (see the Theorem of Sect. 5.6).

Some further relations between these models – and especially the introduction of a group structure on a cubic – will be treated in Sections 1.7 and 2.6 of Chapter 3.

Remark 1. The models of type (*b*) and (*c*) are of an algebraic nature. This is reflected in the algebraic dependence of the absolute invariant on the natural parameters of these models: λ, g_2, and g_3.

Remark 2. A Riemann surface of genus 1 is arithmetic if and only if its absolute invariant is an algebraic number.

Remark 3. Each of these models has some natural generalizations: (*a*) complex tori and abelian varieties (see Chap. 3, § 1); (*b*) hyperelliptic Riemann surfaces and, more generally, hyperelliptic varieties; (*c*) cubics, that is, zeros of cubic forms in higher-dimensional projective space (see Manin [1972]).

Chapter 2
Algebraic Curves

In the present chapter an algebraic curve is regarded as an object of external geometry. The main results on projective embeddings – properties and constructions – are discussed in § 3. Of major interest here are the highlights of the enumerative geometry of curves: Castelnuovo's inequality and the formulae of Clebsch and Plücker. The required notions and techniques are introduced in the first two sections. A detailed treatment of the material of this chapter, and proofs, can be found in Arbarello et al. [1985], Griffiths-Harris [1978], Hartshorne [1977], and Walker [1950].

§ 1. Basic Notions

According to Chap. 1, § 6, a compact Riemann surface can be identified with an algebraic subset of $\mathbb{CP}^n$. An exact description of these subsets involves the notion of a (complex) algebraic curve. But we cannot explain this concept without including in our dicussion some more general objects from algebraic geometry, namely, projective and quasi-projective varieties. And some of these, like projective space, will be needed from the very outset. This section is based on some first notions, examples, and results from algebraic geometry. For details the reader is referred to Hartshorne [1977] and Shafarevich [1972]. (See also the survey by V. I. Danilov, 'Algebraic varieties and schemes', in this volume.) We complete the section by comparing some fundamental concepts for Riemann surfaces and for complex algebraic curves.

1.1. Algebraic Varieties; Zariski Topology. We take as ground field an arbitrary algebraically closed field k, which will be fixed for all subsequent definitions and constructions. By $\mathbb{P}^n$ we denote n-dimensional projective space over k. A *projective algebraic variety* V is an algebraic subset of $\mathbb{P}^n$, that is, the zero-set of some homogeneous polynomials f_i, $i \in I$, in the homogeneous coordinates $(x_0 : \ldots : x_n)$ of $\mathbb{P}^n$:

$$V = \{(x_0 : \ldots : x_n) \mid f_i(x_0, \ldots, x_n) = 0, \quad i \in I\}.$$

In view of the homogeneity of the f_i, the relation $f_i(x_0, \ldots, x_n) = 0$ holds regardless of the choice of homogeneous coordinates for a point. As the polynomial ring over a field is noetherian, we may assume without loss of generality that the set of polynomials f_i, $i \in I$, which defines the variety V, is finite.

Example 1. Algebraic varieties of the form $\{f = 0\} \subset \mathbb{P}^n$, where f is a nonzero homogeneous polynomial, are called *hypersurfaces*. The hypersurface corresponding to a linear polynomial f is a *hyperplane*, a linear subspace of codimension 1.

It is readily seen that the union of two algebraic subsets of $\mathbb{P}^n$, and the intersection of any number of them, are algebraic. Hence we can think of algebraic sets as being closed in $\mathbb{P}^n$ and we see that $\mathbb{P}^n$ is equipped with a topology. This induces a topology on every projective algebraic variety, which is called the *Zariski topology.* By *algebraic variety* we mean a *quasi-projective variety,* that is, an open subset of some projective variety. A closed subset of an algebraic variety is said to be a *subvariety.*

Caution: The Zariski topology differs markedly from that of complex and differentiable manifolds. First, it is only weakly separable, almost never Hausdorff. For example, the open subsets of $\mathbb{CP}^1$ in the Zariski topology (apart from the empty set) are the complements of finite subsets. Secondly, this topology is compact, in the following sense. A descending chain $V_0 \supsetneq V_1 \supsetneq \dots$ of subvarieties of a quasi-projective variety always terminates. (It is finite.) But, as we shall see later, a more exact analogue of compactness for complex manifolds is 'being projective'. The compactness referred to above is called *Noether's property,* all the more since it bears some direct relationship with the Noether property for polynomial rings over a field.

Example 2. The complement of a hyperplane in $\mathbb{P}^n$ is n-dimensional affine space $\mathbb{A}^n$. Its subvarieties are called *affine varieties.* An affine variety $V \subset \mathbb{A}^n$ is algebraic, that is, quasi-projective, as an open subset of the closure $\overline{V} \subset \mathbb{P}^n$ in the Zariski topology. If $H \subset \mathbb{A}^n$ is the affine hypersurface defined by the zeros of a polynomial $f(x_1, \dots, x_n)$ of degree d, then its closure $\overline{H} \subset \mathbb{P}^n$ is the hypersurface given by the zeros of the homogeneous polynomial $x_0^d f(x_1/x_0, \dots, x_n/x_0)$.

Example 3. It is easy to check that a product of affine algebraic varieties is an affine variety. For example, $\mathbb{A}^n \times \mathbb{A}^m = \mathbb{A}^{n+m}$. To define the product of projective varieties, one may use the Segre embedding

$$\varphi\colon \mathbb{P}^n \times \mathbb{P}^m \hookrightarrow \mathbb{P}^{nm+n+m}$$
$$\big((x_0 : \dots : x_n), (y_0 : \dots : y_m)\big) \mapsto (w_{ij} = x_i\, y_j).$$

Indeed, its image is algebraic, as it is given by the quadratic equations:

$$w_{ij}\, w_{hl} = w_{hj}\, w_{il} \qquad (i, h = 0, \dots, n; \quad j, l = 0, \dots, m)$$

(cf. Shafarevich [1972]). Given two projective varieties, $V \subset \mathbb{P}^n$ and $W \subset \mathbb{P}^m$, it is easy to verify that $\varphi(V \times W)$ is also projective. It is natural to regard this variety as the *product* of V and W. Similarly, one can define the product of quasi-projective varieties.

1.2. Regular Functions and Mappings. A *rational function* on $\mathbb{P}^n$ is the ratio f/g of two homogeneous polynomials f and $g \neq 0$ of the same degree (unless $f = 0$) in the homogeneous coordinates of $\mathbb{P}^n$. The condition on the degrees ensures that the function has a well-defined value in its natural domain of definition, $\mathbb{P}^n - \{g = 0\}$. Such a function on the open set $\mathbb{P}^n - \{g = 0\}$ is said to be regular. More generally, a function $f \colon V \to k$ on a quasi-projective variety $V \subset \mathbb{P}^n$ is called *regular* if for every point there is a neighbourhood (in the Zariski topology) where it is the restriction of a regular function defined on some open subset of $\mathbb{P}^n$.

Example 1. Let f and g be two regular functions, defined on two open subsets U, respectively $V \subset \mathbb{P}^n$. Suppose they coincide on the intersection $U \cap V$. Then they define a regular function on $U \cup V$. But this construction, as well as the general definition, does not lead to any new regular functions. Indeed, every regular function on an open subset $U \subset \mathbb{P}^n$ is given by a rational function f/g such that $U \subset \mathbb{P}^n - \{g = 0\}$. To prove this fact, essential use is made of Hilbert's Nullstellensatz. It is easy to deduce that the only functions that are regular in $\mathbb{P}^n$ are the constants: $f \equiv c, \;\; c \in k$.

Example 2. Similarly, any function which is regular on $\mathbb{A}^n$ can, in affine coordinates, be given by a polynomial.

A *map* $f \colon V \to W$ of quasi-projective varieties $V \subset \mathbb{P}^n$ and $W \subset \mathbb{P}^m$ is said to be *regular* if it is given locally by regular functions. This means that, in some neighbourhood U of any point $p \in U$, one can write f in coordinate form as $y_1 = f_1(p), \ldots, y_m = f_m(p)$, where $y_1, \ldots, y_m$ are affine coordinates in $\mathbb{P}^m$, and $f_1, \ldots, f_m$ are regular functions on U. Algebraic varieties, together with regular mappings, form a category. The invertible maps of this category are called *isomorphisms.* If $V = W$, an isomorphism is called an *automorphism*. A regular function $g \colon W \to k$ may be viewed as a regular mapping into the affine coordinate space k. The regular function $f^*(g) \overset{\text{def}}{=} g \circ f \colon V \to k$ is called the *pull-back* of g with respect to f.

Example 3. The product $\varphi(V \times W)$ of two quasi-projective varieties V and W (see Example 3 of Sect. 1.1) is in fact a product in the sense of category theory. This can be checked by showing that the two projections are regular, as well as all maps arising from universality. Consider, for example, the projection $\varphi(V \times W) \to V$. On the open set $\varphi(V \times W) \cap \{w_{ij} \neq 0\}$, this map is given as follows: $x_0/x_i = w_{0j}/w_{ij}, \; \ldots, \; x_n/x_i = w_{nj}/w_{ij}$.

Let G be a finite automorphism group acting on an algebraic variety V. The *quotient variety* V/G is defined to be an algebraic variety, together with a quotient mapping $V \to V/G$, which is regular and endowed with the standard universal property. The variety V/G exists and is projective, quasi-projective or affine, as the case may be, if V is (see Mumford [1965]). Locally, V/G can be constructed as the image of some embedding in $\mathbb{A}^n$, which is given by a suitable collection of G-invariant functions.

Example 4. The permutation group $\mathcal{S}_d$ acts naturally on the products $\underbrace{\mathbb{A}^1 \times \ldots \times \mathbb{A}^1}_{d}$ and $\underbrace{\mathbb{P}^1 \times \ldots \times \mathbb{P}^1}_{d}$. This action is regular. Now the elementary symmetric polynomials define an isomorphism $(\underbrace{\mathbb{A}^1 \times \ldots \times \mathbb{A}^1}_{d})/\mathcal{S}_d \xrightarrow{\sim} \mathbb{A}^d$, which extends to an isomorphism $(\underbrace{\mathbb{P}^1 \times \ldots \times \mathbb{P}^1}_{d})/\mathcal{S}_d \xrightarrow{\sim} \mathbb{P}^d$.

Example 5. Let L and $P \subset \mathbb{P}^n$ be two disjoint subspaces whose dimensions add up to $n-1$. The *projection map*

$$\begin{aligned} \pi_L \colon \mathbb{P}^n - L &\to P \\ p &\mapsto \overline{Lp} \cap P, \end{aligned}$$

from L to P, is regular. Here $\overline{Lp}$ denotes the linear span of $L \cup \{p\}$. Indeed, if the homogeneous coordinates $(x_0 : \ldots : x_n)$ are chosen so that $L = \{x_m = \ldots = x_n = 0\}$ and $P = \{x_0 = \ldots = x_{m-1} = 0\}$, then we have $\pi_L(x_0 : \ldots : x_n) = (x_m : \ldots : x_n)$.

Example 6. An isomorphism of vector spaces $L_1 \xrightarrow{\sim} L_2$ induces an isomorphism of the corresponding projectivizations $\mathbb{P}(L_1) \xrightarrow{\sim} \mathbb{P}(L_2)$. It is easy to see that every isomorphism between projective spaces can be obtained in this way. In particular, every automorphism of $\mathbb{P}^n$ is linear fractional in affine coordinates (cf. Example 3 of Chap. 1, Sect. 1.4). The bidual space $\mathbb{P}^{n\vee\vee}$ is canonically isomorphic to $\mathbb{P}^n$. There are so-called *correlation isomorphisms* $\mathbb{P}^{n\vee} \xrightarrow{\sim} \mathbb{P}^n$, but they are not canonical.

Example 7. Proportional polynomials define the same hypersurface. Hence those hypersurfaces of $\mathbb{P}^n$ which are given by homogeneous polynomials of degree m are in one-to-one correspondence with the points of the projectivization $\mathbb{P}^{\binom{n+m}{m}-1} = \mathbb{P}(L)$, where L is the space of homogeneous polynomials of degree m in the homogeneous coordinates of $\mathbb{P}^n$. In fact, the monomials $x_0^{i_0} \ldots x_n^{i_n}$, $i_0 + \ldots + i_n = m$, form a basis for this space, and the corresponding coefficients $v_{i_0 \ldots i_n}$ are the coordinates. It is easy to check that the mapping $v_m \colon \mathbb{P}^n \to \mathbb{P}^{\binom{n+m}{m}-1}$ defined by $v_{i_0 \ldots i_n} = x_0^{i_0} \ldots x_n^{i_n}$, $i_0 + \ldots + i_n = m$, is regular. It is known as the *Veronese mapping*, and the image $v_m(\mathbb{P}^n)$ as a *Veronese variety*. This is given by the quadratic equations:

$$v_{i_0 \ldots i_n} v_{j_0 \ldots j_n} = v_{k_0 \ldots k_n} v_{l_0 \ldots l_n},$$

where $i_0 + j_0 = k_0 + l_0, \ldots, i_n + j_n = k_n + l_n$. Moreover, the Veronese mapping is an isomorphism onto its image. If $H \subset \mathbb{P}^n$ is the hypersurface associated with a homogeneous polynomial $f = \sum a_{i_0 \ldots i_n} x_0^{i_0} \ldots x_n^{i_n}$ of degree m, then the image $v_m(H)$ is the intersection of $v_m(\mathbb{P}^n)$ with the hyperplane $\{\sum a_{i_0 \ldots i_n} v_{i_0 \ldots i_n} = 0\}$ in $\mathbb{P}^{\binom{n+m}{m}-1}$. Thus any hypersurface H is isomorphic to a projective variety defined by some quadratic equations. A similar statement holds for any projective variety.

1.3. The Image of a Projective Variety is Closed. The fact that the image is closed under continuous mappings is one of the characteristic properties of compact sets. The analogous property is distinctive of projective varieties.

Example. By definition every quasi-projective variety U has an open inclusion $U \subset V$ in some projective variety V. Obviously, the image is closed only if $U = V$ is projective.

Theorem (cf. Mumford [1976]). *The image of a projective variety under a regular mapping is closed.*

Corollary. *A regular function on a connected projective variety is constant.*

Cf. Corollary 2 of Chap. 1, Sect. 2.5, and see Example 1 of Sect. 1.2 in this chapter.

1.4. Irreducibility; Dimension. An *algebraic variety* is said to be *irreducible* if it cannot be represented as a union of two nonempty proper subvarieties. By the noetherian property of the Zariski topology, every quasi-projective variety can be written as a finite union of irreducible subvarieties. This decomposition is unique. This explains why irreducible varieties are of particular interest in the study of algebraic geometry. An algebraic variety V is of dimension n if the maximal length of a descending chain of irreducible subvarieties

$$V \supset V_0 \supsetneq V_1 \supsetneq \dots \supsetneq V_n \neq \varnothing$$

is $n+1$. The dimension of V is denoted by $\dim V$.

Example 1. $\dim \mathbb{P}^n = \dim \mathbb{A}^n = n$. So, the algebro-geometric dimension is equal to the linear one. A maximal descending chain can be composed of linear subspaces: $\mathbb{P}^n \supset \mathbb{P}^{n-1} \supset \dots \supset \mathbb{P}^0$, respectively, $\mathbb{A}^n \supset \mathbb{A}^{n-1} \supset \dots \supset \mathbb{A}^0$.

Example 2. An irreducible zero-dimensional variety is a point.

Example 3. A hypersurface $H \subset \mathbb{P}^n$ and each of its irreducible components have dimension $n-1$. The converse is also true: a subvariety $H \subset \mathbb{P}^n$, each irreducible component of which is of dimension $n-1$, is a hypersurface. A hypersurface H splits into irreducible components in the same way as its defining polynomial f splits into irreducible factors. If $f = \prod f_i^{p_i}$, where all the f_i are irreducible, then the decomposition into irreducible components is $H = \bigcup H_i$, where $H_i = \{f_i = 0\}$. The irreducible polynomials f_i are determined from H uniquely up to a nonzero constant multiple. The total degree $d = \sum \deg f_i$ is called the *degree of the hypersurface* H. Geometrically, the degree may be defined as the maximum number of points in the intersection of H with a straight line not contained in it. A hypersurface of degree 1 is a hyperplane. The irreducible hypersurfaces of degree 2, 3, 4, etc., are called *quadrics, cubics, quartics,* etc. If they lie in $\mathbb{P}^2$, they are said to be *plane.* But a plane quadric is usually called a *conic.*

One says that some property *holds at a generic point* if it does on a nonempty (Zariski-)open subset. (In the irreducible case, such a set is everywhere dense and the complement is of smaller dimension.) For example, a generic straight line intersects a hypersurface of degree d in d distinct points. Here the lines are regarded as the points of the Grassmann variety of straight lines (see Griffiths-Harris [1978]).

A variety of dimension 2 is called a *surface.* A *curve* is a variety of dimension 1.

1.5. Algebraic Curves. Usually, by an *algebraic curve* one means either a connected or an irreducible quasi-projective or projective variety of dimension 1 (thus excluding arithmetic curves and one-dimensional schemes). In what follows, unless otherwise specified, a curve means an irreducible projective algebraic variety of dimension 1. Moreover we shall be concerned mainly with nonsingular curves (see Sect. 1.6 below).

Example. A hypersurface in $\mathbb{P}^2$ is called a *plane curve.* Any two conics are isomorphic. More precisely, there is a homogeneous coordinate system in which the equation of the conic is $x_0^2 = x_1 x_2$ (or $x_0^2 + x_1^2 + x_2^2 = 0$ if $\operatorname{char} k \neq 2$).

Remark. A curve $C \subset \mathbb{P}^3$ is said to be a *space curve.* Obviously, any space curve is given by at least two equations. It is not known, however, whether it can always be given, as a set, by exactly two equations.

1.6. Singular and Nonsingular Points on Varieties. A first approximation to a variety is its tangent space. Let p be a point of a quasi-projective variety $V \subset \mathbb{P}^n$. Choose an affine coordinate system in $\mathbb{P}^n$ such that $p = (0, \ldots, 0)$. Then any polynomial f that vanishes on V takes the form

$$f(x_1, \ldots, x_n) = \sum_{i=1}^{n} a_i x_i + \text{terms of order} \geq 2.$$

The linear part $\sum a_i x_i$ is the differential of f at p. It can be expressed in terms of (formal) partial derivatives

$$df(p) \stackrel{\text{def}}{=} \sum_{i=1}^{n} a_i x_i = \sum_{i=1}^{n} \frac{\partial f}{\partial x_i}(p)\, x_i .$$

The differentials of all the functions defining V form a subspace

$$T_p = \{df(p) = 0 \mid f(V) = 0\},$$

which is called the *embedded tangent space* to V at the point p. In projective space it is convenient to work with the closure

$$\overline{T}_p = \{df(p) = 0 \mid f \text{ is a homogeneous polynomial vanishing on } V\}.$$

A point $p \in V$ is said to be *nonsingular* if $\dim T_p = \dim_p V$, where $\dim_p V$ is the *dimension of V at the point p*, that is, the maximal dimension of an irreducible component of V through p. The variety V is said to be *nonsingular* if all of its points have this property. The set of singular points of V is denoted by $\operatorname{Sing} V$.

Example 1. $\mathbb{A}^n$ and $\mathbb{P}^n$ are nonsingular. Further, $T_p = \mathbb{A}^n$, respectively $\overline{T}_p = \mathbb{P}^n$, at every point p.

Example 2. A conic is always nonsingular. If it is given by $x_0^2 = x_1 x_2$ then the tangent line $\overline{T}_p$ at a point $p = (a_0 : a_1 : a_2)$ is given by the equation $2a_0 x_0 = a_2 x_1 + a_1 x_2$. If $\operatorname{char} k = 2$, note that every tangent $\overline{T}_p = \{a_2 x_1 + a_1 x_2 = 0\}$ passes through the point $(1 : 0 : 0)$. A nonsingular curve $C \subset \mathbb{P}^n$ is said to be *strange* if all of its tangents pass through some common point. (It is strange because an object always 'moves' along the curve in the same direction, namely towards the intersection of the tangents.) It turns out that the straight line and the conic in characteristic 2 are the only (nonsingular) strange curves (Hartshorne [1977]).

Example 3. A generic plane curve $C \subset \mathbb{P}^2$ of degree d is nonsingular.

Example 4. Let C be a nonsingular algebraic curve. The permutation group S_d acts naturally on the product $C^d = \underbrace{C \times \ldots \times C}_{d}$. The quotient variety $C_d \overset{\text{def}}{=} C^d / S_d$ is called the *d-fold symmetric product* (or *power*) of C. By using the main theorem on symmetric polynomials, it is not difficult to establish that C_d is nonsingular (cf. Example 4 of Sect. 1.2). Similarly, any symmetric power of a Riemann surface is a complex manifold.

Proposition. (a) $\dim T_p \geq \dim_p V$;

(b) $\operatorname{Sing} V$ *is a subvariety of V, and a generic point of each irreducible component of V is nonsingular;*

(c) *The intersection points of the irreducible components of V are singular.*

Nonsingular algebraic curves are in many respects analogous to Riemann surfaces.

Lemma. *Given a nonsingular point p on a curve C, there exists a regular function t in a neighbourhood of that point, such that any other function f which is regular at p can be uniquely expressed as*

$$f = t^{\nu} g, \tag{1}$$

where the function g is also regular at p and $g(p) \neq 0$. (Cf. formula (1) of Chapter 1.)

If $C \subset \mathbb{A}^n$ and $p = 0$, then one can take for t the restriction of a linear form defining a hyperplane through p which is transversal to T_p (that is, which does not contain it).

Definition 1. A function t satisfying the Lemma is called a *local parameter* at p.

Definition 2. The number ν that appears in (1) is independent of the choice of a local parameter t. It is called the *multiplicity* or the *order (of vanishing)* of f at the point p. It is denoted by $\operatorname{ord}_p f$.

Fix a local parameter t at p. Then, given a function f which is regular at p, there is a uniquely determined formal power series $\sum_{i \geq 0} a_i t^i \in k[[t]]$ (Taylor series) with the property that, for every $j \geq 0$, we have:

$$\operatorname{ord}_p\left(f - \sum_{i=0}^{j} a_i t^i\right) \geq j + 1.$$

Obviously, $a_0 = f(0)$, $a_1 = \dfrac{f - a_0}{t}(0)$, etc.

Example 5. If x is an affine coordinate on $\mathbb{A}^1$, then $t = x - x(p)$ is a local parameter at $p \in \mathbb{A}^1$. The function $1/(1-t)$ is regular in a neighbourhood of p and its Taylor series is simply $\sum_{i \geq 0} t^i$. Indeed,

$$\frac{1}{1-t} - \sum_{i=0}^{j} t^i = \frac{t^{j+1}}{1-t}.$$

Locally a regular function f is uniquely determined by its Taylor series or, equivalently, it vanishes in a neighbourhood of p when the Taylor series does. (This circumstance makes regular and analytic functions differ in a very essential manner from differentiable functions, which can have an identically zero Taylor series.) Thus f can be identified with its Taylor series. The formal equality $f = \sum_{i \geq 0} a_i t^i$ has much significance and content (cf. Sect. 1.9 below). Example 5 would then be written as $1/(1-t) = \sum_{i \geq 0} t^i$.

1.7. Rational Functions, Mappings and Varieties. Throughout this section all varieties are supposed irreducible. The reason for this assumption will become clear a little later. By a *rational function* on an algebraic variety V, we mean a regular function $f \colon U \to k$, given on some nonempty open subset $U \subset V$. Two such functions are considered equal if they coincide on some nonempty open subset. As we see, this notion differs notably from the set-theoretic one of a function. To begin with, a rational function need not be defined on the whole of V, but only at a generic point. And then it is an equivalence class of functions. (Something like the Lebesgue functions of calculus.) The latter circumstance is relatively harmless, since we could replace the equivalence class by the choice of a representative with the largest possible domain of definition. This is called the *domain of regularity* of the rational function.

Example 1. We consider projective space $\mathbb{P}^n$, with homogeneous coordinates $(x_0 : \ldots : x_n)$. By definition (see the beginning of Sect. 1.2), a rational function on $\mathbb{P}^n$ is given as a ratio f/g of homogeneous polynomials in $x_0, \ldots, x_n$ of the same degree. Therefore rational functions on $\mathbb{P}^n$ form a field, $k(x_1/x_0, \ldots, x_n/x_0)$. If f and g are coprime polynomials (so that f/g is an irreducible fraction), then the set $\{g(x_0, \ldots x_n) \neq 0\}$ is the domain of regularity of the function f/g.

In the general case, a rational function on a variety $V \subset \mathbb{P}^n$ is the restriction of a rational function on $\mathbb{P}^n$ which is defined at a generic point of V. It easily follows that the rational functions on any (irreducible!) variety V form a field, which is denoted by $k(V)$. This field is finitely generated over k. Indeed, one can take as generators the restrictions of the functions x_i/x_0 (provided that V is not contained in the hyperplane $\{x_0 = 0\}$). In particular, the transcendence degree of $k(V)$ over k is finite.

Example 2. Let $H \subset \mathbb{A}^n$ be a hypersurface, given by an irreducible polynomial $f(x_1, \ldots, x_n)$. Suppose the variable x_n effectively appears in f. Then the functions $x_1, \ldots, x_{n-1}$ are algebraically independent in $k(H)$. So, $k(H)$ is obtained by extending the field $k(x_1, \ldots, x_{n-1})$ by the function x_n, which is connected with $x_1, \ldots, x_{n-1}$ by the algebraic relation $f(x_1, \ldots, x_n) = 0$. Hence the transcendence degree of $k(H)$ is equal to $n - 1 = \dim H$.

Conversely, if a field $k(x_1, \ldots, x_n)$ is finitely generated over k, then it is the rational function field of some algebraic variety. For instance, that of the affine variety in $\mathbb{A}^n$ whose defining equations $f(x_1, \ldots, x_n) = 0$ are all algebraic relations between the elements $x_i \in k(x_1, \ldots, x_n)$. Such an algebraic variety is called a *model of the field* $k(x_1, \ldots, x_n)$. As a model of $k(x_1, \ldots, x_n)$ one may even take a hypersurface (affine or projective, at will). Indeed, any field of transcendence degree m which is finitely generated over k, can be generated by $m + 1$ elements. If $\operatorname{char} k = 0$, this follows immediately from the primitive element theorem. The case where $\operatorname{char} k > 0$ requires some more refined arguments, like choosing a separating transcendence basis. This relies on the fact that the field k, being algebraically closed, is perfect (cf. Zariski-Samuel [1958]). A geometric version of this statement is given in Example 6 below.

On an affine variety every rational function is a ratio of regular functions. This does not apply to projective varieties, where regular functions are always constant. Nevertheless, locally (that is, in a neighbourhood of every point) a rational function is a ratio of regular functions. Hence, by Sect. 1.6, we have the following lemma.

Lemma. *Let p be a nonsingular point on a curve C and let t be a local parameter at p. Then every rational function $f \neq 0 \in k(C)$ can be written in* *a unique way as*

$$f = t^\nu g,$$

where $\nu \in \mathbb{Z}$, the function g is regular at p, and $g(p) \neq 0$.

The integer ν is independent of the choice of a local parameter t. It is called the *order* of f at p and is denoted by $\operatorname{ord}_p f$. If $\operatorname{ord}_p f < 0$ then f is said to have a *pole of multiplicity* or *of order* $-\operatorname{ord}_p f$ at p. Now, since g is regular at p, we can consider its Taylor expansion for t and multiply it by t^ν. We thus obtain the Laurent expansion for f in terms of the parameter t:

$$f = \sum_{i \geq \operatorname{ord}_p f} a_i t^i, \qquad a_{\operatorname{ord}_p f} \neq 0.$$

A *rational map* of V into W is a regular mapping $f\colon U \to W$, where U is a nonempty open subset of V. Two rational maps from V to W are regarded as equal if they coincide on some nonempty open subset. A rational map f from V to W is usually denoted by $f\colon V \dashrightarrow W$. (The broken arrow points to the fact that f is not defined everywhere.) Like rational functions, a rational map has a natural domain of definition, its *domain of regularity,* whose complement consists of points of indeterminacy. The image of a rational map can be thought of as the image of its domain of regularity. Rational functions are rational maps into k.

Example 3. The projection map $\pi_L\colon \mathbb{P}^n \dashrightarrow P$ (see Example 5 of Sect. 1.2) is rational. Its indeterminacy points reduce to the centre of projection L. Let $V \subset \mathbb{P}^n$ be any subvariety which is not contained in L. Then the projection map $\pi_L\colon V \dashrightarrow P$ is also rational. Its points of indeterminacy are included in the intersection $L \cap V$, but do not always coincide with it.

Theorem (on regularity). *A rational map $f\colon C \dashrightarrow V$ from a curve C to a projective variety V is regular at every nonsingular point of C. In particular, it is regular when C is a nonsingular curve.*

Without loss of generality, $V = \mathbb{P}^n$ and $f(C)$ is not contained in the hyperplane $\{x_0 = 0\}$. Then $f = (1 : f_1 : \ldots : f_n)$, where $f_i = f^*(x_i/x_0) \in k(C)$. Then indeterminacies can be cleared away as in Chap. 1, Sect. 5.8. The theorem allows us to regard a rational function on a nonsingular curve C as a regular mapping $C \to \mathbb{P}^1$ (cf. Chap. 1, Sect. 2.2).

Example 4. Let $C \subset \mathbb{P}^n$ be a curve and let p be a nonsingular point on it. Then the projection map π_p, with centre p, defines a regular mapping on C. An easy verification shows that $\pi_p(p) = \overline{T}_p \cap P$. In this sense the tangent line $\overline{T}_p$ to C at p appears as the limit of all secants $\overline{pq}$, $q \in C$.

The composition of two rational maps, $f\colon V \dashrightarrow W$ and $g\colon W \dashrightarrow U$, is well-defined as a rational map, provided the image of f contains an open subset of W. When such is the case, we say that f is a *dominant map.* If a rational map $f\colon V \dashrightarrow W$ is invertible, in the sense that there exists a rational map $g\colon W \dashrightarrow V$ with $g \circ f = id_V$ and $f \circ g = id_W$ as rational maps, then f is said to be *birational*. Further, V and W are *birationally equivalent* if there exists a birational map $f\colon V \dashrightarrow W$. A dominant rational map $f\colon V \dashrightarrow W$ defines a k-extension of fields, $f^*\colon k(W) \hookrightarrow k(V)$. Conversely, every k-extension $\varphi\colon k(W) \hookrightarrow k(V)$ is given by a dominant rational map, $f = (1 : f_1 : \ldots : f_n)\colon V \dashrightarrow W \subset \mathbb{P}^n$, where $f_i = \varphi(x_i/x_0) \in k(V)$.

Remark 1. The functor

$$V \rightsquigarrow k(V),$$
$$f\colon V \dashrightarrow W \rightsquigarrow f^*\colon k(W) \hookrightarrow k(V)$$

defines a duality between the category of algebraic varieties with dominant rational maps and the category of finitely generated fields over k with k-extensions as morphisms.

Thus the classification of algebraic varieties up to birational equivalence is tantamount to that of finitely generated fields over k up to k-isomorphism. The simplest example of such a field is the purely transcendental extension $k(x_1, \ldots, x_n)$ of k, by algebraically independent elements $x_1, \ldots, x_n$. An algebraic variety with this field of rational functions is also said to be *rational*. For example, $\mathbb{P}^n$, $\mathbb{A}^n$, $\mathbb{P}^n \times \mathbb{P}^m$, as well as quadrics, are rational varieties. The rationality of a variety has some important applications, not only in calculus (see indefinite integration in Remark 2 of Chap. 1, Sect. 4.8) and hence in physics, but also in arithmetic. Indeed, if $f = (1 : f_1 : \ldots : f_m)\colon \mathbb{P}^n \dashrightarrow V \subset \mathbb{P}^m$ is a birational dominant map into V, given by rational functions $f_1, \ldots, f_m$ with coefficients in $\mathbb{Q}$, then 'almost all' rational points $(x_0 : \ldots : x_m)$, $x_i \in \mathbb{Q}$, on V are the images of rational points of $\mathbb{P}^n$ under f. In particular, such are the points $(1 : f_1(y_0, \ldots, y_n) : \ldots : f_m(y_0, \ldots, y_n))$, $y_i \in \mathbb{Q}$. This underlies the description of rational and integer solutions for many known diophantine equations.

Obviously, the dimension and the transcendence degree of a field of rational functions are invariant under birational maps. From this, as in Example 2, one derives the following result.

Theorem. *The transcendence degree over k of the field of rational functions of an algebraic variety is equal to its dimension. In particular, algebraic curves are all irreducible algebraic varieties whose field of rational functions is of transcendence degree* 1 *over k.*

Corollary. *A nonsingular projective rational curve is isomorphic to $\mathbb{P}^1$.*

Let $f\colon V \dashrightarrow W$ be a rational, dominant map of algebraic varieties of the same dimension. Such, for example, is a nonconstant mapping of irreducible curves. By the theorem, $f^*\colon k(W) \hookrightarrow k(V)$ is a finite k-extension. The degree of the extension f^* is called the *degree of the map* f and is denoted by $\deg f$. There are notions and results in the geometry of algebraic curves that have no substantial analogue for Riemann surfaces. Many of them have to do with separability. A *map* f is said to be *separable, inseparable,* or *purely inseparable* if f^* has the respective property. In characteristic 0, all maps are separable. The geometric meaning of degree for a map is that, in the separable case, the fibre $f^{-1}(p)$ over a generic point $p \in W$ consists of exactly $\deg f$ points. But, in the inseparable case, the number of points in $f^{-1}(p)$ is less than $\deg f$ for every $p \in W$, and it is even equal to 1 if f is purely inseparable. A more precise statement for curves is given in Sect. 2.1 below. Clearly, f is birational if and only if $\deg f = 1$.

Example 5. The projection map $\pi\colon C \to P$ of a nonsingular curve $C \subset \mathbb{P}^n$ from a point $p \in \mathbb{P}^n - C$ is birational precisely when for a generic point $q \in C$ the straight line $\overline{pq}$ is not tangent to C and does not intersect C at any other point. Further, π is an isomorphic embedding if the preceding condition holds for every point $q \in C$. It is easy to show by counting parameters that, for $n \geq 4$, projection from a generic point $p \in \mathbb{P}^n$ yields a regular embedding of C into $P = \mathbb{P}^{n-1}$. Therefore every nonsingular projective curve can be embedded in $\mathbb{P}^3$. By making use of successive projections from generic points, one can construct a birational map from any projective curve $C \subset \mathbb{P}^n$ to a possibly singular curve in $\mathbb{P}^2$. The centre of projection is chosen at every stage off the surface spanned by the tangent and the chords passing through some nonsingular point of C.

Example 6. One can prove in a similar way that every algebraic variety is birationally equivalent to a hypersurface.

Let V be a projective algebraic variety. A regular mapping $f\colon W \to V$ of degree 1, where W is a nonsingular projective variety, is called a *resolution of the singularities,* or a *desingularization,* of V. The fact that W is projective precludes trivial 'desingularizations' $W \subset V$, with W an open subset composed of nonsingular points of V. The existence of desingularizations for algebraic varieties of dimension ≥ 2 is rather delicate. The case of curves is easier and amounts to the following statement.

Theorem (on the model). *A finitely generated field of transcendence degree* 1 *over* k *has a model formed of a nonsingular projective curve, which is defined uniquely up to isomorphism.*

Uniqueness follows directly from the regularity theorem. One of the simplest approaches to the proof of existence involves the notion of normalization for algebraic varieties. In the case of curves this coincides with desingularization (see the survey by V. I. Danilov in this volume). For $k = \mathbb{C}$, one can also use the analytic desingularization (see Corollary 4 of Chap. 1, Sect. 2.11), together with the algebraic nature of compact Riemann surfaces.

Example 7. The affine curve $y^2 = f(x)$, where $\operatorname{char} k \neq 2$ and $f(x)$ is a polynomial of degree n without multiple roots, is a nonsingular model of the field $k(x, \sqrt{f})$. However, for $n \geq 4$ its closure in $\mathbb{P}^2$ is singular at the only extra point $(0:1:0)$. A nonsingular projective model C of $k(x,\sqrt{f})$ is called a *hyperelliptic curve*. The embedding $k(x) \hookrightarrow k(x,\sqrt{f})$ induces a *hyperelliptic projection* $\gamma\colon C \to \mathbb{P}^1$, of degree 2. This map γ is separable. Moreover, $k(x) \hookrightarrow k(x,\sqrt{f})$ is a Galois extension, whose only nontrivial automorphism sends $\sqrt{f}$ into $-\sqrt{f}$. By the regularity theorem, this defines a *hyperelliptic involution* $j\colon C \to C$, which exchanges the two points in every fibre $\gamma^{-1}(p)$, $p \in \mathbb{P}^1$.

As a matter of fact, it would be more accurate to say that a hyperelliptic curve is a nonsingular projective curve C with a separable projection $\gamma\colon C \to \mathbb{P}^1$ of degree 2. If $\operatorname{char} k \neq 2$, this does not make any difference. But for $\operatorname{char} k = 2$ we must examine the desingularization of the closure of a curve with equation $y^2 + y\,g(x) = f(x)$, where $f(x)$ and $g(x)$ are polynomials in x. (The method of completing the square does not work in characteristic 2 !)

Remark 2. If follows from Remark 1 that there is a duality between the category of nonsingular projective curves with nonconstant mappings and the category of finitely generated fields over k of transcendence degree 1 with k-extensions as morphisms. In particular, the theory of nonsingular projective curves can be presented entirely in the language of field theory (see Chevalley [1951] and Lang [1982]).

Example 8. Let C be a nonsingular projective curve, on which there is a regular action by a finite group of automorphisms G. Then G acts also on the rational function field $k(C)$. It is easy to check that the subfield of invariant functions,

$$k(C)^G = \{f \in k(C) \mid g^*(f) = f, \quad g \in G\},$$

is finitely generated over k and of transcendence degree 1. A nonsingular projective model of $k(C)^G$ is the *quotient curve* C/G. The quotient map $C \to C/G$ is induced by the inclusion $k(C)^G \subset k(C)$. Hence it is separable and of degree equal to the order of G. So it is a *Galois mapping*: G acts transitively on the fibres of the quotient.

Example 9. Let $C \subset \mathbb{P}^n$ be a nonsingular curve and $p = \operatorname{char} k > 0$. The curve C is a model of the field $k(C)$. A change in the action of constants on rational functions, however surprising it may at first seem, can lead to a new curve. For instance, suppose we replace the inclusion $k \subset k(C)$ by its composition with the Frobenius map $k \xrightarrow[\sim]{a^p} k \subset k(C)$, which is an isomorphism since k is perfect. Then, as a nonsingular projective model we get a curve $C_p \subset \mathbb{P}^n$. The rational functions on C_p and on C are the same, but in the former case k acts via multiplication by the p-th powers of constants. Correspondingly, the equations of C_p can be obtained from those of C by taking p-th roots of the coefficients. And raising the coordinates of $\mathbb{P}^n$ to the p-th power yields a regular homeomorphism $\varphi\colon C_p \to C$, which is called the *Frobenius mapping*. Now, though φ is one-to-one, we have $\deg\varphi = p$. The Frobenius map is associated with the purely inseparable extension $k(C) \overset{f^p}{\hookrightarrow} k(C_p) = k(C)$, of degree p. The latter isomorphism $f \mapsto f$ is defined not over k, but over $k \xrightarrow[\sim]{a^{1/p}} k$. The Frobenius mapping and its powers φ^m play an important role in the arithmetic of algebraic varieties over finite fields. This is due to the fact that, if a curve $C \subset \mathbb{P}^n$ is defined over $\mathbb{F}_p$ (that is, given by equations with coefficients in $\mathbb{F}_p$) then $C_p = C$ and $\varphi^m \colon C \to C$ is an endomorphism. Its fixed points are the points of C defined over $\mathbb{F}_{p^m}$: $x_i^{p^m} = x_i \iff x_i \in \mathbb{F}_{p^m}$. These points are counted by the zeta-function of the curve C, which is an analogue of the Riemann ζ-function (see Weil [1949]).

Theorem (on decomposition). *A nonconstant mapping of nonsingular projective curves can be written as the composition of a separable mapping and some Frobenius endomorphisms.*

The proof results from field theory (see Hartshorne [1977]).

1.8. Differentials. Let K be some field over k. A mapping $D \colon K \to L$ into a K-vector space L is called a *differentiation* if

$$D(f+g) = D(f) + D(g) \quad \text{and} \quad D(fg) = f\,D(g) + g\,D(f).$$

If $D(k) = 0$ we talk about a differentiation *over* k. We say that a differentiation $d \colon K \to K^1$ over k is *universal* if, for every differentiation $D \colon K \to L$ over k, there exists a unique K-linear map $f \colon K^1 \to L$ such that $D = f \circ d$. In other words, the following triangle is commutative:

$$\begin{array}{ccc} K & \xrightarrow{\ d\ } & K^1 \\ & D \searrow \quad \swarrow f & \\ & L & \end{array}$$

Obviously, a universal differentiation is defined uniquely up to isomorphism. If K is finitely generated over k, it is not difficult to establish that such a universal differentiation exists and to check that $\dim_K K^1 = \operatorname{tr.deg} K/k$.

Theorem. *For every irreducible algebraic variety V there exists a universal differentiation $d \colon k(V) \to k(V)^1$, and $\dim_{k(V)} k(V)^1 = \dim V$.*

The elements of $k(V)^1$ are called *rational differentials* of degree 1 on V. A differential of the form df is the *differential of the function* $f \in k(V)$.

Example 1. If $x_1, \ldots, x_n$ are affine coordinates on $\mathbb{A}^n$, then $dx_1, \ldots, dx_n$ is a basis for the vector space $k(\mathbb{A}^n)^1$ over $k(\mathbb{A}^n)$. Further, $df = \sum \dfrac{\partial f}{\partial x_i} dx_i$, where $\dfrac{\partial f}{\partial x_i}$ is the formal partial derivative of the rational function $f \in k(\mathbb{A}^n)$ with respect to x_i.

Example 2. For a hypersurface $H \subset \mathbb{A}^n$ given by the zeros of an irreducible polynomial f, the vector space $k(H)^1$ is generated over $k(H)$ by the differentials $dx_1, \ldots, dx_n$, with the single relation $df = \frac{\partial f}{\partial x_i} dx_i = 0$ on H.

Let p be a nonsingular point of a curve C, and let t be a local parameter at p. It is easy to see that $dt \neq 0$. Hence every rational differential ω of degree 1 on C has the form $\omega = f\,dt$, where $f \in k(C)$. If f is regular at p then we say that ω is regular at p. A differential which is regular everywhere on a nonsingular curve C is said to be *regular*. These differentials form a k-vector space, which is denoted by Ω_C, or simply Ω. Rational differentials of degree ≥ 2 are defined as tensor (symmetric) products of differentials of degree 1. The regularity of these differentials is defined locally, just as for differentials of degree 1. The pull-back of rational functions under a dominant rational mapping defines the pull-back for rational differentials. In turn, a regular map defines the pull-back for regular differentials. In what follows, unless stated to the contrary, a differential means a differential of degree 1.

Let $f\colon C \to B$ be a nonconstant regular mapping of nonsingular curves. The pull-back of differentials (of any degree) is trivial exactly if f is inseparable. (This is dual to the triviality of the tangent map at all points, a pathological fact from the real or complex analytic viewpoint.) By contrast, in the separable case, and in particular when $\operatorname{char} k = 0$, the pull-back of differentials (of any degree) is injective: $f^*\omega = 0$ if and only if $\omega = 0$. Moreover, if f is a Galois mapping with group G, then $f^*(k(B)) = k(C)^G$ and $f^*(k(B)^1) = (k(C)^1)^G$. Hence every invariant rational function or differential is the pull-back of a rational function or differential. The corresponding statement for regular differentials in positive characteristic is false in general (cf. Corollary 2 of Chap. 1, Sect. 4.8 and see Example 4 below). On the other hand, for a purely inseparable map f (like the Frobenius mapping φ), there is an isomorphism

$$k(B)^1 \xrightarrow{\sim} k(C)^1, \quad g\,dt \mapsto f^*(g)^{1/\deg f}\, df^*(t)^{1/\deg f},$$

which induces an isomorphism $\Omega_B \xrightarrow{\sim} \Omega_C$ in the projective case. However, these isomorphisms are defined not over k but over $k \xrightarrow[\sim]{a^{1/\deg f}} k$ (cf. Example 9 of Sect. 1.7).

Example 3. The regular differentials on a nonsingular plane curve $C \subset \mathbb{P}^2$ can be described in the same way as the holomorphic differentials in Example 2 of Chap. 1, Sect. 4.8. Therefore $\dim \Omega = \frac{1}{2}(d-1)(d-2)$, where d is the degree of C.

Example 4. Let $C \subset \mathbb{P}^2$ be the closure of the affine curve $y^2 + y = x^3 + x$. It is a nonsingular cubic (if we assume that $\operatorname{char} k \neq 7, 13$). The hyperelliptic projection $\gamma\colon (x, y) \mapsto x$ is a Galois mapping of degree 2. The space Ω of regular differentials is one-dimensional and is generated by the differential

$$\omega = \frac{dy}{3x^2 + x} = \frac{dx}{2y + 1}.$$

If char $k = 2$, it is worth noting that the differential $\omega = dx = \gamma^* dx$ is invariant under the hyperelliptic involution which exchanges the points in each fibre of γ. It is the pull-back of the non-regular differential dx on $\mathbb{P}^1$. The reason for the appearance of such examples will become clear in Sect. 2.4.

Let C be a nonsingular curve and let $\omega \neq 0$ be a rational differential on C. Then at every point $p \in C$ there is a well-defined notion of *order*

$$\operatorname{ord}_p \omega = \operatorname{ord}_p f,$$

where $\omega = f\,dt$ and t is a local parameter at p. If $\operatorname{ord}_p \omega \geq 0$ then ω is regular at p and has a *zero of order* $\operatorname{ord}_p \omega$. Now if $\operatorname{ord}_p \omega < 0$, we say that ω has a *pole of order* $-\operatorname{ord}_p \omega$ at p.

1.9. Comparison Theorems. Now we make the assumption that $k = \mathbb{C}$. A nonsingular complex algebraic curve $C \subset \mathbb{CP}^n$ is a complex submanifold of dimension 1. Therefore it has an *associated Riemann surface* C^{an}. Connectedness follows from the irreducibility as in the construction of the Riemann surface of an algebraic function (cf. Corollary 5 of Chap. 1, Sect. 2.11). Further the surface C^{an} is compact precisely when C is projective. The results of Sections 6.4 and 6.5 in Chapter 1 lead to the following theorem.

Theorem. *Every compact Riemann surface is associated with a complex algebraic curve, which is unique up to isomorphism.*

It is not surprising, therefore, that many notions involving Riemann surfaces can be made algebraic.

Comparison theorems. *Let C and B be nonsingular complex projective curves, with associated Riemann surfaces C^{an} and B^{an}. The following assertions hold:*

1. *A mapping $f\colon C \to B$ is regular if and only if $f^{an} \stackrel{\text{def}}{=} f\colon C^{an} \to B^{an}$ is holomorphic.*
2. $\operatorname{Aut} C^{an} = \operatorname{Aut} C$*; whence f^{an} is normal if and only if f is.*
3. $\mathcal{M}(C^{an}) = \mathbb{C}(C)$*; that is, a function is rational if and only if it is meromorphic. Similarly, for differentials we have:*
4. $\mathcal{M}^1(C^{an}) = \mathbb{C}(C)^1$.
5. *The orders* $\operatorname{ord}_p f$ *and* $\operatorname{ord}_p \omega$ *of f, respectively ω, at a point $p \in C$ are the same, no matter whether we consider meromorphic or rational functions, respectively, differentials.*
6. *For a differential of any degree on C, being holomorphic is the same as being regular; in particular, $\Omega_{C^{an}} = \Omega_C$.*

(To be continued in Sect. 2.6.)

The key point here is assertion 3, which is a restatement of the rationality theorem of Chap. 1, Sect. 6.5. By the inverse function theorem over $\mathbb{C}$, a local parameter for C is also a local parameter for the Riemann surface C^{an} (at the same point). Hence we obtain assertion 5. (A formal power series expansion in terms of a local parameter acquires the usual meaning on C^{an}.)

Assertion 6 paves the way for the algebraization of the genus (see Sect. 2.5 below). However, a number of topological and analytic notions from the theory of Riemann surfaces (and, more generally, of complex manifolds) have no direct algebraic analogues, which points to their transcendental nature. Such, for example, are the universal covering surface, the fundamental group, homology, integration, etc. However, some algebraic substitutes do exist for most of them. Thus we have the algebraic fundamental group, étale topology, l-adic or crystalline cohomology (cf. Milne [1980]), etc.

Remark. With every nonsingular complex algebraic variety V one can associate a complex manifold V^{an} of the same dimension (see, for example, the Theorem of Chap. 3, Sect. 1.2). (One associates a complex analytic space if the variety is singular.) For the relevant comparison theorems, like the identity of 'regular' and 'holomorphic' for mappings of projective varieties, we refer to Serre [1956].

1.10. Lefschetz Principle. It is quite natural to use transcendental methods in the study of complex algebraic varieties, since they can be regarded as complex analytic spaces, or manifolds in the nonsingular case (cf. Griffiths-Harris [1978]). With the Lefschetz principle, the validity of these methods extends to algebraic varieties over an arbitrary algebraically closed field k of characteristic 0. More precisely, any statement about algebraic varieties and mappings that depends on no more than a finite number (or even a continuum) of constants and is true over $\mathbb{C}$, is true over an arbitrary algebraically closed field k of characteristic 0. This is due to the fact that any finitely generated field over $\mathbb{Q}$ (or even with a continuum of generators) can be viewed as a subfield of $\mathbb{C}$. Nevertheless, every particular application of the Lefschetz principle requires some meaningful additions according to the situation.

Example 1. Let $f\colon C \to \mathbb{A}^1$ be a regular mapping of algebraic curves, such that each fibre consists of n distinct points. If $\operatorname{char} k = 0$ then f has degree 1. Hence $n = 1$ and f is an isomorphism. If $k = \mathbb{C}$, f is in fact a finite (n-sheeted) unramified covering of Riemann surfaces. Now the complex space associated with the line $\mathbb{A}^1$ over $\mathbb{C}$ is the Gaussian plane. This being simply connected, $n = 1$ and f is an isomorphism. The case where $\operatorname{char} k = 0$ then follows from the Lefschetz principle. Indeed, f is the restriction of a rational function to the curve $C \subset \mathbb{P}^n$. Further, defining C and f requires only a finite number of constants, namely, the coefficients of a system of equations for the closure $\overline{C} \subset \mathbb{P}^n$, those of f, and the coordinates of the points in the complementary set $\overline{C} - C$. A careful justification, however, has to rely on some additional information on the multiplicities of mappings between curves (see Sect. 2.1). For instance, we must be confident that no new branch points show up as the ground field is extended. This can easily be deduced from Hilbert's Nullstellensatz.

The mapping f is clearly the algebraic analogue of an unramified covering of the affine line $\mathbb{A}^1$. Our result therefore asserts that the algebraic fundamental group of $\mathbb{A}^1$ is trivial, or that $\mathbb{A}^1$ is simply connected provided $\operatorname{char} k = 0$. This is false when $\operatorname{char} k > 0$ (cf. Example 4 of Sect. 1.8 for $\operatorname{char} k = 2$).

Example 2. As with Riemann surfaces, expanding functions in terms of some local parameter (cf. Sect. 1.8) enables one to define the principal part of a rational function or differential. In particular, there is a well-defined notion of residue. Moreover, the sum of all residues of a rational differential on a nonsingular projective curve vanishes. In characteristic 0, this result can be proved by means of the Lefschetz principle. The general case follows from this by the 'extension principle for algebraic identities' (Serre [1959], Chap. II, no. 13).

Catchword: Apply the Lefschetz principle if a result is true only in characteristic 0.

Example 2 shows that this rule should not be interpreted too strictly.

§ 2. Riemann-Roch Formula

We discuss here the Riemann-Roch formula for algebraic curves and some parallel notions (cf. Chap. 1, § 6). The Riemann-Roch formula is one of the most important instruments of study for algebraic curves. The section concludes with a few elementary applications.

2.1. Multiplicity of a Mapping; Ramification. Let $f\colon C \to B$ be a nonconstant regular mapping of nonsingular algebraic curves.

Definition 1. The *multiplicity* of f at a point $p \in C$ is the number

$$\operatorname{mult}_p f \stackrel{\text{def}}{=} \operatorname{ord}_p f^*(t),$$

where t is a local parameter of B at $f(p)$.

Definition 2. If $\operatorname{mult}_p f \geq 2$, the point $p \in C$ is called a *ramification point*, and $f(p) \in B$ a *branch point*, of f. Then $\operatorname{mult}_p f$ is the *multiplicity* or *order of ramification*. Further, a ramification point of order 2 is said to be *simple*.

Example 1. If $f\colon C \to \mathbb{P}^1$ is a rational function, then $\operatorname{mult}_p f = \operatorname{ord}_p f$ when $f(p) = 0$ and $\operatorname{mult}_p f = -\operatorname{ord}_p f$ when $f(p) = \infty$.

Example 2. If $p = \operatorname{char} k > 0$ and $\varphi\colon C_p \to C$ is the Frobenius mapping, then every point $q \in C$ is a branch point of φ of order p. So, by the decomposition theorem of Sect. 1.7, an inseparable map is ramified everywhere. By contrast, a separable mapping has only a finite number of branch points.

Theorem. *Let $f\colon C \to B$ be a nonconstant mapping of nonsingular projective curves. Then the number of points in each fibre $f^{-1}(q)$, $q \in B$, counting multiplicities, is equal to the degree of the mapping, that is,* $\deg f = \sum_{f(p)=q} \operatorname{mult}_p f$.

Corollary. *The number of points in a fibre* $f^{-1}(q)$ *does not exceed* $\deg f$. *It is equal to* $\deg f$ *if and only if* f *is unramified over* q.

The assumption that the curves are projective is clearly essential in the theorem. But the statement remains true in the non-projective case, provided we remove the entire fibre $f^{-1}(q)$ each time a point $q \in B$ is removed. The resulting *mappings* on curves are said to be *finite*. The notion of finiteness may be given a purely algebraic meaning, which underlies the proof of the theorem.

2.2. Divisors. Let C be a nonsingular curve.

Definition 1. A *divisor* D on C is a formal finite sum

$$D = \sum a_i p_i,$$

with $a_i \in \mathbb{Z}$ and $p_i \in C$.

Definition 2. The set of all divisors on an algebraic curve C form a group under addition, known as the *divisor group* $\operatorname{Div} C$.

Definition 3. A divisor $D = \sum a_i p_i$ is said to be *effective* if all $a_i \geq 0$. We write $D \geq 0$ and, more generally, $D_1 \geq D_2$ if the divisor $D_1 - D_2$ is effective.

Definition 4. The number $\deg D \stackrel{\text{def}}{=} \sum a_i$ is called the *degree* of the divisor $D = \sum a_i p_i$.

The degree map $\deg\colon \operatorname{Div} C \twoheadrightarrow \mathbb{Z}$ is an epimorphism.

Example 1. Let $f\colon C \to B$ be a nonconstant mapping of nonsingular curves. Then, for every point $p \in B$, the *fibre* over p defines a divisor

$$f^*p \stackrel{\text{def}}{=} \sum_{f(q)=p} \operatorname{mult}_q f \cdot q.$$

This correspondence extends to a pull-back homomorphism

$$f^*\colon \operatorname{Div} B \to \operatorname{Div} C, \qquad f^*(\textstyle\sum a_i p_i) = \sum a_i f^* p_i.$$

If C and B are projective curves (or if f is a finite mapping) then, by the Theorem of Sect. 2.1, $\deg f^*p = \deg f$ and $\deg f^*D = \deg f \cdot \deg D$ for all $p \in B$ and $D \in \operatorname{Div} B$.

Example 2. A nonzero rational function f on a nonsingular curve C defines a divisor

$$(f) \stackrel{\text{def}}{=} \sum \operatorname{ord}_p f . p.$$

This is called the *divisor of* f. It is indeed a divisor, since every nonzero rational function has a finite number of zeros and poles. On considering f as a mapping $f\colon C \to \mathbb{P}^1$, we get a decomposition $(f) = (f)_0 - (f)_\infty$, where

$(f)_0 \stackrel{\text{def}}{=} f^*0$ is known as the *divisor of zeros,* and $(f)_\infty \stackrel{\text{def}}{=} (1/f)_0 = f^*\infty$ as the *divisor of poles,* of f. The divisor of a rational function on C is said to be *principal.* These divisors form a subgroup of $\operatorname{Div} C$. We say that two divisors, D_1 and D_2, are *linearly equivalent* if their difference $D_1 - D_2$ is principal. The fact that D_1 and D_2 are linearly equivalent is denoted by $D_1 \sim D_2$. If C is a projective curve then every principal divisor has degree zero: $\deg(f) = \deg f^*(0 - \infty) = 0$. Linearly equivalent divisors have the same degree. In particular, if a divisor is linearly equivalent to an effective divisor, its degree is ≥ 0. For instance, a polynomial $f(x)$ is a rational function on $\mathbb{P}^1$ and $(f(x)) = \sum a_i\, p_i - d\,\infty$, where the sum is taken over the roots p_i, with multiplicity a_i, and $d = \deg f$. In this case, saying that the divisor (f) has degree 0 means that the degree of the polynomial is equal to the number of its roots, counted with multiplicities. Every divisor D on $\mathbb{P}^1$ of degree 0 is principal, and every divisor on $\mathbb{P}^1$ of degree d is linearly equivalent to $d.p$, where p is any point of $\mathbb{P}^1$. In the general situation, the classification of divisors of degree 0 on an algebraic curve modulo principal ones leads up to Jacobians (see Chap. 3, Sect. 2.3).

Example 3. A nonzero rational differential ω on a nonsingular curve C defines a divisor

$$(\omega) \stackrel{\text{def}}{=} \sum \operatorname{ord}_p \omega . p.$$

A divisor of this form is called *canonical* and denoted by K_C, or simply K. By the Theorem of Sect. 1.8, all canonical divisors form a complete linear equivalence class, known as the *canonical class.*

A *prime divisor* on an irreducible variety V is a subvariety $D \subset V$ of codimension 1. A *divisor* is a finite linear combination $\sum a_i D_i$ of prime divisors D_i, with integer coefficients. Effective divisors, and the ordering, are defined in the same way as for curves. One can also define the divisor of a rational function (cf. Shafarevich [1972]).

Example 4. The prime divisors in $\mathbb{A}^n$ and in $\mathbb{P}^n$ are the irreducible hypersurfaces. A divisor is any finite combination with integer coefficients. A hypersurface can be identified with an effective divisor $\sum D_i$, where the D_i are distinct prime divisors. The degree of a divisor $\sum a_i D_i$ in $\mathbb{P}^n$ is given by the rule

$$\deg D = \sum a_i \cdot \deg D_i$$

(cf. Example 3 in Sect. 1.4). Further, every divisor $\sum a_i D_i$ in $\mathbb{A}^n$ is principal: $(f) = \sum a_i D_i$, where $f = \prod f_i^{a_i}$ and f_i is an irreducible polynomial defining D_i. Similarly, any effective divisor $\sum a_i D_i$ in $\mathbb{P}^n$ of degree d can be obtained as the divisor of zeros of some homogeneous polynomial $f = \prod f_i^{a_i}$ of degree d. All principal divisors in $\mathbb{P}^n$ have degree 0.

Example 5. To begin with, let $C \subset \mathbb{A}^n$ be a nonsingular affine curve, and let $D = \sum a_i D_i$ be a divisor in $\mathbb{A}^n$ whose *support* $\operatorname{Supp} D \stackrel{\text{def}}{=} \bigcup_{a_i \neq 0} D_i$ does not contain C. The restriction of D to C is given by the relation

$$D|_C \stackrel{\text{def}}{=} (f|_C),$$

where f is a rational function on $\mathbb{A}^n$ such that $(f) = D$. It is well-defined, since f is determined uniquely up to multiplication by a nonzero constant. With the help of affine charts one can define the restriction $D|_C$ to a nonsingular projective curve $C \subset \mathbb{P}^n$ of any divisor in $\mathbb{P}^n$ with $\operatorname{Supp} D \not\supset C$. The restriction of an effective divisor D of degree d is known as a *hypersurface section divisor*. For $d = 1$, one talks about a *hyperplane section divisor* $H|_C$, where $H \subset \mathbb{P}^n$ is a hyperplane.

All hyperplane section divisors (respectively, all sections by hypersurfaces of a given degree) are linearly equivalent and therefore have the same degree. The *degree of a curve* $C \subset \mathbb{P}^n$ is defined to be the degree of any hyperplane section divisor $H|_C$: $\deg C \stackrel{\text{def}}{=} \deg(H|_C)$. A generic hyperplane meets the curve C in $\deg C$ points. Accordingly, the divisor of a hypersurface section of degree d is of degree $d \cdot \deg C$. On replacing the curve C by a nonconstant mapping $\varphi\colon C \to \mathbb{P}^n$ (or $\to \mathbb{A}^n$), we can define the pull-back $\varphi^* D$ of any divisor D whose support does not contain the image $\varphi(C)$: $\varphi^* D|_{\varphi^{-1}U} \stackrel{\text{def}}{=} (\varphi^* f)|_{\varphi^{-1}U}$ for any (affine) open subset $U \subset \mathbb{P}^n$ and any rational function f on $\mathbb{P}^n$ such that $(f)|_U = D|_U$, where

$$\sum a_i D_i\Big|_U \stackrel{\text{def}}{=} \sum_{D_i \cap U \neq \varnothing} a_i D_i.$$

In particular, the *degree* of a possibly singular projective curve $B \subset \mathbb{P}^n$ is the degree of the divisor $\varphi^* H$, where $\varphi\colon C \to B \subset \mathbb{P}^n$ is a desingularization of B and H is a hyperplane in $\mathbb{P}^n$ not containing B.

Proposition. *Let $f\colon C \to \mathbb{P}^n$ be a nonconstant mapping of a projective curve C, and let $H \subset \mathbb{P}^n$ be a hyperplane not containing $f(C)$. Then $\deg f^* H = \deg f \cdot \deg f(C)$.*

Example 6. If $C \subset \mathbb{P}^2$ is a nonsingular curve of degree d, then the geometric version of the formula in Example 2 of Chap. 1, Sect. 4.8 can be rephrased as $K \sim (d-3)L|_C$ for a generic line $L \subset \mathbb{P}^2$. Hence $\deg K = d(d-3)$. The above relations are connected with the *adjunction formula* (Hartshorne [1977]).

2.3. Intersection of Plane Curves. In view of Example 5 of Sect. 2.2, the number of intersection points of any two distinct irreducible curves $C_1 \neq C_2 \subset \mathbb{P}^2$ does not exceed $\deg C_1 \cdot \deg C_2$. More generally, we have:

Theorem (Bézout). *If $C_1, C_2 \subset \mathbb{P}^2$ are two (possibly reducible) curves with no common irreducible component, then they may intersect in no more than $\deg C_1 \cdot \deg C_2$ points.*

A more classical version of this result is as follows.

Corollary 1. *Let $f, g \in k[x,y]$ be two coprime polynomials, of degree n and m, respectively. Then the system $f(x,y) = g(x,y) = 0$ has at most $n \cdot m$ solutions.*

Example 1. Requiring that a conic should pass through an assigned point represents one linear condition on its coefficients. Therefore, through five points lying in general position in $\mathbb{P}^2$ (that is, no three of them are collinear; cf. Sect. 3.3), there is at least one conic. By Bézout's theorem that conic is unique.

Example 2 (Theorem of Pascal). Let $L_1, \ldots, L_6$ be the extended sides of a hexagon inscribed in a conic Q. Then the two cubics $C_1 = L_1 + L_3 + L_5$ and $C_2 = L_2 + L_4 + L_6$ intersect in the nine points p_{12}, p_{23}, p_{34}, p_{45}, p_{56}, p_{61}, p_{14}, p_{25}, and p_{36}, where $p_{ij} = L_i \cap L_j$ (see Fig. 19). Now, the linear conditions for a curve of degree 3 to pass through the first eight points are independent. Indeed, a curve of degree 3 is given by a cubic form with ten coefficients. Further, requiring that the curve should pass through the first eight points and a generic point of Q yields a unique cubic, namely $C_3 = Q + L$, where L is the line through p_{14} and p_{25}. In fact, it follows from Bézout's theorem that a curve of degree 3 which meets a conic in 7 or more points contains it entirely. However, by construction the condition that a cubic should pass through p_{36} depends linearly on the conditions for passing through the first 8 points. Hence every curve of degree 3 passing through these eight points passes through p_{36}. In particular, the curve C_3, and so also the line L, pass through p_{36}, which proves Pascal's theorem.

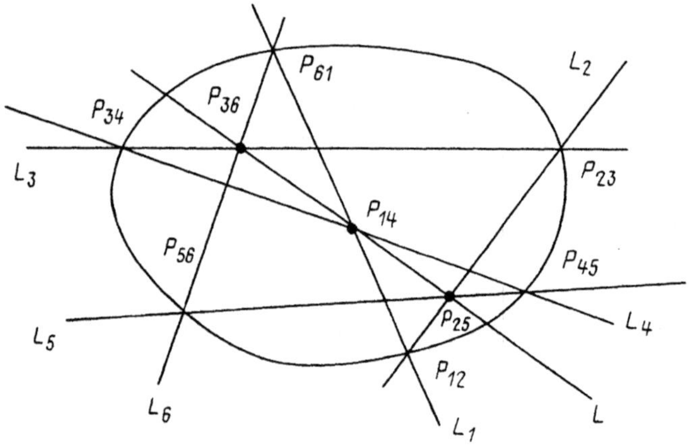

Fig. 19

Corollary 2. *A complex, possibly reducible, curve $C \subset \mathbb{CP}^2$ is connected.*

Bézout's theorem can be turned into an equality: two curves, C_1 and C_2, have $\deg C_1 \cdot \deg C_2$ common points, provided these are counted with suitable multiplicities (cf. the Theorem of Sect. 2.1, and Sect. 3.11 below). Moreover, a point $p \in C_1 \cap C_2$ has multiplicity 1 precisely when C_1 and C_2 are nonsingular at p and their tangents at p meet only in the point p. This explains the following example.

Example 3. A plane cubic $C \subset \mathbb{P}^2$ has no more than one singular point. If we project such a cubic from its singular point, we obtain a birational equivalence with $\mathbb{P}^1$. Hence every singular cubic is rational.

Other applications of Bézout's theorem can be found in Walker [1950]. Some generalizations are given in Shafarevich's book [1972], and further on in Danilov's survey.

2.4. The Hurwitz Formulae. Let $f\colon C \to B$ be a separable mapping of nonsingular curves.

Definition 1. The *ramification index* of a mapping f at a point $p \in C$ is the number $r_p(f) \overset{\text{def}}{=} \operatorname{ord}_p(f^*dt)$, where t is a local parameter at $f(p)$. Then $R = \sum r_p(f).p$ is the *ramification divisor.*

The following two assertions are not hard to prove (cf. Hartshorne [1977]). The second one states when the ramification index can be computed in the usual way, in terms of multiplicity (cf. Chap. 1, Sect. 2.4).

Proposition 1. *If $\omega \neq 0$ is a rational differential on B, then*

$$(f^*\omega) = f^*(\omega) + R.$$

Proposition 2. $r_p(f) = \operatorname{mult}_p f - 1$ *if and only if* $\operatorname{mult}_p f$ *is not divisible by* $\operatorname{char} k$. *In particular, this is always true when* $\operatorname{char} k = 0$.

Remark. If $\operatorname{char} k$ divides $\operatorname{mult}_p f$ then $r_p(f) \geq \operatorname{mult}_p f$.

Definition 2. We say that ramification is *weak*, or *tame*, at a point p if $\operatorname{mult}_p f$ is not divisible by $\operatorname{char} k$. If $\operatorname{char} k = 0$ then ramification is tame at every point. In this sense, mappings of Riemann surfaces are always tame.

Proposition 1 yields the following results (cf. Chap. 1, Sect. 4.8).

Hurwitz formula for canonical divisors. $K_C \sim f^*K_B + R.$

Hurwitz formula for the degree of a canonical divisor. *If C and B are two projective curves then* $\deg K_C = \deg f \cdot \deg K_B + \deg R$.

Example. If $\gamma\colon C \to \mathbb{P}^1$ is a hyperelliptic projection, with ramification points $p_1, \ldots, p_r$, then $K_C \sim \left(\sum\limits_{i=1}^{r-2} p_i\right) - p_{r-1} - p_r$ and $\deg K_C = r - 4$, provided $\operatorname{char} k \neq 2$. In that case, ramification is tame at all points p_i. If, now, $\operatorname{char} k$ is equal to 2 then ramification is *wild* (that is, not tame) at every point

p_i. Further, $\deg K_C \geq 2r - 4$. In particular, a separable mapping $\gamma\colon \mathbb{P}^1 \to \mathbb{P}^1$ of degree two in characteristic 2 has one ramification point of order and index 2.

2.5. Function Spaces and Spaces of Differentials Associated with Divisors. Let $D = \sum a_i p_i$ be a divisor on a nonsingular projective curve C. By analogy with § 6 of Chapter 1 we set:

$$\begin{aligned} L(D) &= \{f \in k(C) \mid f \equiv 0 \text{ or } (f) + D \geq 0\} \\ &= \{f \equiv 0 \text{ or } \operatorname{ord}_{p_i} f \geq -a_i \text{ for all } p_i\}; \end{aligned}$$

$$\begin{aligned} \Omega(D) &= \{\omega \in k(C)^1 \mid \omega = 0 \text{ or } (\omega) + D \geq 0\} \\ &= \{\omega = 0 \text{ or } \operatorname{ord}_{p_i} \omega \geq -a_i \text{ for all } p_i\}. \end{aligned}$$

$L(D)$ and $\Omega(D)$ are vector spaces over k. They are called, respectively, the *rational function space* and the *space of rational differentials* associated with D. As in Chap. 1, § 6, one can prove the following results.

Proposition 1. *If D_1 and D_2 are two linearly equivalent divisors then the spaces $L(D_1)$ and $L(D_2)$ (respectively, $\Omega(D_1)$ and $\Omega(D_2)$) are isomorphic.*

Proposition 2. *$\Omega(D)$ is isomorphic to $L(K + D)$.*

Corollary 1. *$L(D) \neq 0$ if and only if D is linearly equivalent to an effective divisor.*

Corollary 2. *If* $\deg D < 0$ *then* $L(D) = 0$.

Corollary 3. *If $\deg D > \deg K$ then $\Omega(-D) = 0$.*

Theorem. *$L(D)$ and $\Omega(D)$ are finite-dimensional spaces. More precisely,* $\dim L(D) \leq \deg D + 1$, *provided* $\deg D \geq -1$.

We set $l(D) \overset{\text{def}}{=} \dim L(D)$ and $i(D) \overset{\text{def}}{=} \dim \Omega(-D) = l(K - D)$. The number $i(D)$ is called the *irregularity,* or the *index of speciality,* of the divisor D.

Definition. The dimension of the space Ω of regular differentials on a curve C is called its *genus*. It is denoted by $g(C)$, or simply g.

Example. $g(\mathbb{P}^1) = 0$. If x is an affine coordinate on $\mathbb{P}^1$, then $L(d\,\infty) = \{$space of polynomials in x of degree $\leq d\}$. Hence, for every divisor D on $\mathbb{P}^1$ of degree $d \geq -1$, we have $l(D) = l(d\,\infty) = d + 1$. Notice that, in case C is not isomorphic to $\mathbb{P}^1$, then for every divisor D of degree ≥ 1 we have the stronger inequality $l(D) \leq \deg D$ (cf. Example 1 of Chap. 1, Sect. 6.1).

2.6. Comparison Theorems (Continued)

Let C be a nonsingular complex projective curve. From the results of Sect. 1.9 we get:

7. $\operatorname{Div} C^{an} = \operatorname{Div} C$.

8. *The divisor of a function (respectively, of a differential) is the same whether we consider it to be meromorphic or rational. Therefore the property of being principal, and linear equivalence, have the same meaning on C and on C^{an}.*

9. *The spaces $L(D)$ and $\Omega(D)$, and the numbers $l(D)$ and $i(D)$, are the same for C and for C^{an}.*

10. *The canonical divisors on C and on C^{an} coincide.*

11. $g(C) = g(C^{an})$.

2.7. Riemann-Roch Formula. Let C be a nonsingular projective curve.

Theorem (Riemann, Roch). $l(D) - l(K - D) = \deg D - g + 1$.

All versions of the Riemann-Roch formula in Chap. 1, § 6, when properly understood, remain true for an algebraic curve C.

Example 1. If $C = \mathbb{P}^1$ then, by the Example of Sect. 2.5 and the relation $\deg K_{\mathbb{P}^1} = -2$, we obtain the following special case of the theorem:

$$l(D) - l(K_{\mathbb{P}^1} - D) = \deg D + 1.$$

Example 2. The theorem is also obvious for $D = 0$, since $l(0) = 1$ and $l(K) = g$.

2.8. Approaches to the Proof. For $k = \mathbb{C}$, the comparison theorems reduce the Riemann-Roch formula for a curve C to the corresponding formula for the Riemann surface C^{an}. Thus, by the Lefschetz Principle, the analytic proof of the Riemann-Roch formula given in Chap. 1, § 6 is sufficient for algebraic curves in characteristic 0. To be complete, one needs to check only that $l(D)$ does not depend on an extension of the ground field k.

There is also a cohomological proof of the Riemann-Roch formula. It can be carried out by purely algebraic methods, valid in any characteristic. This approach is somewhat simpler for curves than for Riemann surfaces (cf. Remark 2 of Chap. 1, Sect. 6.2). This not only because the Zariski topology on a Riemann surface is much simpler than the classical one, but also because there is a fundamental difference between rational and meromorphic functions, even on a Zariski open subset $U \subset S$. Indeed, meromorphic functions may have some essential singularities at the points of the complement $S - U$. The adelic interpretation of cohomology (see Serre [1959]) is in fact connected with the Mittag-Leffler problems on an algebraic curve. These can be stated and solved in much the same way as for Riemann surfaces (cf. Example 2 of Sect. 1.10).

2.9. First Applications. The Riemann-Roch formula for a canonical divisor $D = K$ yields:

Riemann-Hurwitz formula. $\deg K = 2g - 2$.

Corollary (Hurwitz formula for the genus). *Let $f\colon C \to B$ be a separable mapping of nonsingular projective curves. Then*

$$g(C) = \deg f \cdot g(B) + \tfrac{1}{2}\deg R - \deg f + 1,$$

where R is the ramification divisor (cf. Sect. 3.6 of Chap. 1).

On the other hand, by Sect. 1.8 we have

Proposition 1. *The genus is invariant under a purely inseparable mapping $f\colon C \to B$, that is, $g(C) = g(B)$.*

Example 1. Let $C \subset \mathbb{P}^2$ be a nonsingular plane curve. By Example 6 of Sect. 2.2, the Riemann-Hurwitz formula for C comes down to the formula for the genus of a plane curve (see Example 3 of Sect. 1.8).

Example 2. Let $f\colon C \to B$ be a nonconstant regular mapping of nonsingular projective curves. Then $g(C) \geq g(B)$. Indeed, by Proposition 1 and the decomposition theorem of Sect. 1.7, we may assume that f is separable. So there is a monomorphism $f^*\colon \Omega_B \hookrightarrow \Omega_C$, whence the conclusion. A more precise result is given by Hurwitz's formula for the genus: the equation $g(C) = g(B)$ is possible only if either f is purely inseparable, or $\deg f = 1$, or $g(C) = 0$, or $g(C) = 1$ and f is unramified.

Most applications that follow are based on ideas discussed in Chap. 1, § 6.

Riemann's formula.

$$l(D) = \deg D - g + 1, \quad \textit{provided} \quad \deg D \geq \deg K + 1 = 2g - 1.$$

Let D be a divisor on a nonsingular projective curve C and suppose $l(D) \geq 1$. Choosing a basis $f_0, \dots, f_n$ for $L(D)$ one can define a regular mapping

$$\varphi_D\colon C \to \mathbb{P}^n$$
$$p \mapsto (f_0(p) : \dots : f_n(p)),$$

where $n = l(D) - 1$. That φ_D is rational, is obvious. Then one appeals to the regularity theorem (see Sect. 1.7).

Definition. $\varphi_D\colon C \to \mathbb{P}^n$ is the *mapping associated with* the divisor D.

Theorem 1. *φ_D is an embedding if* $\deg D \geq 2g + 1$.

Example 3. If C is a curve of genus 0, then for every point $p \in C$ there is an isomorphism $\varphi_p\colon C \xrightarrow{\sim} \mathbb{P}^1$.

Corollary 1. *A nonsingular projective curve is rational, that is, isomorphic to $\mathbb{P}^1$, if and only if its genus is equal to zero.*

Corollary 2 (Lüroth's theorem). *If $L \supset k$ is a subfield of a purely transcendental extension $k(x)$ of k, then L is also purely transcendental over k.*

In the nontrivial case where $L \neq k$, we have tr. $\deg L/k = 1$. Hence L is isomorphic to the rational function field of a nonsingular projective curve C. The inclusion $L \subset k(x)$ corresponds to some regular nonconstant mapping $\mathbb{P}^1 \to C$. By Example 2 above, $g(C) = 0$. Thus $C \simeq \mathbb{P}^1$ and $L \simeq k(C) \simeq k(x)$.

Example 4. If C is a curve of genus 1 then any point $p \in C$ gives an embedding $\varphi_{3p}\colon C \hookrightarrow \mathbb{P}^2$. Its image $\varphi_{3p}(C)$ has degree 3, since $\deg K = 0$.

From this one readily obtains:

Corollary 3. *Each of the following properties is characteristic of a curve of genus* 1*:*
(a) $\deg K = 0$;
(b) $K \sim 0$;
(c) *C is isomorphic to a plane cubic.*

Remark 1. One more characterization of curves of genus 1 is given in Sect. 1.2 of Chap. 3.

From the proof of Lüroth's theorem we derive:

Corollary 4. *A regular mapping from $\mathbb{P}^1$, and a fortiori from $\mathbb{P}^n$, to a curve C of genus ≥ 1 is constant.*

If we apply this to a nonsingular plane cubic C, we see that its ninth intersection point with a curve of degree 3 is uniquely determined by the remaining eight points of intersection. This is the clue to understanding Pascal's theorem, since a generic cubic through the first eight points (see Example 2 of Sect. 2.3) is nonsingular.

Example 5. Let C be a curve of genus $g \geq 2$. Then for every $d \geq 1$ we have the *pluricanonical mapping:*

$$\varkappa_d\colon C \longrightarrow \mathbb{P}^n$$
$$p \mapsto (\omega_0(p) : \ldots : \omega_n(p)),$$

where $\omega_0, \ldots, \omega_n$ is a basis for the space Ω^d of regular differentials of degree d. It is easy to verify that this map is isomorphic to φ_{dK}. In particular, $\varkappa_d$ is an embedding, providing $d \geq 3$. But it is more canonical than φ_{dK}. For instance, every automorphism of $\varkappa_d(C)$ induces an automorphism of Ω^d and of $\mathbb{P}^n$. The latter automorphism extends the given one. Thus the group $\operatorname{Aut} C \simeq \operatorname{Aut} \varkappa_d(C)$, for $d \geq 3$, is the group of all linear-fractional transformations (see Example 6 of Sect. 1.2) that leave $\varkappa_d(C)$ invariant. Hence it has a natural structure of algebraic variety together with a regular action of $\operatorname{Aut} C$.

Theorem 2. *The group of automorphisms* $\operatorname{Aut} C$ *of a curve C of genus ≥ 2 is finite.*

By homogeneity the variety $\operatorname{Aut} C$ is smooth (cf. Chap. 3, Sect. 1.1) and all of its components have the same dimension. It remains to check that its dimension is zero. Indeed, by the Noether property, the number of components is finite. Hence it suffices to find the dimension of its tangent space at any point, for example at the point id, which corresponds to the identity automorphism.

Lemma. $\dim T_{id} = l(-K)$.

Indeed, from the infinitesimal viewpoint an automorphism of C is a vector field on C. In other words, the tangent space to $\operatorname{Aut} C$ at id can be naturally identified with the space of regular vector fields on C. Now, the latter space is isomorphic to $L(-K)$.

Example 6. Suppose, as before, that C is a curve of genus $g \geq 2$. Assume further that $\operatorname{char} k = 0$. The degree of the quotient map $f\colon C \to C/\operatorname{Aut} C$ is equal to the order n of $\operatorname{Aut} C$. Denote by $p_1, \ldots, p_s$ a maximal set of ramification points of f lying over distinct points of $C/\operatorname{Aut} C$. Further, let r_i be the order of ramification $\operatorname{mult}_{p_i} f$. The fibre $f^{-1}f(p_i)$ consists of n/r_i ramification points of order r_i. Thus, by Hurwitz's genus formula, we have

$$(2g-2)/n = 2g(C/\operatorname{Aut} C) - 2 + \sum_{i=1}^{s}(1 - 1/r_i).$$

Now, if $g' \geq 0$, $s \geq 0$, and $r_i \geq 2$ (for $i = 1, \ldots, s$) are integers satisfying the inequality

$$2g' - 2 + \sum_{i=1}^{s}(1 - 1/r_i) > 0,$$

it is easy to show that the minimum value of the left-hand side is $1/42$. In this way we have obtained a theorem of Hurwitz: a curve of genus $g \geq 2$ over a field of characteristic 0 has at most $84(g-1)$ automorphisms.

Remark 2. The automorphisms of any nonsingular projective curve C form a group with a natural algebraic variety structure. The dimension of this variety can again be computed from the Lemma. If $g = 0$ then $C \simeq \mathbb{P}^1$ and the dimension of the group of linear-fractional transformations is equal to $l(-K_{\mathbb{P}^1}) = 3$. If now $g = 1$ then C is a plane cubic. Then each point p of C defines a hyperelliptic involution $C \to C$, which exchanges the points forming a pair $L \cap C - \{p\}$, where L is a generic straight line through p (see Fig. 20). This situation is discussed in greater detail in Sect. 1.7 of Chapter 3.

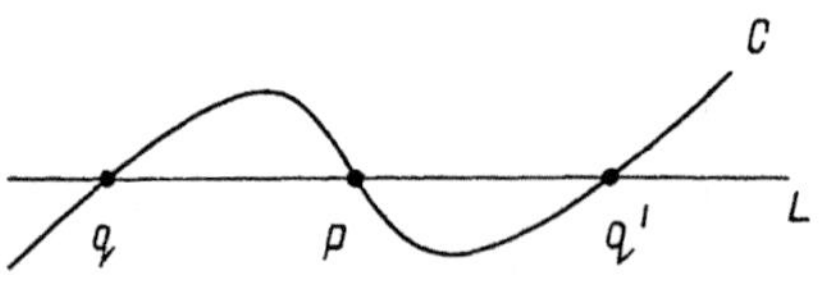

Fig. 20. A hyperelliptic involution on a cubic: $q \leftrightarrow q'$

Example 7. A curve which is the image of a canonical mapping is said to be *canonical*. Unless C is hyperelliptic, the *canonical map* $\varkappa \stackrel{\text{def}}{=} \varkappa_1 : C \to \mathbb{P}^{g-1}$ is an embedding. Further, the canonical map of a hyperelliptic curve C can be written as the composition of a hyperelliptic projection $\gamma\colon C \to \mathbb{P}^1$ and the Veronese mapping $v_{g-1}\colon \mathbb{P}^1 \to \mathbb{P}^{g-1}$ (cf. Example 3 of Chap. 1, Sect. 6.1). It follows that a hyperelliptic curve C of genus ≥ 2 has only one hyperelliptic projection. We further observe that every curve of genus 2 is hyperelliptic. As we shall see below, a generic curve of genus $g \geq 3$ is not hyperelliptic.

The behaviour of $l(D)$ when $0 \leq \deg D \leq 2g-2$ strongly depends on the curve C and on the linear equivalence class of the divisor D. Still, some information can be recovered from the Riemann-Roch formula.

Proposition 2. $l(D) \geq \min\{0, \deg D - g + 1\}$. *In particular,* $l(D) \geq 1$ *if* $\deg D \geq g$.

Proposition 3. $l(D) = 1$ *for a generic effective divisor* D *of degree* $d \leq g$.

The effective divisors of degree d on a curve C can be identified with the points of its symmetric power C_d. By a generic effective divisor of degree d, we understand a generic point of C_d. The proof of Proposition 3 uses the obvious fact that $l(D-p) \geq l(D) - 1$ if $l(D) \geq 1$. Further, equality holds for a generic point $p \in C$. It follows that $l(K-D) = g - \deg D$ for a generic divisor D of degree $d \leq g$.

Remark 3. The Riemann-Hurwitz formula can also be used (cf. Hartshorne [1977]) to substantiate the description of strange curves (see Example 2 of Sect. 1.6).

2.10. Riemann Count. Just like Riemann surfaces, the set $\mathcal{M}_g$ of isomorphism classes of nonsingular projective curves of genus g has a natural algebraic variety structure. This structure is uniquely determined by the following universal property. Every *algebraic family* $f\colon V \to B$ of curves of genus g – that is, every regular mapping f whose fibres $f^{-1}(b)$ are curves of genus g – defines a regular mapping of the base $B \to \mathcal{M}_g$, which sends a point $b \in B$ into the isomorphism class of $f^{-1}(b)$. The variety $\mathcal{M}_g$ is called a *coarse moduli space* for curves of genus g.

Theorem. $\mathcal{M}_g$ *is a quasi-projective irreducible algebraic variety, and its dimension is* $3g-3$ *provided* $g \geq 2$.

The case $g = 0$, where $\mathcal{M}_0$ is a point, is obvious. Further, $\mathcal{M}_1 \simeq k$ (see Chap. 3, Sect. 1.7, and cf. Chap. 1, Sect. 5.10). To prove the existence of a coarse moduli space $\mathcal{M}_g$ when $g \geq 2$, one needs to set up a rather deep and elaborate technique. This is discussed at length, together with the properties of $\mathcal{M}_g$, in the survey on the theory of moduli (see also Deligne-Mumford [1969]). Here we shall mention only how to compute the dimension of $\mathcal{M}_g$ when $g \geq 2$. For this parameter count, it is convenient to assume that $k = \mathbb{C}$.

Then, by the comparison theorems and Example 2 of Chap. 1, Sect. 3.6, an arbitrary collection of $2n+2g-2$ points in $\mathbb{P}^1$ corresponds to some finite (nonempty) collection of algebraic curves C of genus g, equipped with a mapping f of degree n onto $\mathbb{P}^1$, having a simple ramification point above each of the chosen $2n+2g-2$ points of $\mathbb{P}^1$. To find the dimension of $\mathcal{M}_g$, it remains to discover how many such collections correspond to one algebraic curve C of genus g. To this effect, we fix an affine coordinate on $\mathbb{P}^1$. Then our mappings can be identified with rational functions f on C such that $\deg(f)_\infty = n$. For $n > 2g$, these functions are determined by the choice of a divisor D of degree n and a generic element of the space $L(D)$, which has dimension $n-g+1$. Further, the mapping is defined by the rational function up to an automorphism of C. But, when $g \geq 2$, the automorphism group is finite. Hence

$$\dim \mathcal{M}_g = (2n+2g-2) - (n+n-g+1) = 3g-3.$$

From the existence of curves of genus $g \geq 3$ with a trivial automorphism group and from the irreducibility of $\mathcal{M}_g$ we obtain:

Corollary. *A generic curve of genus $g \geq 3$ has a trivial automorphism group. In particular, it is not hyperelliptic* (cf. the Corollary of Chap. 1, Sect. 5.10).

Actually, if $\operatorname{char} k \neq 2$, it is easy to check that hyperelliptic curves form an irreducible subvariety of dimension $2g-1$ in $\mathcal{M}_g$.

Remark 1. For $g \geq 3$, the variety $\mathcal{M}_g$ is singular. Its singular points correspond to curves having a nontrivial automorphism group.

Remark 2. For $g \geq 1$, the variety $\mathcal{M}_g$ is not complete. A natural completion $\overline{\mathcal{M}}_g$ is described in Deligne-Mumford [1969].

Remark 3. If $k = \mathbb{C}$ then $\mathcal{M}_g^{an}$ is a coarse moduli space for Riemann surfaces of genus g. The fact mentioned in Chap. 1, Sect. 5.10, that there are no independent parameters, means that $\mathcal{M}_g$ is not rational if $g \geq 40$ (see Harris [1984] and Harris-Mumford [1982]).

§ 3. Geometry of Projective Curves

In this section we discuss the properties of projective embeddings of curves: the relationship between external and intrinsic numerical invariants, and the structure of equations. The final part is devoted to plane curves with simple singularities.

3.1. Linear Systems. Let D be a divisor on a nonsingular projective curve C. The set of effective divisors

$$|D| \stackrel{\text{def}}{=} \{D' \in \operatorname{Div} C \mid D' \geq 0 \text{ and } D' \sim D\}$$

has a natural structure as a projective space. Indeed,

$$|D| = \{(f) + D \mid 0 \neq f \in L(D)\}$$

and

$$(f) + D = (g) + D \Leftrightarrow g = \lambda f, \ \lambda \in k^\times .$$

Hence $|D|$ is the projectivization of $L(D)$, that is, $|D| = \mathbb{P}(L(D))$. It is easy to see that the projective space structure on $|D|$ is unchanged if we replace D by a linearly equivalent divisor: $|D| = |D'|$ if $D \sim D'$. By construction, $\dim |D| = l(D) - 1$.

Definition. A *linear system*, or *series*, of divisors on a curve C is a projective subspace L of $|D|$. A linear system of the form $|D|$ is said to be *complete.*

A linear system is made up of effective divisors linearly equivalent to one another. One can therefore define the *degree,* $\deg L$, of a nonempty linear system to be the degree of any one of its members. Linear systems of degree d and dimension n have been traditionally denoted by g^n_d.

Example 1. If $C \subset \mathbb{P}^n$ then the hyperplane section divisors (see Example 5 of Sect. 2.2) define on C a linear system $L_C \stackrel{\text{def}}{=} \{H|_C\}$, of degree $\deg C$.

Example 2. Similarly, a mapping $\varphi\colon C \to \mathbb{P}^n$ determines a linear system $L_\varphi \stackrel{\text{def}}{=} \{\varphi^* H\}$. For instance, a hyperelliptic projection $\gamma\colon C \to \mathbb{P}^1$ yields a system $g^1_2 = L_\gamma = \{\gamma^* p \mid p \in \mathbb{P}^1\}$.

Example 3. By the same argument, the pull-backs and the restrictions of effective divisors of degree d in $\mathbb{P}^n$ form linear systems L^d_φ, respectively L^d_C.

A *base point* of a linear system L is a point $p \in C$ that occurs with nonzero multiplicity in every element of L. A linear system containing no base points is said to be *without base points,* or *free.* Such are the linear systems L_C, L_φ, L^d_C, and L^d_φ of the preceding examples. As a general result we have:

Proposition 1. *If $L \neq \varnothing$ then $L = B + L'$, where B is an effective divisor on C and the linear system L' is without base points. Further, this decomposition is unique.*

We say that the divisor B is the *base* of the linear system L. It can be found from the relation $B = \inf_{D \in L} D$, where the infimum is taken with respect to the relation $\geq$ on divisors.

Example 4. The set of divisors $L(-p) = \{D \geq 0 \mid D + p \in L\}$, $p \in C$, is a linear system (cf. Lemma in Sect. 3.2 below).

Proposition 2. *A linear system L is free if and only if* $\dim L(-p) = \dim L - 1$ *for every $p \in C$. In particular, a complete linear system $|D|$ is free if and only if $l(D - p) = l(D) - 1$ for all $p \in C$.*

Hence, by the formulae of Riemann-Roch and of Riemann, we obtain:

Corollary. *A complete linear system* $|D|$ *on a curve* C *of genus* g *is certainly free if one of the following conditions holds:*
(a) $D = K$ *and* $g \geq 1$;
(b) $\deg D \geq 2g$.

3.2. Mappings of Curves into $\mathbb{P}^n$. We wish to describe mappings of the form $\varphi\colon C \to \mathbb{P}^n$, where C is a nonsingular projective curve. A convenient assumption is that the image $\varphi(C)$ is not contained in any linear subspace of $\mathbb{P}^n$. A mapping satisfying this condition is said to be *nondegenerate.* Of course we also say that a curve $C \subset \mathbb{P}^n$ is *nondegenerate* if its inclusion map into $\mathbb{P}^n$ has this property. Let L be a free linear system on C.

Lemma. *The set* $p + L(-p) = \{D \in L \mid D \geq p\}$ *of all effective divisors passing through a point* $p \in C$ *is a hyperplane in* L.

This defines a regular mapping

$$\begin{aligned} \varphi_L\colon C &\to L^\vee \\ p &\mapsto p + L(-p), \end{aligned}$$

where $L^\vee$ is the projective space dual to L.

The regularity can be seen from a representation in coordinates. Fix a divisor $D \in L$. Then $L \subset |D|$; so the system L is the projectivization of a subspace of $L(D)$. Thus φ_L is given by a basis $f_0, \ldots, f_n$ of this subspace: $\varphi_L(p) = (f_0(p) : \ldots : f_n(p))$ (cf. the Remark in Chap. 1, Sect. 6.1). We say that φ_L is the *mapping associated with the free linear system* L. Obviously, φ_L is nondegenerate.

Conversely,

Theorem. *Every nondegenerate mapping* $\varphi\colon C \to \mathbb{P}^n$ *is of the form* φ_L, *where* $L = L_\varphi$ *is a free linear system on* C.

Example 1. For every divisor D the mapping φ_D is isomorphic to $\varphi_{|D-B|}$, where B is the base of the complete linear system $|D|$.

Example 2. If $g \geq 1$, the *canonical* linear system $|K|$ is free. It induces the canonical map $\varkappa = \varphi_{|K|}$. In the same way, the *pluricanonical system* $|dK|$, $d \geq 1$, defines the pluricanonical mapping $\varkappa_d = \varphi_{|dK|}$.

Definition. A mapping $\varphi\colon C \to \mathbb{P}^n$ is said to be *linearly normal* if the linear system L_φ of its hyperplane sections is complete. And we say that a curve $C \subset \mathbb{P}^n$ is *linearly normal* if its inclusion map into $\mathbb{P}^n$ has this property.

Thus the maps φ_D are nondegenerate, linearly normal mappings.

Example 3. The projection map π_p of a nondegenerate curve $C \subset \mathbb{P}^n$ from any point $p \in \mathbb{P}^n - C$ is not linearly normal, for it is induced by the proper subsystem $L_{\pi_p} \subsetneq L$.

Remark. A singular curve $C \subset \mathbb{P}^n$ is said to be nondegenerate, respectively linearly normal, if its desingularization $f\colon \widetilde{C} \to C \subset \mathbb{P}^n$ has this property.

The Proposition of Sect. 2.2 yields the following

Corollary. $\deg L = \deg \varphi_L \cdot \deg \varphi_L(C)$. *In particular,* $\deg \varphi_L(C) = \deg L$ *if* φ_L *is an embedding.*

Example 4. The degree of a canonical, non-hyperelliptic curve $C \subset \mathbb{P}^{g-1}$ is equal to $2g-2$. Conversely, any nondegenerate curve $C \subset \mathbb{P}^{g-1}$ of genus g and degree $2g-2$ is a non-hyperelliptic canonical curve.

Example 5. The image $v_d(\mathbb{P}^1) \subset \mathbb{P}^d$ of the Veronese embedding $v_d = \varphi_{|dp|}$ of degree $d \geq 1$ is a rational, linearly normal curve of degree d. It is known as the *rational normal curve* in $\mathbb{P}^d$. Conversely, every (possibly singular) nondegenerate curve $C \subset \mathbb{P}^d$ of degree d is a rational normal curve. Indeed, if $f\colon \widetilde{C} \to C \subset \mathbb{P}^d$ is its desingularization then L_f is a linear system of degree d and dimension d. So, by the Example in Sect. 2.5, f is the Veronese embedding.

One can use this to show that the canonical image, $\varkappa(C) \subset \mathbb{P}^{g-1}$, of a hyperelliptic curve ($g \geq 2$) is a rational normal curve of degree $g-1$. In fact, the curve $\varkappa(C)$ is nondegenerate and has degree $\frac{1}{2}(2g-2) = g-1$. This approach is useful for proving the existence of a decomposition $\varkappa = v_{g-1} \circ \gamma$ in the hyperelliptic case (see Example 7 in Sect. 2.9) when $\operatorname{char} k = 2$ and there is no description of regular differentials on C comparable to the one in Example 1 of Chap. 1, Sect. 4.8.

Example 6. Let $\pi_p\colon C \to \mathbb{P}^{n-1}$ be the birational projection map from a point $p \in \mathbb{P}^n - C$. Then $\deg \pi_p(C) = \deg C$. Now, if p is a nonsingular point of C then $\deg \pi_p(C) = \deg C - 1$. For instance, if we project a rational normal curve of degree d from a point lying on it, what we get is a rational normal curve of degree $d-1$ (cf. Example 3 above).

3.3. Generic Hyperplane Sections

Proposition. *Let* $C \subset \mathbb{P}^n$ *be a nondegenerate (possibly singular) curve. Then* $\deg C \geq n$, *and equality holds precisely when* C *is a rational normal curve.*

Equivalently,

Corollary. *For any linear system* g^n_d *on a nonsingular curve we have* $n \leq d$, *and equality holds only if the curve is isomorphic to* $\mathbb{P}^1$ *and the system* g^n_d *is complete.*

The latter statement is a reformulation of the Example in Sect. 2.5. Another, more geometric, proof of the Proposition uses the fact that a generic hyperplane section $H|_C$ is a divisor consisting of $\deg C$ points, which span H. This means that the points of $H|_C$ do not lie in any proper projective subspace of H. Otherwise, by the definition of the degree of a curve, the hyperplane passing through all points of the divisor $H|_C$ and one more, arbitrary, point of C would contain C. This would contradict the fact that C is nondegenerate.

Now a stronger statement is true. We say that the points of a subset $M \subset \mathbb{P}^n$ are *in general position* in $\mathbb{P}^n$ if any m of them, for $m \leq n+1$, span a projective subspace of dimension $m-1$:

$$\dim\{\bigcap H \mid \text{the hyperplane } H \text{ contains the given points}\} = m-1.$$

Theorem (on general position). *Let $C \subset \mathbb{P}^n$ be a nonsingular, nondegenerate curve. Then the divisor $H|_C$ of a generic hyperplane section $H \subset \mathbb{P}^n$ consists of* $\deg C$ *points in general position in H.*

It is easy to reduce the general case to that of a space curve $C \subset \mathbb{P}^3$. Then the Theorem states that a generic secant of C is simple. That is, the line through two generic points $p, q \in C$ meets C only at p and q, and it is not tangent to C at either of these points. Besides, it is enough to prove the existence of at least one simple secant. Indeed, simple secants form a Zariski open subset in the variety of all secants of C. Further, it is not hard to show that only strange curves (see Hartshorne [1977]) can have no simple secants. Now, there are no nonsingular, nondegenerate strange curves $C \subset \mathbb{P}^3$. This completes the proof. The classification of strange curves in Example 2 of Sect. 1.6 is therefore the clue to the theorem.

Example (Mumford). Suppose $p = \operatorname{char} k > 0$ and $n \geq 3$. Then the mapping

$$\varphi\colon \mathbb{P}^1 \to \mathbb{P}^n$$
$$(x_0 : x_1) \mapsto (x_0^{p^n} : x_0^{p^n - p} x_1^p : \ldots : x_0^{p^n - p^{n-1}} x_1^{p^{n-1}} : x_1^{p^n})$$

is one-to-one at a generic point of $\mathbb{P}^1$, and the theorem fails to hold for the curve $\varphi(\mathbb{P}^1)$. Indeed, given two generic points, $a = x_1/x_0$ and $b = y_1/y_0$ in $\mathbb{A}^1 \subset \mathbb{P}^1$, the line through $\varphi(a)$ and $\varphi(b)$ contains the points $\varphi(\alpha a + \beta b)$, $\alpha, \beta \in \mathbb{F}_p$, $\alpha + \beta = 1$, which are all situated on the curve $\varphi(\mathbb{P}^1)$:

$$\begin{aligned}\varphi(\alpha a + \beta b) &= \big((\alpha a + \beta b)^p, \ldots, (\alpha a + \beta b)^{p^n}\big) \\ &= \alpha(a^p, \ldots, a^{p^n}) + \beta(b^p, \ldots, b^{p^n}) = \alpha\,\varphi(a) + \beta\,\varphi(b)\end{aligned}$$

in the relevant chart $\mathbb{A}^n \subset \mathbb{P}^n$.

Remark. The theorem is true for singular curves in characteristic 0. The proof in this case is somewhat simpler. One shows that the line is the only strange curve (the tangents at generic points all pass through one point), even among singular curves. This follows from the separability of mappings in characteristic 0. The analytic variant of this argument can be found in Griffiths-Harris [1978].

3.4. Geometrical Interpretation of the Riemann-Roch Formula. The presence or absence on a given curve of a linear system g_d^n with prescribed n and d, is deeply connected with its geometry. For instance, a curve C is hyperelliptic when there is a linear system g_2^1 on it. So, a basic question is: for what n and d does there exist a linear system g_d^n on a curve of genus g? By the Corollary of Sect. 3.3, we know that $n \le d$, with equality only if $g = 0$. A more precise inequality will be given in Sect. 3.5. But we already note that there is no loss of generality in restricting to the case of a complete linear system $g_d^n = |D|$, where D is some effective divisor. If, in addition, $l(K - D) = 0$ then $n = \deg D - g$. The most interesting situation, therefore, is that of a system $g_d^n = |D|$ for which $l(K - D) \ge 1$ or, equivalently, $|K - D| \ne \varnothing$. Such a linear system, and the divisors in it, are called *special*. In view of Proposition 3 of Sect. 2.9, $g_d^n \ne \varnothing$ is a special linear system when $d \le g - 1$; and then $n = 0$ in the generic case. These latter special systems, and their divisors, are said to be *ordinary*. The remaining special systems g_d^n, for which $n > \max\{0, d - g\}$, and their divisors, are said to be *exceptional*. The existence of exceptional special systems is settled by the Brill-Noether theorem and its converse (see Arbarello et al. [1985]), which are discussed in another article of this survey. To understand the geometric meaning of the terms 'special' and 'exceptional', let us consider a non-hyperelliptic canonical curve $C \subset \mathbb{P}^{g-1}$. We denote by $\overline{D}$ the linear span of an effective divisor D:

$$\overline{D} \stackrel{\text{def}}{=} \left\{\bigcap H \mid H \subset \mathbb{P}^{g-1} \text{ is a hyperplane such that } H|_C \ge D\right\}.$$

Then we have:

Theorem. $\dim \overline{D} = \deg D - \dim |D| - 1.$

If, in particular, D is the sum of $\deg D$ distinct points, then the dimension of the linear system $|D|$ is equal to the number of linearly independent relations between these points.

The theorem reduces to the Riemann-Roch formula for the effective divisor D. Indeed, C is a canonical curve; so, the elements of the canonical system are the hyperplane section divisors of C. For a special divisor D, it follows that $\overline{D}$ is a proper subspace of $\mathbb{P}^{g-1}$. Now, the map $H \mapsto H|_C - D$, $H \supset \overline{D}$, identifies the space of hyperplanes through $\overline{D}$ with $|K - D|$. Hence

$$\dim \overline{D} = g - 1 - l(K - D) = \deg D - l(D) = \deg D - \dim |D| - 1,$$

where the second equality is the Riemann-Roch formula. Further, if D is non-special then $\overline{D} = \mathbb{P}^{g-1}$ and the result is again a consequence of the Riemann-Roch formula.

Thus a divisor $D \ge 0$ is special exactly when $\overline{D}$ is a proper subspace of $\mathbb{P}^{g-1}$, and D is ordinary when $\dim \overline{D} = \deg D - 1$. For D consisting of distinct points, this means that the points are in general position in $\overline{D}$.

Remark 1. The geometric interpretation of the Riemann-Roch formula remains valid for a hyperelliptic curve C, provided we replace $H|_C$ in the definition of $\overline{D}$ by the divisor $\varkappa^*H$, where $\varkappa\colon C \to \mathbb{P}^{g-1}$ is the canonical mapping.

Remark 2. A proof of the theorem making no direct use of the Riemann-Roch formula is presented in Griffiths-Harris [1978].

Example 1. The canonical linear system $|K|$, and its divisors, are special for $g \geq 1$ and exceptional for $g \geq 2$. There are no other special (or exceptional) linear systems of degree $2g-2$.

Example 2. Let C be a hyperelliptic curve of genus ≥ 2 with projection γ. The linear system $g_2^1 = |\gamma^* p|$, $p \in \mathbb{P}^1$, is special and exceptional.

Example 3. A curve C is said to be *trigonal* if it allows a mapping $C \to \mathbb{P}^1$ of degree 3. This is obviously equivalent to the existence on C of a free linear system g_3^1, which is complete for $g \geq 2$. It is fairly easy to show that on a hyperelliptic curve of genus ≥ 3 we have: $g_3^1 = g_2^1 +$ one base point. Hence a trigonal curve C of genus ≥ 3 is never hyperelliptic. On the other hand, a non-hyperelliptic, canonical curve $C \subset \mathbb{P}^{g-1}$ is trigonal if there is a one-dimensional family of straight lines $\{\overline{D} \mid D \in g_3^1\}$ which meet C in at least three points, counting osculation multiplicities (cf. Sect. 3.11). The *ruled surface* generated by these lines is nonsingular. If $g \geq 5$, it has no other one-dimensional family of lines. (More precisely, if $g \geq 5$ then one more straight line is possible only for $g = 6$.) Therefore any trigonal curve of genus ≥ 5 has a unique mapping $C \to \mathbb{P}^1$ of degree 3 and a unique linear system g_3^1. The possible types of ruled surfaces and the location of C on them can be found in Shokurov [1971]. A simple parameter count shows that a generic curve of genus $g \geq 5$ is not trigonal. By contrast, any non-hyperelliptic curve of genus 3 or 4 is trigonal. Indeed, in genus 3 such a curve is isomorphic to its canonical model, which is a plane quartic $C \subset \mathbb{P}^2$. Now, every map $C \to \mathbb{P}^1$ of degree 3 is the projection from some point $p \in C$. So the corresponding linear systems g_3^1 are of the form $\{L|_C - p \mid L \text{ is a line through } p\}$. The case $g = 4$ is discussed in Example 6 of Sect. 3.10 below. Note, further, that every curve C of genus ≤ 2 is not only hyperelliptic, but also trigonal.

3.5. Clifford's Inequality

Lemma. *For any two effective divisors D and D', we have*

$$\dim|D| + \dim|D'| \leq \dim|D + D'|.$$

Indeed, the fibres of the following regular mapping are finite:

$$\begin{aligned} |D| \times |D'| &\to |D + D'| \\ (D_1, D_2) &\mapsto D_1 + D_2. \end{aligned}$$

Theorem (Clifford). *Let D be a special divisor on a curve C. Then*

$$\dim|D| \leq \tfrac{1}{2}\deg D.$$

Equality holds only if $D=0$, *or* $D=K\geq 0$, *or if* C *is hyperelliptic and* $D\in(\frac{1}{2}\deg D)\,g_2^1=\underbrace{g_2^1+\ldots+g_2^1}_{\frac{1}{2}\deg D}$, *with* $\deg D\leq 2g-2$.

Indeed,

$$\begin{array}{lr} \dim|D|+\dim|K-D|\leq\dim|K|=g-1 & \text{(Lemma)}\\ +\ \underline{\dim|D|-\dim|K-D|=\deg D-g+1} & \text{(Riemann-Roch)}\\ 2\dim|D|\leq\deg D & \end{array}$$

The description of possible cases of equality amounts to the general position theorem for a canonical curve. A more straightforward approach is explained in Hartshorne [1977]. Figure 21 summarizes what is known about possible values of n and d for a complete linear system g_d^n on a curve C of genus g. Moreover, with one additional constraint the Clifford theorem strengthens the Proposition of Sect. 3.3.

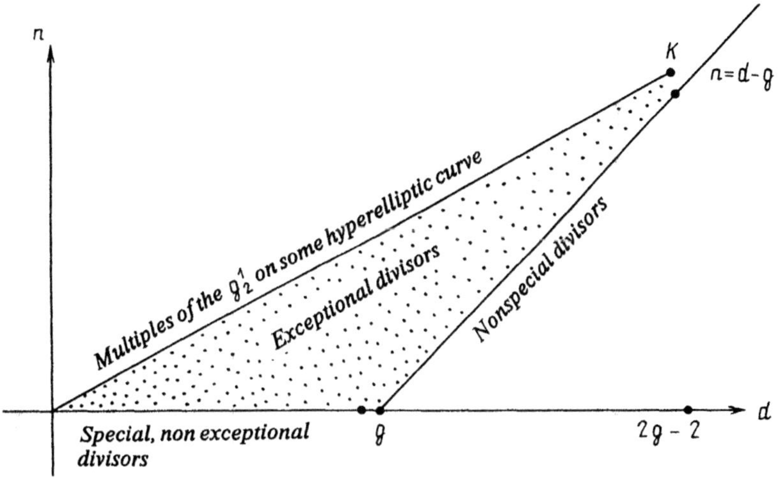

Fig. 21

Corollary. *Let* $C\subset\mathbb{P}^n$ *be a nonsingular nondegenerate curve of degree* $d<2n$ *and genus* g. *Then*

$$d\geq n+g,$$

and equality holds precisely when the curve C *is linear normal.*

Equality in the corollary is attained on any curve of genus $g=d-n$ embedded in $\mathbb{P}^n$ by the mapping associated with an arbitrary divisor of degree d.

Indeed, for such a divisor we have $d \geq 2g+1$. The above inequality can also be written as $g \leq d-n$, which is generalized in the forthcoming section.

3.6. Castelnuovo's Inequality

Let $C \subset \mathbb{P}^n$ be a nonsingular, nondegenerate curve of degree d and genus g. Then

$$g \leq \tfrac{m(m-1)}{2}(n-1) + m\,\varepsilon, \tag{2}$$

where $m = \left[\dfrac{d-1}{n-1}\right]$, $d-1 = m(n-1)+\varepsilon$, *and* $[\ ]$ *is Gauss' integral part.*

Let D be the divisor of a generic hyperplane section of C. By the general position theorem, one can construct sufficiently many linearly independent rational functions in $L(iD)$, and hence bound $l(iD)$ from below. More precisely, we have the estimates:

$$\begin{gathered} l(D) \geq n+1 \\ l(2D) \geq 3(n-1)+3 \\ l(3D) \geq 6(n-1)+4 \\ \cdots\cdots\cdots\cdots\cdots\cdots \\ l(mD) \geq \tfrac{m(m+1)}{2}(n-1)+m+1 \\ \cdots\cdots\cdots\cdots\cdots\cdots\cdots\cdots \\ l((m+i)D) \geq \tfrac{m(m+1)}{2}(n-1)+m+1+i.d \end{gathered}$$

For $i \gg 0$, however, the divisor $(m+i)D$ is non-special. So, by the Riemann-Roch formula,

$$l((m+i)D) = (m+i)d - g + 1.$$

Thus we get the required estimate for the genus.

If the bound is attained, so that equality holds in (2), we say that C is an *extremal curve,* or a *Castelnuovo curve.*

Remark 1. For every pair $d \geq n$, there are some extremal curves $C \subset \mathbb{P}^n$ of degree d. They have an explicit description (see Arbarello et al. [1985] or Griffiths-Harris [1978]).

Remark 2. The Corollary of Sect. 3.5 and the Castelnuovo inequality hold also for a singular curve C, provided it is nondegenerate and $\operatorname{char} k = 0$. However, g means the genus of a desingularization of C. Extremal curves happen to be nonsingular (see Arbarello et al. [1985] or Griffiths-Harris [1978]).

Example 1. As we already know, the case $d = n$ is possible only for the rational normal curve. This curve is extremal.

Example 2. For $n < d < 2n$ one gets the estimate for the genus obtained at the end of Sect. 3.5: $g \leq d-n$. In this case, C is extremal if and only if it is linearly normal.

Example 3. If $n = 2$ then $m = d - 1$, $\varepsilon = 0$, and $g \leq \frac{1}{2}(d-1)(d-2)$. Hence, by the genus formula (see Example 3 in Sect. 1.8), every nonsingular plane curve of degree $d \geq 2$ is extremal.

Example 4. If $d = 2n$ then $m = 2$, $\varepsilon = 1$, and $g \leq n + 1$. Further, equality holds precisely when C is a non-hyperelliptic canonical curve. Hence every canonical curve is extremal.

3.7. Space Curves. In 1882 the Berlin Academy of Sciences proposed to award the valued Steiner prize for the best work on the classification of space curves. The prize was shared by two geometers: Max Noether ([1882]) and G-H. Halphen ([1882]). Their achievements in this field of mathematics have remained unsurpassed until very recently.

Let $C \subset \mathbb{P}^3$ be a nonsingular space curve of degree d and genus g. The first question that arises is how to describe all possible g and d. Only the nondegenerate case is of interest. Otherwise, C is a plane curve with $g = \frac{1}{2}(d-1)(d-2)$.

Theorem (Castelnuovo). *Let $C \subset \mathbb{P}^3$ be a nonsingular, nondegenerate curve. Then $d \geq 3$ and*

$$g \leq \begin{cases} \frac{1}{4}d^2 - d + 1, & \text{if } d \text{ is even}, \\ \frac{1}{4}(d^2 - 1) - d + 1, & \text{if } d \text{ is odd}. \end{cases}$$

Equality is attained for every $d \geq 3$, and a curve for which equality holds is extremal and lies on a quadric in $\mathbb{P}^3$.

For $n = 3$, this is more precise than the inequality of Sect. 3.6. Moreover, if C lies on a quadric then it is known that d and g are of the form $d = a + b$ and $g = (a-1)(b-1)$, for some $a, b \geq 1$. Otherwise,

$$g \leq \tfrac{1}{6}d(d-3) + 1. \tag{3}$$

Hence, for fixed d, there is a gap for the genus above $\frac{1}{6}d(d-3)$. This fact was known to Halphen. He also laid claim to the converse statement: there exists a nonsingular space curve with given nonnegative d and g satisfying (3). We now have rigorous proofs of these facts (cf. Hartshorne [1981–82]). We shall content ourselves with a weaker statement:

Proposition (Halphen). *Any curve C of genus $g \leq d - 3$ can be embedded in $\mathbb{P}^3$ as a curve of degree d.*

One can check that the linear system $|D|$, for a generic effective divisor D of degree $d \geq g + 3$, is free and defines an embedding of C in $\mathbb{P}^{d-g}$. Subsequent projections from generic points yield an embedding in $\mathbb{P}^3$ (cf. Example 5 in Sect. 1.7).

Using the construction of the Chow variety or the Hilbert scheme, one can show that curves of given degree d and genus g in $\mathbb{P}^3$ are parametrized by a finite union of quasi-projective varieties (see Mumford [1965]). Thus we are faced with a more delicate question: to determine the number and the dimensions of these varieties, for all possible d and g. Noether and Halphen provided an exhaustive answer for curves of low degrees (typically, up to degree 20). In the general case the question is as yet open. We illustrate the difficulties that can arise.

Example 1. The only curve of degree 1 in $\mathbb{P}^3$ is the line.

Example 2. The only curve of degree 2 in $\mathbb{P}^3$ is the conic.

Example 3. If $d = 3$ then $C \subset \mathbb{P}^3$ is either a plane cubic with $g = 1$, or a rational normal curve of degree 3.

Example 4. If $d = 4$ then $C \subset \mathbb{P}^3$ is either a plane quartic with $g = 3$, or a rational curve of degree 4, or a curve of genus 1. Curves of this last type are extremal.

The first three examples are supported by the Proposition of Sect. 3.3. For Example 4, one also needs the Corollary in Sect. 3.5. A much more complicated example is as follows.

Example 5. Let $Q \subset \mathbb{P}^3$ be a quadric of rank 4, and let L be one of its generating lines. The intersection of Q with a general quintic through L consists of the line L and a nonsingular curve of degree 9. This curve is extremal and has genus 12 ($a = 5$, $b = 5 - 1 = 4$). It so happens that there are no curves $C \subset \mathbb{P}^3$ with $d = 9$ and $g = 11$, but there are some for any $g \leq 10$. Moreover, there exist two families of curves of degree 9 and genus 10 in $\mathbb{P}^3$. A generic element of the first family is the intersection of two generic cubics in $\mathbb{P}^3$. To describe a generic element of the second family, consider three generators, L_1, L_2, and L_3, belonging to one of the two systems of lines on a quadric $Q \subset \mathbb{P}^3$ of rank 4. Besides these lines, a generic sextic through L_1, L_2, L_3 cuts out on Q a curve of degree 9 and genus 10, which lies in the second family ($a = 6$, $b = 6 - 3 = 3$).

More about this, and some further examples, can be found in Hartshorne [1977].

3.8. Projective Normality. Let $C \subset \mathbb{P}^n$ be a nonsingular projective curve.

Definition 1. A curve C is said to be *m-normal* ($m \geq 1$) if the linear system L_C^m on C is complete.

Definition 2. C is said to be *projectively normal* if it is m-normal for every $m \geq 1$.

Obviously, 1-normal is the same as linearly normal. If a curve $C \subset \mathbb{P}^n$ is linearly normal, and D is a generic hyperplane section, then m-normality stands for the surjectivity of the natural map

$$\operatorname{Sym}^m L(D) \to L(mD). \tag{4}$$

This map takes a k-polynomial of degree m in the rational functions belonging to $L(D)$ and carries it into the associated rational function of $L(mD)$. Thus Sym^m is the symmetric tensor product of degree m over k.

Theorem. *An extremal curve is projectively normal.*

For extremal curves, the inequalities of Sect. 3.6 are in fact equalities. On the other hand, the functions constructed in $L(mD)$ during the proof of Castelnuovo's inequality are in the image of (4). Projective normality follows.

Example 1. A rational normal curve is projectively normal.

Example 2. The image of an embedding $\varphi_D \colon C \to \mathbb{P}^{d-g}$, given by a divisor D of degree $d \geq 2g+1$, is a projectively normal curve. Indeed, it is extremal according to Examples 1 and 2 of Sect. 3.6.

Example 3. Every nonsingular plane curve $C \subset \mathbb{P}^2$ is projectively normal. In particular, the linear system $g_d^2 = L_C$ on C is complete ($d = \deg C$). Moreover, from the geometric interpretation of the Riemann-Roch formula, this g_d^2 is readily seen to be the only linear system of the form g_d^2 on the plane curve C, provided $d \geq 4$. A plane curve of genus ≥ 3 can therefore be embedded in $\mathbb{P}^2$ in a unique way (up to an isomorphism of $\mathbb{P}^2$).

Example 4. A canonical curve is projectively normal. As we know, in the non-hyperelliptic case this means that, for every $m \geq 1$, the map

$$\mathrm{Sym}^m L(K) \to L(mK) \tag{5}$$

is surjective. Now, $L(mK)$ is naturally isomorphic to Ω^m, whence we obtain the famous

Theorem (Noether). *Let C be a non-hyperelliptic curve. Then the natural maps*

$$\mathrm{Sym}^m \Omega \to \Omega^m$$

are epimorphic for all $m \geq 1$. In other words, the algebra of regular differentials $\Omega^\bullet = \bigoplus_{m \geq 1} \Omega^m$ on a non-hyperelliptic curve, is generated by the differentials of degree 1 (in Ω).

By contrast, for a hyperelliptic curve of genus ≥ 2, the maps (5) are not epimorphic for $m \geq 2$. The algebra $\Omega^\bullet$ of regular differentials on a hyperelliptic curve of genus ≥ 3 is generated by differentials of degree 1 and 2.

3.9. The Ideal of a Curve; Intersections of Quadrics. Let C be a projective curve in projective space $\mathbb{P}^n$, with homogeneous coordinates $(x_0 : \ldots : x_n)$. The set of polynomials

$$I(C) = \bigl\{ f \in k[x_0, \ldots, x_n] \bigm| f(\alpha_0, \ldots, \alpha_n) = 0 \text{ for all } (\alpha_0 : \ldots : \alpha_n) \in C \bigr\}$$

is a homogeneous ideal in $k[x_0, \dots, x_n]$. It is called the *ideal of the curve C*. Homogeneity means that

$$I(C) = \bigoplus_{m \geq 1} I_m(C),$$

where $I_m(C)$ is the vector space of all homogeneous polynomials of degree m in $x_0, \dots, x_n$ that vanish on C. Clearly, $0 \neq f \in I_m(C)$ if and only if the hypersurface $\{f = 0\}$ contains or – to put it in another way – passes through C. By definition, any curve can be described as an intersection

$$C = \bigcap \{f_i = 0\}, \tag{6}$$

where the f_i are homogeneous polynomials in $I(C)$. If C is irreducible, we can assume that they, too, are irreducible. By Hilbert's basis theorem, we may further suppose that the number of hypersurfaces $\{f_i = 0\}$ in (6) is finite. Indeed, to achieve this the polynomials f_i need only generate the ideal $I(C)$. When this last condition is met, we talk about a *scheme-theoretic intersection*. If $C \subset \mathbb{P}^n$ is a nondegenerate curve then $I_1(C) = 0$. Hence a scheme-theoretic intersection (6) contains only hypersurfaces $\{f_i = 0\}$ of degree ≥ 2. Sometimes, in the irreducible case, hypersurfaces of degree 2 – or quadrics – are sufficient to define a scheme-theoretic intersection.

Proposition. *The image of an embedding $\varphi_D \colon C \to \mathbb{P}^n$, given by a divisor D of degree $\geq 2g+2$, is a scheme-theoretic intersection of quadrics.*

What the proof uses, in essence, is the general position theorem and a formula giving the dimensions of the spaces $I_m(C)$. This can be more conveniently explained in a more general situation, when $C \subset \mathbb{P}^n$ is a nondegenerate projectively normal curve. By the linear normality of $C \subset \mathbb{P}^n$, the subspace of linear polynomials in $k[x_0, \dots, x_n]$ can be identified with $L(D)$, where D is a hyperplane section of C. This identification extends naturally to

$$k[x_0, \dots, x_n] = \operatorname{Sym} L(D) \stackrel{\text{def}}{=} \bigoplus_{m \geq 0} \operatorname{Sym}^m L(D).$$

Now, there is a natural homomorphism of graded k-algebras

$$\operatorname{Sym} L(D) \to \bigoplus_{m \geq 0} L(mD),$$

whose kernel – under this identification – is the ideal of the curve C. Hence, by projective normality, $I_m(C)$ is the kernel of the epimorphism

$$\operatorname{Sym}^m L(D) \twoheadrightarrow L(mD).$$

But $\dim \operatorname{Sym}^m L(D) = \binom{n+m}{m}$ (cf. Example 7 of Sect. 1.2), whence we have the following lemma.

Lemma. $\dim I_m(C) = \binom{n+m}{m} - l(mD).$

Example 1. A rational normal curve is a scheme-theoretic intersection of quadrics.

The canonical image of a hyperelliptic curve, being a rational normal curve, is a scheme-theoretic intersection of quadrics. By contrast, in the non-hyperelliptic case a canonical curve $C \subset \mathbb{P}^{g-1}$ need not even be an intersection of quadrics.

Example 2. A trigonal, non-hyperelliptic, canonical curve $C \subset \mathbb{P}^{g-1}$ is not an intersection of quadrics. Indeed, any quadric through C contains the straight lines $\overline{D}$, for $D \in g_3^1$, and hence also the ruled surface they generate (see Example 3 of Sect. 3.4).

Example 3. Let $C \subset \mathbb{P}^5$ be the canonical image of a plane quintic $q \subset \mathbb{P}^2$. Now, a plane curve is projectively normal. So, from the description of the canonical divisor of a plane curve (see Example 6 of Sect. 2.2) we see that $|K_q| = L_C^2$. Therefore C is the Veronese image $v_2(q)$. Further, a simple dimension count shows that $I_2(C) = I_2(v_2(\mathbb{P}^2))$. Hence the intersection of all quadrics through C is nothing but the Veronese surface $v_2(\mathbb{P}^2) \subset \mathbb{P}^5$.

As a matter of fact, this deals with all exceptions.

Theorem (Enriques, Babbage, Petri). *A non-hyperelliptic and non-trigonal canonical curve $C \subset \mathbb{P}^{g-1}$ which is not isomorphic to a plane quintic (in particular, if $g \neq 6$) is a scheme-theoretic intersection of quadrics.*

The geometric part of the theorem, namely, that quadrics through such a canonical curve cut out precisely that curve, is due to Enriques ([1919]) and Babbage ([1939]). The more precise analysis of the ideal of a canonical curve – called Petri's analysis – goes back to the work of Petri ([1922]). For a present-day exposition of this approach the reader is referred to Arbarello-Sernesi [1978], Mumford [1975], and Saint-Donat [1973]. Another approach has been developed independently by the author (Shokurov [1971]).

As we know, in the non-hyperelliptic case, the algebra of regular differentials $\Omega^\bullet$ is generated by the regular differentials of degree 1 (see Noether's theorem in Sect. 3.8). Concerning the structure of this algebra, the above theorem implies that it is in fact determined by some quadratic relations between those differentials, provided the curve is non-trigonal and not isomorphic to a plane quintic. In these two exceptional cases, it is determined by some relations of degree 2 and 3. Geometrically, this means that a non-hyperelliptic canonical curve $C \subset \mathbb{P}^{g-1}$ is a scheme-theoretic intersection of quadrics and cubics.

Remark 1. The scheme-theoretic intersection (6) is defined for an arbitrary subvariety $V \subset \mathbb{P}^n$ in the same way as for a curve. According to Example 7 in Sect. 1.2, any variety V can be given, in a suitable projective embedding, as a scheme-theoretic intersection of quadrics. Moreover, quadrics of rank ≤ 4 suffice. This is also true for the curves embedded as in the Proposition and in the Theorem. The proof in the case of canonical curves is by no means trivial. It has been obtained recently by M. Green (cf. Arbarello et al. [1985]). The surfaces in Examples 2 and 3 are scheme-theoretic intersections of quadrics. One easily deduces that a generic non-hyperelliptic canonical curve $C \subset \mathbb{P}^{g-1}$, of genus $g \geq 5$, is a scheme-theoretic intersection of quadrics of rank ≤ 4.

Remark 2. The theorem has some important applications in the theory of surfaces and three-dimensional algebraic varieties. It is used, for example, in the only successful approach to date to proving the existence of lines on certain three-dimensional Fano varieties (Shokurov [1979]).

3.10. Complete Intersections. If a curve $C \subset \mathbb{P}^n$ is given as an intersection of hypersurfaces, say, $C = \bigcap H_i$, then by dimension considerations the number of these hypersurfaces is $\geq n-1$. A scheme-theoretic *intersection* (6) is said to be *complete* if it involves exactly $n-1$ hypersurfaces $H_i = \{f_i = 0\}$.

Example 1. A straight line $L \subset \mathbb{P}^3$ is the complete intersection of two planes. It can also be thought of as a generator of some quadric $Q \subset \mathbb{P}^3$ of rank 3, and can thus be viewed as the intersection of the quadric with its tangent plane along L (see Fig. 22). However, this intersection is neither scheme-theoretic nor complete. Similarly, the rational normal curve $C \subset \mathbb{P}^3$ of degree 3 can be given as the intersection of a quadric of rank 3 with a cubic tangent to it along C, the curve C passing through the only singular point of the quadric. This is neither a scheme-theoretic nor a complete intersection, as can be seen from the following necessary condition.

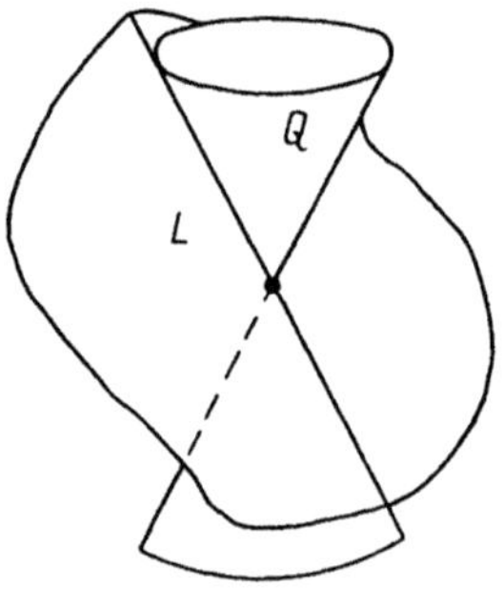

Fig. 22

Theorem. *If a curve $C = \bigcap\{f_i = 0\} \subset \mathbb{P}^n$ is a complete intersection then its degree is equal to the product of the degrees of the polynomials f_i.*

The proof uses the definition of degree and the fact that the hypersurfaces $\{f_i = 0\}$ are nonsingular at all nonsingular points of C, where they meet transversally. In other words, the intersection of their embedded tangent spaces at a nonsingular point $p \in C$ agrees with the embedded tangent space to C at p. (If C is nonsingular, this has to be true at every point.) The argument concludes with Bézout's theorem in projective space (cf. Shafarevich [1972]).

Example 2. It follows from the Proposition of Sect. 3.9 that a nondegenerate, projectively normal curve of degree 4 and genus 1 in $\mathbb{P}^3$ is the complete intersection of two quadrics. It is slightly more difficult to prove the converse, which states that every nonsingular complete intersection of two quadrics in $\mathbb{P}^3$ is of that form. In particular, a nonsingular rational curve of degree 4 in $\mathbb{P}^3$ is not a complete intersection, although $4 = 2 \cdot 2$.

Whether a curve $C \subset \mathbb{P}^n$ is a complete intersection, can be detected by means of the following procedure. For the sake of simplicity, suppose it is irreducible and nondegenerate. First of all, a description of C as a complete intersection should include a maximal collection of linearly independent quadrics $\{f_i = 0\}$ through C, that is, quadrics corresponding to a basis (f_i) of $I_2(C)$. Then it should include a maximal collection of linearly independent cubics $\{f_i = 0\}$ through C that cannot be expressed in terms of quadrics, that is, cubics corresponding to polynomials f_i of degree 3 that are not generated by $I_2(C)$ in $I_3(C)$, and so forth. Then C is a complete intersection if and only if we end up with precisely $n-1$ hypersurfaces $\{f_i = 0\}$. In particular, we see that giving a curve as a complete intersection is in some sense unique.

Example 3. With the Lemma of Sect. 3.9, it is easy to show that $\frac{1}{2}d(d-1)$ linearly independent quadrics pass through the rational normal curve $C \subset \mathbb{P}^d$ of degree d. So, this curve is a complete intersection only if $d \leq 2$.

Example 4. Similarly, for a non-hyperelliptic canonical curve $C \subset \mathbb{P}^{g-1}$, we have:

$$\dim I_2(C) = \binom{g+1}{2} - l(2K) = \tfrac{1}{2}(g-2)(g-3).$$

Hence this curve is not a complete intersection if $g \geq 6$.

Example 5. On the other hand, by the Theorem of Sect. 3.9, a non-hyperelliptic, non-trigonal canonical curve $C \subset \mathbb{P}^4$ of genus 5 is the complete intersection of three quadrics. The converse also holds true for a nonsingular complete intersection of three quadrics in $\mathbb{P}^4$ (see Griffiths-Harris [1978]). By contrast, a non-hyperelliptic, trigonal canonical curve $C \subset \mathbb{P}^4$ of genus 5 is not a complete intersection.

Example 6. Let $C \subset \mathbb{P}^3$ be a non-hyperelliptic, canonical curve of genus 4. There is a unique quadric $Q \subset \mathbb{P}^3$ containing it. Again, the Lemma of Sect. 3.9 yields: $\dim I_3(C) = 5$. Hence there exists a cubic $Q' \subset \mathbb{P}^3$ through C that does not contain Q. An easy verification shows that $Q' \cap Q = C$, the intersection being complete. Conversely, a nonsingular, complete intersection of a quadric and a cubic in $\mathbb{P}^3$ is a non-hyperelliptic canonical curve. Further, if the rank of Q is equal to 4, we note that there are two (complete, free) linear systems g^1_3 on C: they are cut out by the two families of line generators on Q. There are no other systems of type g^1_3 on C. If the rank of Q is equal to 3, then there is only one g^1_3 on C: it is cut out by the unique family of line generators on Q. Thus every non-hyperelliptic curve of genus 4 is trigonal.

Example 7. A non-hyperelliptic, canonical curve $C \subset \mathbb{P}^2$ of genus 3 is a quartic. It is trivially a complete intersection.

3.11. The Simplest Singularities of Curves. Let $f\colon \widetilde{C} \to C \subset \mathbb{P}^2$ be the desingularization of a plane curve C. The intersection multiplicity of C with a line $L \subset \mathbb{P}^2$ at a point $p \in C$ is the total multiplicity $\sum_{f(p_i)=p} a_i$ of the pull-back $f^*L = \sum a_i p_i$ over p. Further, $p \in C$ is called a *double point* if a generic line through p meets C at p with multiplicity 2. Two cases arise.

1. The fibre $f^{-1}(p)$ consists of two points, q_1 and q_2. Each of them corresponds to a tangent line $\overline{T}_j$, with $f^*\overline{T}_j = \sum a_i p_i$, where $a_i \geq 2$ for $p_i = q_j$. If $\overline{T}_1 \cap \overline{T}_2 = \{p\}$ then we say that p is an *ordinary double point.* Intuitively, one may think that there are two nonsingular, transversal branches of C through such a point (see Fig. 23).

2. The fibre $f^{-1}(p)$ consists of one point q. There is a unique tangent line $\overline{T}$ through p, which meets C at p with multiplicity ≥ 3. If $\overline{T}$ intersects C at p with multiplicity 3 then p is called a *cuspidal point,* or a *cusp* (see Fig. 23). Analytically, this amounts to saying that there exist affine coordinates x, y in $\mathbb{P}^2$ such that

$$f^*(x) = t^2 + \text{terms of order} \geq 3,$$
$$f^*(y) = t^3 + \text{terms of order} \geq 4,$$

where t is a local parameter at q.

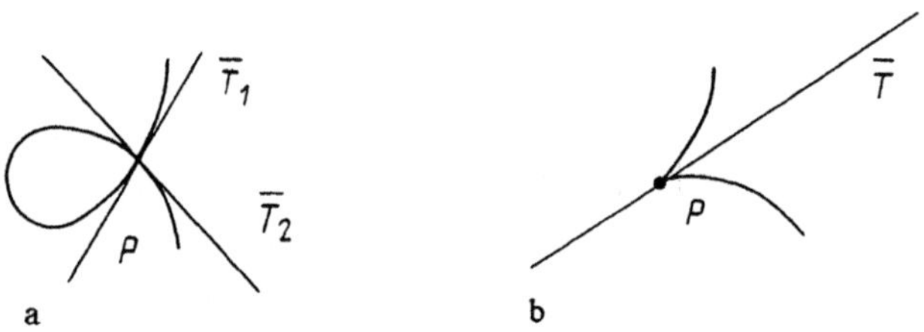

Fig. 23. The most elementary double points: (a) ordinary; (b) cuspidal

Example. Let $C \subset \mathbb{P}^3$ be a nonsingular space curve. The image $\pi(C)$ of the projection $\pi\colon C \to \mathbb{P}^2$ from a generic point of $\mathbb{P}^3$ has only ordinary double points for singularities (cf. Griffiths-Harris [1978]). If the curve C is nondegenerate then, by Example 3 of Sect. 3.8, such singularities do indeed show up on $\pi(C)$. Moreover, it follows from Example 5 of Sect. 1.7 that a generic projection $\pi: C \to \mathbb{P}^2$ of a nonsingular curve $C \subset \mathbb{P}^n$ is birational onto its image $\pi(C)$, which has only ordinary double points for singularities. The curve $\pi(C)$ has the same degree as $C \subset \mathbb{P}^n$.

3.12. The Clebsch Formula

Definition. The *geometric genus* of an irreducible curve C is the genus $g(\widetilde{C})$ of its desingularization $\widetilde{C} \to C$.

Let $C \subset \mathbb{P}^2$ be an irreducible plane curve of degree d and geometric genus g with only the simplest kinds of double points for singularities: δ ordinary double points and $\varkappa$ cusps.

Theorem (Clebsch; cf. Griffiths-Harris [1978]). $g = \frac{1}{2}(d-1)(d-2) - \delta - \varkappa$.

The proof rests on the method of adjunction. In particular, one has the following formula for a canonical divisor of $\widetilde{C}$:

$$K_{\widetilde{C}} \sim (d-3)\, f^*L - \left(\sum (q_1^i + q_2^i) + 2\sum q^j\right),$$

where L is a generic line in $\mathbb{P}^2$ (cf. Example 6 of Sect. 2.2), q_1^i and q_2^i are the inverse images of an ordinary double point, and q^j is that of a cusp. On computing the degree of $K_{\widetilde{C}}$, we obtain Clebsch's formula.

Example 1. Let $C \subset \mathbb{P}^2$ be an irreducible, singular cubic. It has a unique double point, which is either ordinary or cuspidal. Hence its geometric genus is equal to 0. This follows also from the fact that it is rational (see Example 3 of Sect. 2.3).

Example 2. Projecting a non-hyperelliptic, canonical curve of genus g into $\mathbb{P}^2$, one can obtain a birationally isomorphic curve $C \subset \mathbb{P}^2$ of degree $2g-2$, with only ordinary double points. Since its geometric genus is equal to g, this curve has $\frac{1}{2}(2g-3)(2g-4) - g$ ordinary double points.

Example 1 is supplemented by the following corollary.

Corollary. *The number of ordinary double and cuspidal points on an irreducible curve $C \subset \mathbb{P}^2$ of degree d does not exceed $\frac{1}{2}(d-1)(d-2)$. It is equal to $\frac{1}{2}(d-1)(d-2)$ if C is rational.*

Remark. Severi conjectured that there exist irreducible curves $C \subset \mathbb{P}^2$ of degree d with δ ordinary double points, for any preassigned numbers d and $\delta \leq \frac{1}{2}(d-1(d-2)$, and that they form an irreducible family . This has recently been proved (Harris [1986]). This confirms, in particular, that the moduli space $\mathcal{M}_g$ of curves of genus g is irreducible.

3.13. Dual Curves. The well-known duality between the lines and the points of dual projective planes $\mathbb{P}^2$ and $\mathbb{P}^{2^\vee}$ has a natural extension to curves. Let $C \subset \mathbb{P}^2$ be a nondegenerate curve.

Definition. The curve $C^\vee \subset \mathbb{P}^{2^\vee}$, whose generic points are the tangents to $C \subset \mathbb{P}^2$, is called the *dual curve* to C.

The terminology is explained by the following

Biduality theorem. *In characteristic* 0, *we have* $C^{\vee\vee} = C$.

For $k = \mathbb{C}$ there is a fairly intuitive proof. The tangent line $\overline{T}_p \in \mathbb{P}^{2^\vee}$ at a nonsingular point $p \in C$ is the limit of the secants $\overline{pq}$ as $q \to p$ (see Fig. 24). Similarly, the point in $\mathbb{P}^2$ that corresponds to the tangent to $C^\vee \subset \mathbb{P}^{2^\vee}$ at a nonsingular point $\overline{T}_p$ is the limit of the intersection points of the tangents $\overline{T}_p$ and $\overline{T}_q$ as $q \to p$. Of course, this is p (see Fig. 25).

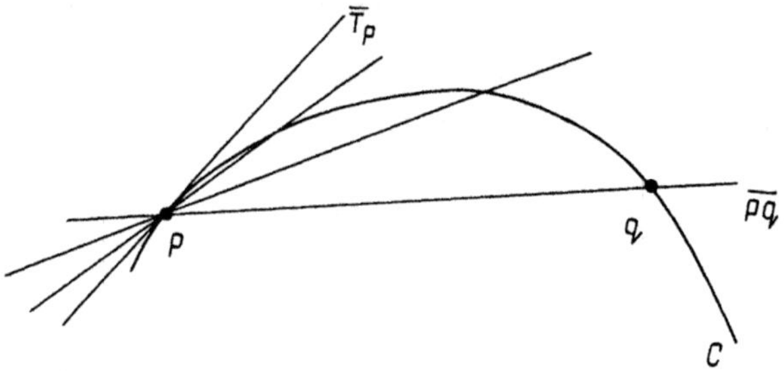

Fig. 24

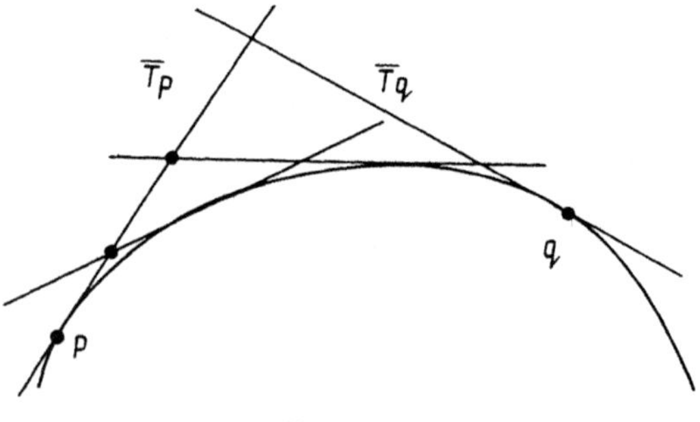

Fig. 25

The biduality theorem is false in general if $\operatorname{char} k > 0$. In fact, the *polar mapping*

$$\tau \colon C \dashrightarrow C^\vee$$
$$p \mapsto \overline{T}_p$$

may be inseparable. On the other hand, if τ is separable then it is birational onto its image and the biduality theorem holds for C (see Kleiman [1977]).

Example 1. When $\operatorname{char} k \neq 2$, the dual of a conic is a conic. In particular, the conic dual to the conic of Example 2 in Sect. 1.6 is defined by the equation $a_0^2 = 4a_1a_2$. In characteristic 2, the dual of a conic is a straight line, and the polar mapping τ is purely inseparable of degree 2.

Example 2. The dual curve to $C = \{x_0^p x_1 + x_1^p x_2 + x_2^p x_0 = 0\} \subset \mathbb{P}^2$, over a field of positive characteristic p, is isomorphic to C, and the mapping $\tau\colon C \to C^\vee$ is purely inseparable (of degree p). Nevertheless, $C^{\vee\vee} = C$.

3.14. Plücker Formula for the Class

Definition. The degree of the dual curve $C^\vee \subset \mathbb{P}^{2^\vee}$ is called the *class* of $C \subset \mathbb{P}^2$.

In addition to the assumptions of Sect. 3.12, we shall suppose that $\operatorname{char} k = 0$. Then the class $d^\vee$ of a curve $C \subset \mathbb{P}^2$ can be computed by the following formula.

Theorem (Plücker). $d^\vee = d(d-1) - 2\delta - 3\varkappa$.

The Hurwitz genus formula, as applied to the composition of a desingularization $f\colon \widetilde{C} \to C$ followed by a generic projection onto a line, leads to a special case of the general Plücker formulae (cf. Arbarello et al. [1985] and Griffiths-Harris [1978]):

$$\varkappa = -d^\vee + 2d + 2g - 2.$$

Here the class is interpreted as the number of tangents to C through a generic point of $\mathbb{P}^2$. Then it suffices to express the geometric genus g by means of Clebsch's formula to obtain what is required. Another proof, due to Plücker, which by-passes the notion of genus, can be found in Griffiths-Harris [1978].

Remark. The theorem is in fact true for any characteristic $\neq 2$, provided the class is correctly defined, as $d^\vee = \deg \tau \cdot \deg C^\vee$ (cf. Example 2 in Sect. 3.13).

3.15. Correspondence of Branches; Dual Formulae. Even if a curve $C \subset \mathbb{P}^2$ has no singularities, the dual curve $C^\vee \subset \mathbb{P}^{2^\vee}$ may have some. As in Sect. 3.14, the characteristic of the ground field k is assumed to be zero.

Example 1. A line L which is tangent to a curve $C \subset \mathbb{P}^2$ in at least two points, is a singular point on the dual curve $C^\vee \subset \mathbb{P}^{2^\vee}$. It is known as a multiple tangent. If a multiple tangent has exactly two tangency points on C and the intersection multiplicity at each of them is equal to 2, then we say it is a *simple bitangent*. A simple bitangent corresponds to an ordinary double point on $C^\vee \subset \mathbb{P}^{2^\vee}$.

Example 2. If the tangent $\overline{T}_p$ at a nonsingular point $p \in C$ intersects C at p with multiplicity ≥ 3, it is a singular point on $C^\vee \subset \mathbb{P}^{2^\vee}$. Then p is called an *inflection point.* An inflection point p is said to be *ordinary* if $\bar{T}$ meets C at p with multiplicity 3 and is not tangent to C at any other point. The tangent at an ordinary inflection point corresponds to a cuspidal point on $C^\vee \subset \mathbb{P}^{2\vee}$.

Example 3. The curve in Example 2 of Sect. 3.13 has the remarkable property that all of its points are inflection points.

We retain the assumptions of Sect. 3.14, but we shall also require that the dual curve should satisfy them. This means that it should have only the simplest kinds of double points for singularities. According to the above examples, this amounts to saying that $C \subset \mathbb{P}^2$ has only the simplest kinds of double points for singularities, only ordinary inflection points, and moreover that any multiple tangent to C is a simple bitangent. Let b be the number of bitangents to C, and let f be the number of inflection points. The corresponding numerical invariants for $C^\vee \subset \mathbb{P}^{2\vee}$ are labelled with the duality sign $^\vee$. For instance, $\delta^\vee$ is the number of ordinary double points on $C^\vee$. In view of the biduality theorem, we have the following

Proposition. $b = \delta^\vee$, $f = \varkappa^\vee$ *and* $b^\vee = \delta$, $f^\vee = \varkappa$.

Since the geometric genus of $C^\vee$ is equal to the geometric genus g of C, the formulae for the genus and the class of $C^\vee \subset \mathbb{P}^{2\vee}$ lead to the following formulae of Clebsch and Plücker.

Theorem (Clebsch, Plücker).

$$g = \tfrac{1}{2}(d^\vee - 1)(d^\vee - 2) - b - f,$$
$$d = d^\vee(d^\vee - 1) - 2b - 3f.$$

Example 4. Let $C \subset \mathbb{P}^2$ be a nonsingular cubic, so that $d = 3$ and $\varkappa = \delta = 0$. By the genus and class formulae, $g = 1$ and $d^\vee = 6$. Obviously, $b = 0$ and the inflection points of the cubic are all ordinary. Therefore the theorem applies to C, each of its formulae giving $f = 9$. Hence C has 9 inflection points. This fact is also true in positive characteristic $\neq 3$, but for quite different reasons (see Chap. 3, Sect. 2.6).

Example 5. A generic quartic in $\mathbb{P}^2$ has 24 inflection points and 28 bitangents. Indeed, for such a curve, $d = 4$, $\varkappa = \delta = 0$, $g = 3$, $d^\vee = 4 \cdot (4 - 1) = 12$, $3 = \frac{1}{2} 11 \cdot 10 - b - f$ and $4 = 12 \cdot 11 - 2b - 3f$. Hence $f = 24$ and $b = 28$.

Chapter 3
Jacobians and Abelian Varieties

These varieties play an important role in the theory of algebraic curves, even though – formally speaking – they are outside its one-dimensional scope. In fact, their geometry is no more complicated than that of curves. For instance, the Jacobian of a complex algebraic curve C can be thought of as a complex torus. The lattice is given by the period matrix of regular differentials on C (cf. Example 2 of Sect. 1.3). This torus is algebraic, since it is associated with a (non-singular) algebraic subvariety of $\mathbb{CP}^n$ (cf. Chap. 2, Sect. 1.9). In view of Abel's theorem, the points on the torus can be identified with the linear equivalence classes of divisors of degree 0 on C. This presentation of Jacobians is adapted for applications and holds over any ground field k. It is developed in § 2. As we see, a Jacobian has two algebraic structures at once: it is both a variety and a group. This brings us to the subject of algebraic groups and abelian varieties.

§ 1. Abelian Varieties

A typical example is any complex algebraic torus $\mathbb{C}^n/\Lambda$. The main topic of this section is a restatement of the condition that $\mathbb{C}^n/\Lambda$ is algebraic, in terms of the lattice Λ. This leads to an important additional structure of abelian varieties, known as a polarization. The section concludes with a discussion of one-dimensional abelian varieties, that is, elliptic curves.

1.1. Algebraic Groups. An *algebraic group* is an algebraic variety G, together with a regular multiplication operation $G \times G \to G$, $(g, h) \mapsto g \cdot h$, on its points, and a regular inverse operation $G \to G$, $g \mapsto g^{-1}$. The additive terminology and notation is customary in the commutative case.

Example 1. A finite-dimensional vector space over k is an algebraic group with respect to addition.

Example 2. The group $\mathrm{GL}(n, k)$ of invertible $n \times n$-matrices with elements in k is a multiplicative algebraic group. Similarly, there is a natural algebraic group structure on the group $\mathrm{Aut}\,\mathbb{P}^n$ of automorphisms of projective space $\mathbb{P}^n$. This group is isomorphic to $\mathbb{P}\mathrm{GL}(n+1, k)$.

Remark. In the same sense, the automorphism group $\mathrm{Aut}\,V$ of any algebraic variety V is algebraic.

Since the action of an algebraic group G on itself is transitive and regular, we obtain the following

Proposition. *(The underlying variety of) any algebraic group G is nonsingular.*

1.2. Abelian Varieties

Definition 1. A commutative algebraic group A on an irreducible projective variety is called an *abelian variety.*

Remark 1. As a matter of fact, the commutativity condition is unnecessary. Indeed an algebraic group on an irreducible projective variety is always commutative (see Mumford [1970]).

Remark 2. A regular mapping of abelian varieties which preserves 0 is a homomorphism (cf. Mumford [1970]). So, the group structure of an abelian variety is uniquely determined by specifying 0.

Definition 2. An abelian variety of dimension 1 is an *elliptic curve.*

Example. Given an elliptic curve C, the infinite group of translation automorphisms $q \mapsto q + p$, with $p, q \in C$, acts without fixed points for $p \neq 0$. It follows that $g(C) = 1$. Conversely, if p is a point on a curve C of genus 1, there is a unique elliptic curve structure on C with $0 = p$. The sum $p_3 = p_1 + p_2$, for $p_1, p_2 \in C$, is defined to be the only element of the linear system $|p_1 + p_2 - p|$, which is zero-dimensional by Riemann's formula (cf. Sect. 2.6).

The relative ease with which one obtains many results on abelian varieties (as in the above Remarks) over $k = \mathbb{C}$ (and, by the Lefschetz principle, in characteristic 0) is explained by the following

Theorem. *With any complex abelian variety A of dimension n, we can associate an n-dimensional complex torus.*

Indeed, A^{an} is a compact, connected, complex Lie group of dimension n (cf. Chap. 2, Sect. 1.9). It is proved in Lie group theory that any such group is a complex torus $\mathbb{C}^n/\Lambda$, the space $\mathbb{C}^n$ being canonically identified with the tangent space T to A at 0, and the quotient mapping $\mathbb{C}^n \to \mathbb{C}^n/\Lambda$ with the exponential map $\exp\colon T \to A^{an}$ (cf. Mumford [1970]). By the comparison theorems (see Serre [1956]), the algebraic variety structure on A can be uniquely reconstructed from the analytic structure of the torus $\mathbb{C}^n/\Lambda$. But by far not every complex torus of dimension ≥ 2 is algebraic.

1.3. Algebraic Complex Tori; Polarized Tori. A general criterion of when a compact complex manifold is algebraic (more precisely, projective), is provided by Kodaira's theorem (see Griffiths-Harris [1978]). The case of complex tori, though nontrivial, is somewhat simpler (cf. Hartshorne [1981–82] and Mumford [1970]).

Example 1. Let $\mathbb{C}^2/\Lambda$ be a two-dimensional complex torus, with associated lattice $\Lambda = \mathbb{Z}e_1 + \mathbb{Z}e_2 + \mathbb{Z}e_3 + \mathbb{Z}e_4$. As it is homeomorphic to the torus $(\mathbb{R}/\mathbb{Z})^4$, its two-dimensional homology group is generated by the six cycles

e_{ij}, $1 \le i < j \le 4$, which are the images of the planes $\mathbb{R}e_i + \mathbb{R}e_j$ in $\mathbb{C}^2/\Lambda$. If $\mathbb{C}^2/\Lambda$ is algebraic, there is an algebraic curve C on it. We may assume that this curve is nonsingular, like a generic hyperplane section of the torus with respect to some projective embedding $\mathbb{C}^2/\Lambda \subset \mathbb{P}^n$. Now the Riemann surface C^{an} is homologous, as a two-dimensional topological cycle, to

$$\sum_{1\le i<j\le 4} a_{ij}\, e_{ij}, \qquad a_{ij} \in \mathbb{Z}.$$

Further, the holomorphic differential $dz_1 \wedge dz_2$ is invariant under translation. So it induces a holomorphic differential ω on $\mathbb{C}^2/\Lambda$. Thus it follows from the Stokes theorem that

$$\int_{C^{an}} \omega = \sum a_{ij} \int_{e_{ij}} \omega,$$

where the $\int_{e_{ij}} \omega$ are easy to compute: if $e_i = (\Pi_{1i}, \Pi_{2i})$ then

$$\int_{e_{ij}} \omega = \Pi_{1i}\Pi_{2j} - \Pi_{1j}\Pi_{2i}.$$

On the other hand, $\int_{C^{an}} \omega = 0$, since there are no exterior holomorphic differentials of degree 2 on C^{an} ($dz \wedge dz = 0$). As a result, we obtain the relation

$$\sum a_{ij}\,(\Pi_{1i}\Pi_{2j} - \Pi_{1j}\Pi_{2i}) = 0. \tag{1}$$

However, this will yield a nontrivial condition on the numbers Π_{1i}, Π_{2i} only if the a_{ij} are not all zero. To this end one can check that $\int_{C^{an}} \eta > 0$, where η is the differential induced on $\mathbb{C}^2/\Lambda$ by the form $\sqrt{-1}\,(dz_1 \wedge d\bar{z}_1 + dz_2 \wedge d\bar{z}_2)$. This is an easy consequence of the relation $\sqrt{-1}\, dz \wedge d\bar{z} = 2\, dx \wedge dy$ (cf. the proof of Riemann's second bilinear relation in Chap. 1, Sect. 4.7). Now, it is not hard to select the numbers Π_{1i}, Π_{2i} in such a way that the products $\Pi_{1i}\Pi_{2j}$ are linearly independent over $\mathbb{Z}$ (cf. Shafarevich [1972]). This contradicts (1). Hence the corresponding torus $\mathbb{C}^2/\Lambda$ is not algebraic!

We write the coordinates of the vectors e_i in the form of a matrix

$$\Pi = (\Pi_{ji}),$$

and we consider the skew-symmetric 4×4-matrix A with elements a_{ij} for $i < j$. Then (1) can be rewritten as

$$\Pi A \Pi^t = 0, \tag{2}$$

where t designates the transpose of a matrix. One can check as above that $\int_{C^{an}} \omega \ge 0$ for differentials of the form

$$\omega = \sqrt{-1}\,(\lambda_1\bar{\lambda}_1 dz_1 \wedge d\bar{z}_1 + \lambda_1\bar{\lambda}_2 dz_1 \wedge d\bar{z}_2 + \bar{\lambda}_1\lambda_2 dz_2 \wedge d\bar{z}_1 + \lambda_2\bar{\lambda}_2 dz_2 \wedge d\bar{z}_2).$$

Moreover, the integral vanishes only when $\lambda_1 = \lambda_2 = 0$ if C is a hyperplane section. This can be rephrased as positive definiteness:

$$\sqrt{-1}\,\Pi A \overline{\Pi}^t > 0. \tag{3}$$

In the case of a general torus $\mathbb{C}^n/\Lambda$, with lattice $\Lambda = \mathbb{Z}e_1 + \ldots + \mathbb{Z}e_{2n}$, we denote by Π the $n \times 2n$-matrix whose columns are the coordinates of the vectors e_i. We say that Π is the *period matrix* of the torus $\mathbb{C}^n/\Lambda$, since $\Pi_{ji} = \int e_i dz_j$.

Frobenius conditions. *The complex torus $\mathbb{C}^n/\Lambda$ is algebraic if and only if there is a skew-symmetric $2n \times 2n$-matrix A, with integer coefficients, such that*

$$\Pi A \Pi^t = 0 \qquad \textit{and} \qquad \sqrt{-1}\,\Pi A \overline{\Pi}^t > 0.$$

The 'only if' part is established in the same way as conditions (2) and (3) in the above Example. We take as a 'curve' C a section of $\mathbb{C}^n/\Lambda$ by $(n-1)$ generic hyperplanes, divided by $(n-1)!$ Readers who are familiar with algebraic topology may prefer the following, more functorial, explanation. Let $D \subset \mathbb{C}^n/\Lambda$ be an effective divisor which is a multiple of a hyperplane section. It is called a *polarization divisor* and defines a topological cycle of dimension $2n-2$. By Poincaré duality, this cycle corresponds to some two-dimensional integer cohomology class on the torus $\mathbb{C}^n/\Lambda$, which is known as the *fundamental class* of the divisor D. Now, such a cohomology class on $\mathbb{C}^n/\Lambda$ can be identified with an integral, skew-symmetric bilinear form E on the lattice Λ. Furthermore the form E, being the fundamental class of a divisor D, is the imaginary part of a unique hermitian form H on $\mathbb{C}^n$. In coordinate notation this yields the first condition of Frobenius. Now, since we assume that D is a polarization divisor, H is also positive definite. In coordinate notation this yields the second Frobenius condition. As a matter of fact, if we denote by E the matrix of the form E in the base $e_1, \ldots, e_{2n}$ then $A = -(\det E).\, E^{-1}$.

Definition 1. The matrix A or, in a more invariant manner, the corresponding skew-symmetric bilinear form E on Λ (or its Hermitian form H on $\mathbb{C}^n$), is called a *polarization* of the torus $\mathbb{C}^n/\Lambda$.

Definition 2. A *principal polarization* is one for which $\det A = 1$ or, equivalently, the bilinear form E is unimodular.

Definition 3. A torus with a fixed polarization is said to be *polarized* (and *principally polarized* if the polarization is principal).

Hence the Frobenius conditions are polarizability conditions on the torus. That they are also sufficient, results from the explicit construction of a projective embedding of a polarized torus, which is given by means of theta functions with characteristics (see Mumford [1970]). In the next section we discuss at length the case of a principal polarization, which is especially important for what follows.

By changing the basis $e_1, \dots, e_{2n}$ of the lattice Λ and the coordinate system of $\mathbb{C}^n$, we can modify the period matrix Π.

Lemma. *A torus $\mathbb{C}^n/\Lambda$ is principally polarizable if and only if there is an integral basis $e_1, \dots, e_{2n}$ of Λ and a coordinate system of $\mathbb{C}^n$ such that the period matrix is of the form*

$$\Pi = (I, Z),$$

where I is the identity $n \times n$-matrix and Z is a Siegel $n \times n$-matrix. *This means that the following conditions are fulfilled:*

$$Z^t = Z \qquad \text{and} \qquad \operatorname{Im} Z \overset{\text{def}}{=} \frac{1}{2\sqrt{-1}} (Z - \overline{Z}^t) > 0. \tag{4}$$

The basis $e_1, \dots, e_{2n}$ for Λ is chosen so that

$$A = \begin{pmatrix} 0 & I \\ -I & 0 \end{pmatrix}.$$

Then, in view of the second Frobenius condition, the vectors $e_1, \dots, e_n$ are linearly independent over $\mathbb{C}$. They form a suitable coordinate system of $\mathbb{C}^n$. Then the Frobenius conditions are expressed by (4). We say that the matrix Π is *normalized.* The polarization is, as it were, lost when the Frobenius conditions are rewritten in the form (4). However, it is present in the choice of the basis for Λ and the coordinates of $\mathbb{C}^n$:

$$E(e_i, e_{n+j}) = \delta_{ij} \quad \text{and} \quad E(e_i, e_j) = E(e_{n+i}, e_{n+j}) = 0 \quad \text{for } 1 \leq i, j \leq n.$$

Example 2. Let S be a compact Riemann surface of genus g. Every cycle $c \in H_1(S, \mathbb{Z})$ defines a $\mathbb{C}$-linear integration mapping $\int_c \omega$ on the space of regular differentials Ω. Thus we get a homomorphism

$$H_1(S, \mathbb{Z}) \to \Omega^\vee$$
$$c \mapsto \int_c$$

By Corollary 2 of Chap. 1, Sect. 4.13, this homomorphism is injective and its image is a lattice of maximal rank $\Lambda \subset \Omega^\vee$. The intersection pairing on $H_1(S, \mathbb{Z})$ (cf. Chap. 1, Sect. 3.8) induces on Λ a skew-symmetric, unimodular bilinear form E. We claim that E is a principal polarization of the g-dimensional torus $\Omega^\vee/\Lambda$. This can be checked by using a standard basis $a_1, \dots, a_g, b_1, \dots, b_g$ of the group $\Lambda = H_1(S, \mathbb{Z})$ and a coordinate system on $\Omega^\vee$ corresponding to an arbitrary base $\omega_1, \dots, \omega_g$ of Ω. In this coordinate system the rows of the period matrix Π are the A- and the B-periods of the forms ω_j: $\Pi_{ji} = \Pi_i^j$. Further, the Riemann bilinear relations (see Sect. 4.7 of Chap. 1) turn into the Frobenius relations for the chosen polarization $A = \begin{pmatrix} 0 & I \\ -I & 0 \end{pmatrix}$. To normalize the period matrix, it is necessary to choose a normalized basis $\omega_1, \dots, \omega_g$ for Ω such that

$$\int_{a_i} \omega_j = \delta_{ij}.$$

The Siegel matrix Z will then consist of B-periods:

$$z_{ji} = \int_{b_i} \omega_j.$$

Note that the first relation of (4) then leads to the following variant of the first Riemann relation

$$\int_{b_i} \omega_j = \int_{b_j} \omega_i.$$

Definition 4. The principally polarized torus $\Omega^\vee/\Lambda$ is called the *Jacobian* of the Riemann surface S. We denote it by $J(S)$.

The normalized period matrices (I, Z) can be identified with the Siegel matrices Z, which make up what is known as the *Siegel upper half-plane*

$$\mathbb{H}_n \overset{\text{def}}{=} \{n \times n\text{-matrix } Z \mid Z^t = Z \text{ and } \operatorname{Im} Z > 0\}.$$

Note that $\mathbb{H}_1 = \mathbb{H}$ is the ordinary upper half-plane. By the Lemma, any principally polarized torus is given by a normalized period matrix (I, Z), and hence by a point $Z \in \mathbb{H}_n$. Two polarized tori are *isomorphic* if they are isomorphic as Lie groups, by an isomorphism preserving the polarization.

Proposition. *Two Siegel matrices define isomorphic principally polarized tori if and only if one of them can be obtained from the other by a transformation of the form*

$$Z \mapsto (AZ + B)(CZ + D)^{-1}, \tag{5}$$

where A, B, C, and D are four integral $n \times n$-matrices satisfying the relation

$$\begin{pmatrix} A & B \\ C & D \end{pmatrix}^t \begin{pmatrix} 0 & I \\ -I & 0 \end{pmatrix} \begin{pmatrix} A & B \\ C & D \end{pmatrix} = \begin{pmatrix} 0 & I \\ -I & 0 \end{pmatrix}. \tag{6}$$

The proof uses the fact that an isomorphism of tori can be lifted to a linear isomorphism of the covering spaces $\mathbb{C}^n$, inducing an isomorphism of the associated lattices (cf. Chap. 1, Sect. 5.6). Now, the property of preserving the polarization is expressed by relation (6) (cf. Siegel [1948–49]). The transformations (5) form the so-called *Siegel modular group* Γ_n, which acts discretely on $\mathbb{H}_n$. Thus principally polarized tori of dimension n are parametrized up to isomorphism by the complex analytic quotient space $\mathbb{H}_n/\Gamma_n$. More precisely, $\mathbb{H}_n/\Gamma_n$ is a coarse moduli space for them (cf. Chap. 1, Sect. 5.10). As a special case, $\mathbb{H}_1/\Gamma_1 = \mathbb{H}/\Gamma$ is a moduli space for complex elliptic curves.

1.4. Theta Function and Riemann Theta Divisor. To establish that a polarized torus $\mathbb{C}^n/\Lambda$ is algebraic, we proceed in two steps. First we construct a projective embedding $f\colon \mathbb{C}^n/\Lambda \hookrightarrow \mathbb{CP}^m$. Then we use the following theorem.

Chow's theorem (cf. Mumford [1976]). *Any compact complex submanifold (and even any complex analytic subspace) of $\mathbb{CP}^m$ is a projective complex algebraic variety.*

The one-dimensional version of this theorem has been discussed in Sect. 6.5 of Chapter 1. To construct the embedding, we choose a collection of holomorphic functions $f_i(u)$ in $\mathbb{C}^n$ which are also automorphic:

$$f_i(u+e) = \mu_e(u) f_i(u),$$

where $e \in \Lambda$ and the holomorphic factors $\mu_e(u)$ are independent of the functions f_i. For a suitable choice of the μ_e, which is determined by the polarization, one can find a collection of functions f_i such that

$$u \mapsto (f_0(u) : \ldots : f_m(u))$$

induces the required embedding f (see Mumford [1970] and cf. formula (6) in Chapter 1).

Let us examine this construction in some more detail for the case of a principally polarized torus $\mathbb{C}^n/\Lambda$. By the lemma of Sect. 1.3, we are free to assume that $\mathbb{C}^n/\Lambda$ is given by a normalized period matrix (I, Z), $Z \in \mathbb{H}_n$. Thus the lattice Λ is spanned by the column vectors e_i of that matrix. The *Riemann theta function* is given on $\mathbb{C}^n \times \mathbb{H}_n$ by the Fourier series

$$\vartheta(u, Z) \overset{\text{def}}{=} \sum_{m\in\mathbb{Z}^n} e^{\pi\sqrt{-1}\,\langle m, mZ\rangle + 2\pi\sqrt{-1}\,\langle m,u\rangle},$$

where $\langle m, v\rangle = \sum_{i=1}^{n} m_i v_i$ is the standard inner product in $\mathbb{C}^n$. Since $\operatorname{Im} Z > 0$ is positive definite, the series is absolutely and uniformly convergent on every compact subspace of $\mathbb{C}^n \times \mathbb{H}_n$. The theta function ϑ is therefore holomorphic on $\mathbb{C}^n \times \mathbb{H}_n$. An immediate verification shows:

Proposition 1. $\vartheta(u + e_i, Z) = \vartheta(u, Z)$, *and*
$\vartheta(u + e_{n+i}, Z) = e^{-\pi\sqrt{-1}(2u_i + z_{ii})}\vartheta(u, Z)$, *where* $1 \le i \le n$

Proposition 2. $\vartheta(-u, Z) = \vartheta(u, Z)$.

Proposition 3. *The theta function satisfies the heat equation:*

$$\frac{\partial^2 \vartheta}{\partial u_i \partial u_j} = 2\pi\sqrt{-1}\,(1+\delta_{ij})\,\frac{\partial \vartheta}{\partial z_{ij}}.$$

To check that the functions appearing in the construction of a projective embedding of $\mathbb{C}^n/\Lambda$ are automorphic, it is enough to look at the generators e_i, e_{n+i} of the lattice. From this point of view, Proposition 1 means that the function $\vartheta(u) \overset{\text{def}}{=} \vartheta(u, Z)$ is automorphic with respect to the factors $\mu_i = 1$, $\mu_{n+i} = e^{-\pi\sqrt{-1}\,(2u_i+z_{ii})}$. Now, it is easy to see that $\vartheta(u)$ is, up to proportionality, the only holomorphic function on $\mathbb{C}^n$ admitting the above factors. That is why the construction of an embedding involves the use of functions that are automorphic with respect to multiple factors μ_i^N, μ_{i+n}^N, $N \in \mathbb{N}$. Let L_N denote the space of all such functions.

Example.

$$\vartheta(u+u_1)\ldots\vartheta(u+u_N) \in L_N,$$

where $u_1, \ldots, u_N \in \mathbb{C}^n$ and $u_1 + \ldots + u_N = 0$.

It is a simple matter to compute the dimension of L_N from the Fourier expansions of its functions: $\dim L_N = N^n$. Further the choice of a basis (f_i) for L_N determines a holomorphic mapping

$$f^N : \mathbb{C}^n/\Lambda \to \mathbb{P}^{N^n-1}.$$

This is indeed a mapping, since for $N \geq 2$ the functions of L_N have no common zeros in $\mathbb{C}^n$ (see the Example). The following result is more difficult to prove.

Lefschetz theorem (cf. Mumford [1970]). *f^N is an embedding, provided $N \geq 3$.*

Hence, by virtue of Chow's theorem, the torus $\mathbb{C}^n/\Lambda$ is associated with a complex projective algebraic variety A. By the comparison theorems for mappings (see Serre [1956]), the group structure on $\mathbb{C}^n/\Lambda$ makes A into an abelian variety. Now we notice that, by Proposition 1, the set of zeros of ϑ is invariant under translations by the vectors of Λ. Thus $\mathbb{C}^n/\Lambda$ contains a well-defined closed analytic subset $\Theta = \{u \bmod \Lambda \mid \vartheta(u) = 0\}$, of dimension $n-1$. By Chow's theorem again, Θ is a divisor on A. It is known as a *Riemann theta divisor.*

Theorem 1. *Θ is a polarization divisor, and its fundamental class is the original principal polarization of $\mathbb{C}^n/\Lambda$.*

Theorem 2. $-\Theta = \Theta$, *where* $-\Theta = \{-p \mid p \in \Theta\}$.

The latter Theorem follows immediately from the fact that ϑ is an even function (see Proposition 2). By the construction of f^N, the divisor $N\Theta$ is a hyperplane section of the image $f^N(\mathbb{C}^n/\Lambda)$, and hence a polarization divisor.

The final assertion in Theorem 1 is a nontrivial example illustrating the deep interrelationship between geometry and calculus. To be properly understood, it requires the more invariantive point of view of line bundles and their Chern classes (cf. Griffiths-Harris [1978], Mumford [1970], and Wells [1973]).

Remark. We have included Proposition 3 for completeness. Its significance will be explained in a forthcoming chapter (see also Arbarello et al. [1985]).

1.5. Principally Polarized Abelian Varieties. Given an n-dimensional abelian variety A over any field k, a *polarization divisor* is an effective divisor D which is a multiple of some hyperplane section divisor of $A \subset \mathbb{P}^m$. That is to say, $ND = H|_A$, where H is a general hyperplane in $\mathbb{P}^m$. We set

$$D^n = (H|_A^n)/N^n,$$

where $(H|_A^n)$ is the *degree* of the variety $A \subset \mathbb{P}^m$. By definition, this is the same as the number of points in the intersection of A with n generic hyperplanes (cf. Example 5 of Chap. 2, Sect. 2.2). As in the case of curves, one can define the divisor of a function (see Shafarevich [1972]) and the space

$$L(D) = \{f \in k(A) \mid f \equiv 0 \text{ or } (f) + D \geq 0\}.$$

Proposition. *The following properties of a polarization divisor D on an abelian variety A, are equivalent:*

(a) $D^n = n!$

(b) $\dim L(D) = 1$;

(c) *if $k = \mathbb{C}$ then the fundamental class of D on the complex torus associated with A is unimodular: it is a principal polarization.*

The equivalence of (a) and (b) derives from the formula:

$$\dim L(D) = D^n/n! \tag{7}$$

(see Mumford [1970]). In characteristic 0 this is a special case of Hirzebruch's Riemann-Roch formula, which – by virtue of the Kodaira vanishing theorem – contains no higher cohomological terms. Properties (a) and (c) are equivalent by the topological relation:

$$D^n = \sqrt{\det E} \cdot n!$$

where E is the skew-symmetric matrix of the fundamental class.

Example 1. Let Θ be a Riemann theta divisor on a complex abelian variety $A = \mathbb{C}^n/\Lambda$. The mapping

$$L_N \to L(N\Theta), \quad f \mapsto f/\vartheta^N,$$

is an isomorphism, whence $\dim L(N\Theta) = \dim L_N = N^n = (N\Theta)^n/n!$

Definition 1. In the general case, a *polarization divisor* Θ on an abelian variety A is said to be *principal*, or a principally polarized *(p.p.) divisor*, if it has the properties stated in the Proposition.

By Theorem 1 of Sect. 1.4, a Riemann theta divisor is a p.p. divisor. Every p.p. divisor on an elliptic curve is a point. Further, all p.p. divisors $\Theta = \Sigma a_i D_i$ are reduced, that is, each $a_i = 1$ (cf. Mumford [1970]). Hence we can think of them as being subvarieties $\Theta \subset A$.

Example 2. If $\Theta \subset A$ is a p.p. divisor then, for any point $p \in A$, the translate $\Theta + p \stackrel{\text{def}}{=} \{q + p \mid q \in \Theta\}$ is also a p.p. divisor.

Definition 2. An *abelian variety* A is said to be *principally polarized*, or *p.p.*, if it is given with a p.p. divisor Θ, which is defined up to translation.

Definition 3. Two p.p. abelian varieties are said to be *isomorphic* if there exists an isomorphism that preserves their p.p. divisors up to translation.

A Riemann theta divisor on a complex p.p. abelian variety A depends on the choice of a normalized period matrix for the associated torus. These divisors, however, differ only by a translation (see Mumford [1970]). Hence they correspond to a well-defined polarization on A. We describe them in the forthcoming subsection.

1.6. Points of Finite Order on Abelian Varieties. A surjective regular homomorphism $f\colon A_1 \twoheadrightarrow A_2$ of abelian varieties with the same dimension is called an *isogeny*. Clearly, the kernel $\ker f$ of an isogeny is a finite abelian group. Its order satisfies the inequality $\# \ker f \leq \deg f$. Further, equality holds if f is separable, in particular if $\operatorname{char} k = 0$.

Example 1. Let $D \subset A$ be a (reduced) polarization divisor. Then the group $G = \{p \in A \mid D + p = D\}$ is finite. Taking the quotient modulo G yields an isogeny $f\colon A \to A/G$. Obviously, $D = f^{-1}(D/G)$ and $D^n = \#G \cdot (D/G)^n$. Consequently, if D is a p.p. divisor then, by (7) and the definition of p.p. divisors, we have $\#G = 1$. Thus the following lemma holds.

Lemma. *If $\Theta \subset A$ is a p.p. divisor then $p + \Theta = \Theta$ only if $p = 0$.*

Example 2. Multiplication by $N \in \mathbb{N}$ is an isogeny:

$$\begin{aligned} \mathbf{N}\colon A &\to A \\ p &\mapsto Np \end{aligned}$$

It is separable, provided $\operatorname{char} k$ does not divide N. (The corresponding isogeny for a complex torus $\mathbb{C}^n/\Lambda$ is just multiplication by N on its universal covering $\mathbb{C}^n$; cf. Remark 2 in Chap. 1, Sect. 5.6.) The kernel of this isogeny is denoted by A_N. It consists of all points $p \in A$ of order N, i.e., such that $Np = 0$.

Corollary. *An abelian variety is divisible: for every positive integer N and any point $p \in A$, there exists a point $q \in A$ such that $p = Nq$.*

Theorem 1. *If A is an abelian variety of dimension n then*

$$A_N \simeq (\mathbb{Z}/N\mathbb{Z})^{2n},$$

provided $\operatorname{char} k \nmid N$. *Hence* $\# A_N = N^{2n}$.

The complex case is straightforward: $A = \mathbb{C}^n/\Lambda$ and $A_N = (\Lambda/N)/\Lambda \simeq (\mathbb{Z}/N\mathbb{Z})^{2n}$, where $\Lambda/N = \{e/N \mid e \in \Lambda\}$ is the lattice divided by N. In positive characteristic $p = \operatorname{char} k$, we have the inequality $\# A_p \leq p^n$ (see Mumford [1970]). The degree of the isogeny $\mathbf{N}$ is nonetheless always equal to N^{2n}.

Example 3. Let C be an elliptic curve and assume that $p = \operatorname{char} k > 0$. Then $\# C_p = p$ or 1. In the latter case, C is said to be *supersingular* (cf. Hartshorne [1977]). For $\operatorname{char} k = 2$ the involution $q \mapsto -q$ has 1 or 2 fixed points, according as C is supersingular or not. These points are ramification points of the hyperelliptic projection $C \to C/\{-q \sim q\}$ (cf. the Example of Sect. 2.4, and Example 4 in Chap. 2, Sect. 1.8).

Suppose now that $\operatorname{char} k \neq 2$, and let Θ be a p.p. divisor. A point $p \in A$ such that $-\Theta + 2p = \Theta$ (see Fig. 26) is called a point of symmetry or a *theta characteristic* of Θ.

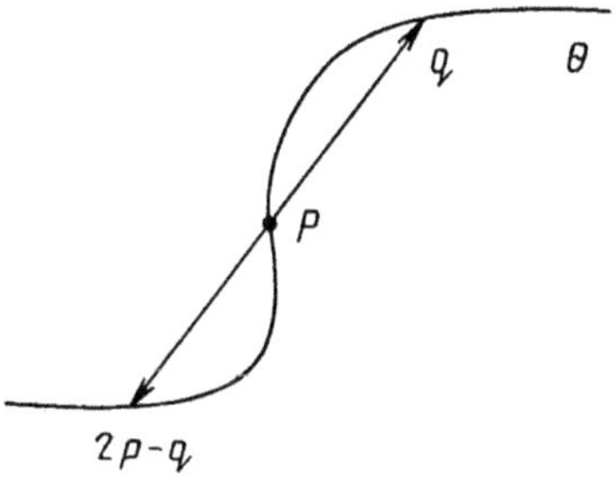

Fig. 26

Theorem 2. *A p.p. divisor Θ on an n-dimensional abelian variety A has 2^{2n} theta characteristics.*

We may assume, up to a translation, that 0 is one of the theta characteristics: $-\Theta = \Theta$. Then, by the Lemma, the theta characteristics form the subgroup A_2 of all points of order 2. Now the existence of at least one theta characteristic in the complex case is guaranteed by the fact that the Riemann theta divisor Θ is symmetric (see Theorem 2 in Sect. 1.4). However, one should not jump to the conclusion that there are 2^{2n} Riemann theta divisors. As a matter of fact, on the functional level symmetries are subdivided into even and odd ones. Indeed, if $f(z) = \sum_{m \geq 0} a_m z^m$ is an even or an odd analytic function then the first nonzero term in its expansion with respect to z is of even, respectively, odd degree. Similarly, in the general case where

k is any field, a theta characteristic is said to be *even* or *odd,* depending on the parity of the multiplicity with which the divisor Θ occurs in it. In other words, one considers the (formal) expansion (with respect to some local parameters) of a function f defining Θ in a neighbourhood of the theta characteristic; and one looks at the parity of the degree of its first nonzero term. From Proposition 2 of Sect. 1.4, we see that 0 is an even theta characteristic for the Riemann divisor Θ. To determine the parity of the other theta characteristics $p \in A_2 = (\Lambda/2)/\Lambda$ we write them as

$$p_\varepsilon^\delta \stackrel{\text{def}}{=} (\varepsilon + \delta Z)/2 \mod \Lambda,$$

where ε and δ are n-vectors consisting of 0's and 1's only (*codes*). Obviously, the divisor $\Theta_\varepsilon^\delta \stackrel{\text{def}}{=} \Theta + p_\varepsilon^\delta$ is the set of zeros of a *theta function with characteristics:*

$$\vartheta\begin{bmatrix}\delta\\ \varepsilon\end{bmatrix}(u,Z) \stackrel{\text{def}}{=} \sum_{m\in\mathbb{Z}^n} e^{\pi\sqrt{-1}\,\langle m+\frac{\delta}{2},(m+\frac{\delta}{2})Z\rangle + 2\pi\sqrt{-1}\,\langle m+\frac{\delta}{2},u+\frac{\varepsilon}{2}\rangle}$$
$$= e^{\pi\sqrt{-1}\,\langle\frac{\delta}{2},\frac{\delta}{2}Z\rangle + 2\pi\sqrt{-1}\,\langle\frac{\delta}{2},u+\frac{\varepsilon}{2}\rangle}\,\vartheta(u+\tfrac{\varepsilon}{2}+\tfrac{\delta}{2}Z,Z);$$
$$\vartheta\begin{bmatrix}0\\ 0\end{bmatrix} = \vartheta.$$

The following relation can be checked directly.

$$\vartheta\begin{bmatrix}\delta\\ \varepsilon\end{bmatrix}(-u,Z) = (-1)^{\langle\delta,\varepsilon\rangle}\,\vartheta\begin{bmatrix}\delta\\ \varepsilon\end{bmatrix}(u,Z). \tag{8}$$

Therefore the theta characteristic p_ε^δ is even if and only if $\langle\delta,\varepsilon\rangle$ is. A simple combinatorial count leads to the following result, which is valid in the general case as well.

Theorem 3. *A p.p. divisor Θ on an n-dimensional abelian variety A has $2^{n-1}(2^n+1)$ even and $2^{n-1}(2^n-1)$ odd theta characteristics.*

Thus a p.p. abelian variety A contains $2^{n-1}(2^n+1)$ polarization divisors Θ for which 0 is an even theta characteristic. In the complex case these are all possible *Riemann theta divisors* for the given polarization. In the general case they are given the same name.

1.7. Elliptic Curves. From Sect. 1.2 we take that an elliptic curve C is a curve of genus 1 with a selected point 0. (This point also uniquely determines a principal polarization.) We assume here that $\operatorname{char} k \neq 2, 3$. Many constructions and statements are then parallel to some over the complex numbers. The involution $p \mapsto -p$ defines a hyperelliptic projection $\gamma\colon C \to C/\{-p \sim p\} = \mathbb{P}^1$. Its four ramification points are the points of order 2 on

C. We choose an affine coordinate on $\mathbb{P}^1$ such that $\gamma(0) = \infty$ and the images of the remaining ramification points are 0, 1, and λ. The parameter λ determines the *absolute invariant* of the curve C (even without the group structure):

$$j(C) \stackrel{\text{def}}{=} \frac{4}{27} \cdot \frac{(\lambda^2 - \lambda + 1)^3}{\lambda^2(1-\lambda)^2}.$$

Theorem 1 (cf. Hartshorne [1977]). *Two elliptic curves (or simply, two curves of genus 1), C and B, are isomorphic if and only if $j(C) = j(B)$* (cf. Chap. 1, Sect. 6.6).

Each embedding $\varphi_{3p}\colon C \to \mathbb{P}^2$ provides an isomorphism onto some cubic. By choosing a suitable basis for $L(3p)$ when $p = 0$, one can bring that cubic to a reasonably simple form (cf. Sect. 2.6). Thus we obtain

Theorem 2. *A curve C of genus 1 is isomorphic to a* cubic in Weierstrass normal form

$$y^2 = 4x^3 - g_2 x - g_3, \qquad g_2, g_3 \in k. \tag{9}$$

Further, $j(C) = g_2^3/\Delta = 1 + 27g_3^2/\Delta$, where $\Delta = g_2^3 - 27g_3^2$.

Let $\operatorname{Aut} C$ be the full automorphism group of the curve C. An automorphism need not preserve the group structure. For instance, we can think of the translation by a point $p \in C\colon\ q \mapsto q + p$. Now, the translations form a normal subgroup of $\operatorname{Aut} C$, which can be identified with C as an elliptic curve. Further, C is the irreducible component of 0, since $\dim \operatorname{Aut} C = l(0) = 1$ (cf. Chap. 2, Sect. 2.9). Hence $\operatorname{Aut} C$ can be viewed as the semidirect product of C with the finite automorphism group $\operatorname{Aut}_0 C$ preserving 0. Moreover, $\operatorname{Aut}_0 C$ is the automorphism group of the elliptic curve C as an algebraic group (see Remark 2 in Sect. 1.2).

Theorem 3 (cf. Hartshorne [1977]). (a) $\operatorname{Aut}_0 C = \{p \mapsto \pm p\} = \mathbb{Z}/2$, *provided $j(C) \neq 0, 1$;*
(b) $\operatorname{Aut}_0 C \simeq \mathbb{Z}/4$ *if $j(C) = 1$;*
(c) $\operatorname{Aut}_0 C \simeq \mathbb{Z}/6$ *if $j(C) = 0$.*

The curves in (*b*) and (*c*) can be given as plane cubics, and their automorphisms described, as in the complex case (see Chap. 1, Sect. 6.6).

Now let $C = \mathbb{C}/(\mathbb{Z} + \tau\mathbb{Z})$ be a complex elliptic curve. The Riemann theta function ϑ has a unique simple zero $1/2 + \tau/2$ in the parallelogram $\{\alpha + \beta\tau \mid 0 \leq \alpha, \beta \leq 1\}$ (see Fig. 27). It corresponds to the unique odd theta characteristic $p_1^1 = \Theta$. This can also be established directly (see Clemens [1980]) by integrating $d \ln \vartheta$ along the frontier of the parallelogram. The remaining three points of order 2, namely,

$$p_0^0 = 0, \quad p_1^0 = 1/2, \quad \text{and} \quad p_0^1 = \tau/2 \mod \mathbb{Z} + \tau\mathbb{Z}$$

are the even theta characteristics. They are nondegenerate, that is, ϑ does not vanish on them. Therefore the even *theta constants*

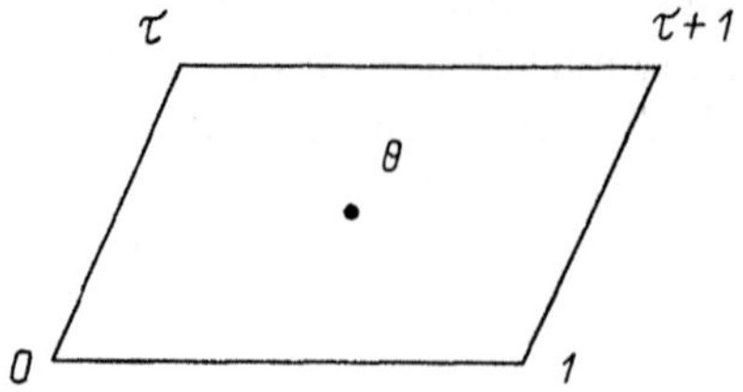

Fig. 27

$$\vartheta\begin{bmatrix}\delta\\ \varepsilon\end{bmatrix}(\tau) \stackrel{\text{def}}{=} \vartheta\begin{bmatrix}\delta\\ \varepsilon\end{bmatrix}(0,\tau), \qquad \delta.\varepsilon = 0,$$

are everywhere nonzero on $\mathbb{H}$.

The theta function with characteristics

$$\vartheta\begin{bmatrix}1\\ 1\end{bmatrix}(z,\tau) = e^{\pi\sqrt{-1}\,\tau/4+\pi\sqrt{-1}\,(z+1/2)}\,\vartheta(z+1/2+\tau/2,\tau)$$

has a zero of order 1 at 0. From the automorphic properties of the Riemann theta function, one easily gets the analogues for $\vartheta\begin{bmatrix}1\\ 1\end{bmatrix}(z,\tau)$:

$$\vartheta\begin{bmatrix}1\\ 1\end{bmatrix}(z+1,\tau) = -\vartheta\begin{bmatrix}1\\ 1\end{bmatrix}(z,\tau)$$

and

$$\vartheta\begin{bmatrix}1\\ 1\end{bmatrix}(z+\tau,\tau) = -e^{-\pi\sqrt{-1}\,(2z+\tau)}\vartheta\begin{bmatrix}1\\ 1\end{bmatrix}(z,\tau).$$

As a consequence, $-\dfrac{d^2}{dz^2}\ln\vartheta\begin{bmatrix}1\\ 1\end{bmatrix}(z,\tau)$ is a doubly periodic function of z (with periods 1 and τ). Furthermore, its Laurent expansion at 0 takes the form $1/z^2 + \text{constant} + \text{terms of even positive degree in } z$. This function is unique up to an additive constant, because $l(2p) = 2$. On comparing it with the Weierstrass $\wp$-function, we obtain:

Proposition. $\wp(z,\mathbb{Z}+\tau\mathbb{Z}) = -\dfrac{d^2}{dz^2}\ln\vartheta\begin{bmatrix}1\\ 1\end{bmatrix}(z,\tau) + c(\tau)$, *where* $c(\tau)$ *is a holomorphic function on* $\mathbb{H}$.

From the automorphic relations for $\vartheta\begin{bmatrix}1\\ 1\end{bmatrix}(z,\tau)$, it also follows that $\vartheta^2\begin{bmatrix}1\\ 1\end{bmatrix}(z,\tau) \in L_2$. In other words, the function $\vartheta^2\begin{bmatrix}1\\ 1\end{bmatrix}(z,\tau)$ has the same multiplicative factors as $\vartheta^2\begin{bmatrix}0\\ 0\end{bmatrix}(z,\tau) = \vartheta^2$. In a similar way, $\vartheta^2\begin{bmatrix}1\\ 0\end{bmatrix}(z,\tau)$

and $\vartheta^2\begin{bmatrix}0\\1\end{bmatrix}(z,\tau) \in L_2$. But $\dim L_2 = 2$; so the functions $\vartheta^2\begin{bmatrix}0\\0\end{bmatrix}(z,\tau)$ and $\vartheta^2\begin{bmatrix}1\\1\end{bmatrix}(z,\tau)$ constitute a basis for L_2. Therefore a map

$$\gamma = \varphi_{2\cdot 0} = f^2\colon \mathbb{C}/(\mathbb{Z}+\tau\mathbb{Z}) \to \mathbb{CP}^1$$

can be defined by

$$z \mod \mathbb{Z}+\tau\mathbb{Z} \mapsto \left(\vartheta^2\begin{bmatrix}0\\0\end{bmatrix}(z,\tau) : \vartheta^2\begin{bmatrix}1\\1\end{bmatrix}(z,\tau)\right).$$

Further it is easy to discover its branch points: $f^2(p_0^0) = \infty$, $f^2(p_1^1) = 0$, $f^2(p_1^0) = \vartheta^2\begin{bmatrix}0\\1\end{bmatrix}(\tau)/\vartheta^2\begin{bmatrix}1\\0\end{bmatrix}(\tau)$, and $f^2(p_0^1) = -\vartheta^2\begin{bmatrix}1\\0\end{bmatrix}(\tau)/\vartheta^2\begin{bmatrix}0\\1\end{bmatrix}(\tau)$. Thus

$$\lambda = -\vartheta^4\begin{bmatrix}1\\0\end{bmatrix}(\tau)/\vartheta^4\begin{bmatrix}0\\1\end{bmatrix}(\tau)$$

for an affine coordinate on $\mathbb{CP}^1$ such that $f^2(p_0^0) = \infty$, $f^2(p_1^1) = 0$, $f^2(p_1^0) = 1$ and $f^2(p_0^1) = \lambda$. From this one can also find an expression for the absolute invariant j in terms of the theta constants $\vartheta\begin{bmatrix}1\\0\end{bmatrix}(\tau)$ and $\vartheta\begin{bmatrix}0\\1\end{bmatrix}(\tau)$.

The modular properties of theta constants are immediate:

$$\vartheta\begin{bmatrix}\delta\\\varepsilon\end{bmatrix}(-1/\tau) = (\tau/\sqrt{-1})^{1/2}(-\sqrt{-1})^{\delta\varepsilon}\,\vartheta\begin{bmatrix}\varepsilon\\\delta\end{bmatrix}(\tau),$$

$$\vartheta\begin{bmatrix}1\\\varepsilon\end{bmatrix}(\tau+1) = e^{\pi\sqrt{-1}/4}\,\vartheta\begin{bmatrix}1\\\varepsilon\end{bmatrix}(\tau),$$

and

$$\vartheta\begin{bmatrix}0\\\varepsilon\end{bmatrix}(\tau+1) = \vartheta\begin{bmatrix}0\\1-\varepsilon\end{bmatrix}(\tau).$$

Further the congruence subgroup $\Gamma_2 = \left\{\begin{pmatrix}a & b\\ c & d\end{pmatrix} \equiv \begin{pmatrix}1 & 0\\ 0 & 1\end{pmatrix} \bmod 2\right\}/\pm I$ is generated by

$$\begin{pmatrix}1&2\\0&1\end{pmatrix} = \begin{pmatrix}1&1\\0&1\end{pmatrix}\begin{pmatrix}1&1\\0&1\end{pmatrix} \quad\text{and}\quad \begin{pmatrix}1&0\\2&1\end{pmatrix} = \begin{pmatrix}0&1\\-1&0\end{pmatrix}\begin{pmatrix}1&-2\\0&1\end{pmatrix}\begin{pmatrix}0&-1\\1&0\end{pmatrix}.$$

It then follows from the modular properties that the fourth powers of theta constants, $\vartheta^4\begin{bmatrix}\delta\\\varepsilon\end{bmatrix}(\tau)$, are automorphic forms of weight 2 with respect to Γ_2. Therefore we have a well-defined mapping $\lambda\colon \mathbb{H}/\Gamma_2 \to \mathbb{C}$, where $\lambda(\tau) = -\vartheta^4\begin{bmatrix}1\\0\end{bmatrix}(\tau)/\vartheta^4\begin{bmatrix}0\\1\end{bmatrix}(\tau)$. As a matter of fact, λ is a coordinate on the Riemann surface $\mathbb{H}/\Gamma_2 \underset{\lambda}{\stackrel{\sim}{\to}} \mathbb{C} - \{0,1\}$ (see Clemens [1980]). We note that

$\lambda(\Gamma_2\, i\infty) = 0$, $\lambda(\Gamma_2\, 0) = \infty$, and $\lambda(\Gamma_2\, 1) = 1$. In much the same way, $\mu(\tau) = \vartheta^4 \begin{bmatrix} 0 \\ 0 \end{bmatrix} (\tau) / \vartheta^4 \begin{bmatrix} 0 \\ 1 \end{bmatrix} (\tau)$ defines a coordinate map μ, with the property that $\mu(\Gamma_2\, i\infty) = 1$, $\mu(\Gamma_2\, 0) = \infty$, and $\mu(\Gamma_2\, 1) = 0$. Hence $\mu = 1 - \lambda$, and we get the *Riemann theta relation:*

$$\vartheta^4 \begin{bmatrix} 0 \\ 1 \end{bmatrix} (\tau) + \vartheta^4 \begin{bmatrix} 1 \\ 0 \end{bmatrix} (\tau) = \vartheta^4 \begin{bmatrix} 0 \\ 0 \end{bmatrix} (\tau).$$

Finally, one can introduce some more general theta constants as the values at 0 of the derivatives of theta functions with characteristics. However, this does not lead to any essentially new constants. This confirms the following beautiful relation, known as *Jacobi's identity:*

$$\frac{\partial}{\partial z} \vartheta \begin{bmatrix} 1 \\ 1 \end{bmatrix} (z, \tau)\Big|_{z=0} = -\pi . \vartheta \begin{bmatrix} 0 \\ 0 \end{bmatrix} (\tau) . \vartheta \begin{bmatrix} 0 \\ 1 \end{bmatrix} (\tau) . \vartheta \begin{bmatrix} 1 \\ 0 \end{bmatrix} (\tau).$$

For a more detailed discussion of this and similar formulae, and about their origin, the reader is referred to the classical textbook by Whittaker & Watson [1902, 1927].

§ 2. Jacobians of Curves and of Riemann Surfaces

In this section, C denotes a nonsingular projective curve of genus g. If $k = \mathbb{C}$ then, according to Example 2 of Sect. 1.3, one can attach to the Riemann surface C^{an} associated with C a p.p. torus $J(C^{an})$, known as the Jacobian of C^{an}. This in turn is associated with a p.p. abelian variety, which is called the Jacobian of C and is denoted by $J(C)$. An obvious question is whether it is possible to construct the variety $J(C)$ and its polarization algebraically from the curve C. In this way we would avoid such transcendental procedures as integration and the consideration of divisors that are defined as the set of zeros of functions given by series on the universal covering. This question has a positive answer, which involves the theorems of Abel and Jacobi, as well as Riemann's theorem on the zeros of theta functions, to be discussed below. We shall also give the definition and explain the simplest properties of the Jacobian of an algebraic curve over an arbitrary algebraically closed field k.

2.1. Principal Divisors on Riemann Surfaces. Let S be a compact Riemann surface of genus g. As we already know, every principal divisor D has degree 0. So, principal divisors lie in the subgroup $\operatorname{Div}^0 S \subset \operatorname{Div} S$ of all divisors of degree 0. On the other hand, every divisor $D = \sum(p_i - q_i)$ of degree 0 defines a $\mathbb{C}$-linear integration mapping

$$\sum \int_{q_i}^{p_i} : \Omega \to \mathbb{C}$$
$$\omega \mapsto \sum \int_{q_i}^{p_i} \omega,$$

where each integral $\int_{q_i}^{p_i}$ is taken over a fixed path joining the points p_i and q_i. Clearly, if these points are taken in a different order, or if the paths connecting them are chosen differently, then the mappings obtained differ by vectors $\int_c$, $c \in H_1(S, \mathbb{Z})$, forming a lattice Λ in $\Omega^\vee$. Hence there is a well-defined holomorphic homomorphism

$$a\colon \operatorname{Div}^0 S \to J(S) = \Omega^\vee/\Lambda$$
$$\Sigma(p_i - q_i) \mapsto \sum \int_{q_i}^{p_i} \quad \operatorname{mod} \Lambda.$$

Example. Let $D = (f)$ be a principal divisor, with $f \in \mathcal{M}(S)$. Then the mapping $\mathbb{CP}^1 \to J(S)$ defined by $\varphi(z) = a((f-z))$ and $\varphi(\infty) = 0$, is readily seen to be holomorphic. Like any other holomorphic mapping of the Riemann sphere into a torus, this map $\varphi\colon \mathbb{CP}^1 \to \mathbb{C}^g/\Lambda$ is constant. Indeed, the differentials dz_i induce holomorphic differentials on the torus $\mathbb{C}^g/\Lambda$, which span the cotangent space at each of its points. But $\varphi^* dz_i = 0$, since $g(\mathbb{CP}^1) = 0$. It follows that $\varphi =$ constant. Now, in our case, $a(D) = \varphi(0) = \varphi(\infty) = 0$. Hence $a(D) = 0$ for any principal divisor D.

Abel's theorem (cf. Griffiths-Harris [1978]). *A divisor $D \in \operatorname{Div}^0 S$ is principal if and only if $a(D) = 0 \in J(S)$.*

2.2. Inversion Problem. The question of finding, for each point $p \in J(S)$, a divisor $D \in \operatorname{Div}^0 S$ such that $a(D) = p$, is called the inversion problem.

Jacobi's inversion theorem. *$a\colon \operatorname{Div}^0 S \to J(S)$ is an epimorphism.*

Therefore the inversion problem can always be solved. In coordinate form, this has the following meaning. Once we fix a basis $\omega_1, \dots, \omega_g \in \Omega$, we can take any vector $(z_1, \dots, z_g) \in \mathbb{C}^g$ and find some points $p_i, q_i \in S$ such that

$$\left(\sum \int_{q_i}^{p_i} \omega_1, \dots, \sum \int_{q_i}^{p_i} \omega_g\right) \equiv (z_1, \dots, z_g) \quad \operatorname{mod} \Lambda.$$

As a matter of fact, we can give ourselves any g points $q_1, \dots, q_g$ and still choose $p_1, \dots, p_g$. This amounts to saying that $f\colon S_g \to J(S)$ is surjective, where f sends an effective divisor Σp_i, regarded as a point of the symmetric product S_g, into the point $a(\Sigma p_i - \Sigma q_i)$. This is an example of a normalized Abel mapping (see Sect. 2.4 below). It is holomorphic, and even regular if we identify the Jacobian $J(S)$ with the corresponding complex abelian variety, and the symmetric product S_g with the g-fold symmetric power of the algebraic curve of the compact Riemann surface S. By Abel's theorem and Proposition 3 of Chap. 2, Sect. 2.9, f is generically injective. But $\dim S_g = \dim J(S)$. Hence, by the Theorem of Chap. 2, Sect. 1.3 – or its complex analogue, the proper mapping theorem –, f is surjective. This is also easy to obtain by checking directly that the Jacobian of f (in the sense of calculus) does not vanish at a generic point of S_g (cf. Griffiths-Harris [1978]). An explicit solution of the inversion problem will be discussed in Sect. 2.5.

From the theorems of Abel and Jacobi we get:

Theorem. *The mapping a defines an isomorphism* $\operatorname{Div}^0 S/{\sim} \xrightarrow{\sim} J(S)$.

Thus the points on the Jacobian $J(S)$ can be identified with the linear equivalence classes of divisors of degree 0 on the Riemann surface S. This is used as a starting point for defining the Jacobian of an algebraic curve.

2.3. Picard Group. The group

$$\operatorname{Pic} C = \operatorname{Div} C/{\sim}$$

of linear equivalence classes of divisors is called the *Picard group* of the algebraic curve C. It is graded naturally as

$$\operatorname{Pic} C = \bigoplus_d \operatorname{Pic}^d C,$$

where $\operatorname{Pic}^d C = \operatorname{Div}^d C/{\sim}$ is the set of classes of degree d. As it happens, every one of these sets $\operatorname{Pic}^d C$ has a unique, natural structure of a projective algebraic variety (in a sense that will be explained below). The group $\operatorname{Pic}^0 C$ then becomes an abelian variety, which is called the *Jacobian* of the curve C. (A more complete definition will be given in Sect. 2.5.)

2.4. Picard Varieties and their Universal Property. The fact that there exists an algebraic variety structure on $\operatorname{Pic}^d C$ is connected with the existence of such a structure on the set of effective divisors of a given degree. Indeed, the effective divisors of degree d on a curve C can be identified with the points of the d-fold symmetric power C_d. In view of Proposition 3 of Chap. 2, Sect. 2.9, the effective divisors $D \in \operatorname{Div}^g C$ such that $l(D) = 1$, define an open subset $U \subset C_g$. Now, $\operatorname{Pic}^d C$ can be made into an algebraic variety by gluing together several copies of this (quasi-projective) variety U. By a noncanonical isomorphism, namely, translation by the class of a divisor $D \in \operatorname{Div}^d C$,

$$\operatorname{Pic}^0 C \xrightarrow{\sim} \operatorname{Pic}^d C, \qquad D'/{\sim} \mapsto (D' + D)/{\sim}, \tag{10}$$

we can restrict ourselves to the case of $\operatorname{Pic}^0 C$. Now we apply (10) for $d = g$ and obtain an embedding of U in $\operatorname{Pic}^0 C$ onto a subset which is open by definition. Then the regularity requirement of the group law provides $\operatorname{Pic}^0 C$

with a well-defined algebraic variety structure. $\operatorname{Pic}^0 C$ becomes a commutative algebraic group, and $\operatorname{Pic}^d C$ a principal homogeneous variety with respect to the natural action of $\operatorname{Pic}^0 C$. Thus, besides being regular and transitive, the action is also principal: $(D' + D)/\sim = D/\sim$ if and only if $D'/\sim = 0$ in $\operatorname{Pic}^0 C$.

Definition 1. The variety $\operatorname{Pic}^d C$ is called the *Picard variety* of degree d of the curve C.

$\operatorname{Pic}^d C$ is an irreducible variety, and its dimension is equal to the genus g of the curve C. A more characteristic property of the Picard variety is that it is universal with respect to algebraic families of divisors. By a *family of divisors* on C with base T, we mean a mapping of the form $f\colon T \to \operatorname{Div} C$, where T is an algebraic variety. For simplicity we shall assume here that T is nonsingular. A family f is said to be algebraic if it is given by a divisor D on the product $C \times T$. In other words, $f(t) = D|_{C\times t}$, $t \in T$. (The restriction $D|_{C\times t}$ is defined as in Example 5 of Chap. 2, Sect. 2.2.) Now the variety $\operatorname{Pic}^d C$ is universal in the following sense.

Theorem. *For every algebraic family $f\colon T \to \operatorname{Div}^d C$ of divisors of degree d, the induced mapping $f/\sim\colon T \to \operatorname{Pic}^d C$, $t \mapsto f(t)/\sim$, is regular.*

Example 1. The algebraic family $C_d \subset \operatorname{Div}^d C$ of effective divisors of degree d determines a regular mapping

$$\mu_d\colon C_d \to \operatorname{Pic}^d C, \qquad D \mapsto D/\sim,$$

called the *Abel mapping*. The fibres $\mu_d^{-1}(D/\sim)$ of the Abel mapping are the complete linear systems $|D|$ of degree d. From Propositions 2 and 3 of Chap. 2, Sect. 2.9 we get:

Proposition. *The Abel mapping (of degree d) is*
(a) *surjective if $d \geq g$, and*
(b) *generically injective if $d \leq g$.*

Thus, in view of the Theorem in Chap. 2, Sect. 1.3, the variety $\operatorname{Pic}^g C = \mu_g(C_g)$ is projective. Of course, this applies to every other Picard variety $\operatorname{Pic}^d C$. As a special case, $\operatorname{Pic}^0 C$ is an abelian variety.

Definition 2. The abelian variety $\operatorname{Pic}^0 C$, also denoted by $J(C)$, is the *Jacobian* of the curve C.

Example 2. If D' is some preassigned divisor of degree d on C, the *normalized Abel mapping*

$$a_d\colon C_d \to J(C), \qquad D \mapsto (D - D')/\sim,$$

is also regular. One often takes as D' a divisor of the form $D' = d.p$, $p \in C$. According to the Example in Sect. 2.5 of Chap. 2, the mapping $a_1\colon C \to J(C)$ is injective when $g \geq 1$. Further, it can be checked that a_1 is an embedding.

Remark 1. The Abel mapping $a_1 \colon C \to J(C)$ is universal in the following sense. For any regular mapping $f \colon C \to A$ into an abelian variety A, there is a unique regular homomorphism $F \colon J(C) \to A$ such that the following diagram is commutative:

$$\begin{array}{ccc} C & \xrightarrow{a_1} & J(C) \\ f \searrow & & \swarrow F \\ & A & \end{array}$$

This is the approach taken in Serre [1959] for the definition and description of the Jacobian of a (possibly singular) curve.

Let us look once more at the complex case. By the Theorem in Sect. 2.2, the Jacobian $J(C)$ can be identified with $J(C^{an}) = \Omega^\vee/\Lambda$. Moreover, one can easily prove the following comparison theorem: $J(C)^{an} = J(C^{an})$. This is supplemented by the traditional integral representation of the Abel mapping $a_d \colon C_d \to J(C)$, normalized by an effective divisor $D' = \Sigma\, q_i$ of degree d:

$$a_d(\sum p_i) = \sum \int_{q_i}^{p_i} \quad \bmod \Lambda$$

or, in coordinate form:

$$a_d(\sum p_i) = \left(\sum \int_{q_i}^{p_i} \omega_1, \ldots, \sum \int_{q_i}^{p_i} \omega_g\right) \quad \bmod \Lambda.$$

Remark 2. For further details on the universal property of the Picard varieties of algebraic curves and of higher-dimensional varieties, see Artin [1969], Grothendieck [1961–62], and Mumford [1966].

2.5. Polarization Divisor of the Jacobian of a Curve; Poincaré Formulae. By the Proposition of Sect. 2.4, the Abel mapping $a_{g-1} \colon C_{g-1} \to J(C)$ is birational onto its image. In fact, $\Theta = a_{g-1}(C_{g-1})$ is an irreducible subvariety of dimension $g-1$, and hence a prime divisor on $J(C)$. Now, by construction, the divisor $\Theta \subset J(C)$ is defined uniquely up to a translation. Moreover, it is a principal polarization divisor. From the definition of principal polarization, this is a consequence of the next theorem for $d = 0$.

Theorem (Poincaré). $a_d(C_d) = \dfrac{1}{(g-d)!}\Theta^{g-d}, \quad 0 \le d \le g.$

Here '=' denotes numerical equivalence. So the formula means that the intersection numbers with any divisors $D_1, \ldots, D_d$ on $J(C)$ are equal:

$$a_d(C_d)\, D_1 \ldots D_d = \frac{1}{(g-d)!}\Theta^{g-d}\, D_1 \ldots D_d$$

In the complex case, '=' may be replaced by homological equivalence. A topological proof of the Theorem in that situation can be found in Arbarello et al. [1985]. Thus the Jacobian $J(C)$ has an entirely natural principal polarization. This conveys a more complete picture of it as a p.p. abelian variety associated with the curve C. In the complex case, one may ask whether this polarization agrees with the one given by the Riemann theta divisor. In fact, they coincide.

Riemann's theorem on the theta divisor. *Let C be a complex curve. The Riemann theta divisor Θ coincides up to a translation with the polarization divisor $a_{g-1}(C_{g-1}) \subset J(C)$. Hence $\Theta = a_{g-1}(C_{g-1}) + \varkappa$ for some point $\varkappa \in J(C)$.*

Once more we see that, in the complex case, $a_{g-1}(C_{g-1})$ is a principal polarization divisor. A purely topological proof of this theorem rests on a computation of the fundamental class of $a_{g-1}(C_{g-1})$ (see Griffiths-Harris [1978]). There is also a more direct proof, which goes back to Riemann (a modern presentation is given in Griffiths-Harris [1978] and Lewittes [1964]). We fix a point $q \in C$ and consider the mapping $a_1 \colon C \to J(C) = \mathbb{C}^g/\Lambda$, $p \mapsto (p-q)/\!\sim\, = \left(\int_q^p \omega_1, \ldots, \int_q^p \omega_g\right) \bmod \Lambda$. Notice that, by the Poincaré formula for $d = 1$, we have:

$$a_1(C) \cdot a_{g-1}(C_{g-1}) = \frac{1}{(g-1)!}\, a_{g-1}(C_{g-1})^g = g.$$

(This is easy to check directly as well!) Therefore each translate $\Theta + t$ of the Riemann divisor by a generic point $t \in J(C)$ must also intersect $a_1(C)$ (which happens to be the curve when $g \geq 1$) in g points, counting multiplicities. This can be shown by considering the differential $d \ln \vartheta\left(\int_q^p \omega_1 - z_1, \ldots, \int_q^p \omega_g - z_g\right)$ in the variable p (where ϑ is the Riemann theta function and $t = (z_1, \ldots, z_g)$ mod Λ), and integrating it along the frontier of a development for C^{an}.

Thus $\Theta + t$ defines on C an effective divisor of degree g, namely, $a_1^*(\Theta + t) = \sum p_i(t)$. Through integration again, one checks that

$$a_g\big(\textstyle\sum p_i(t)\big) + \varkappa = \sum a_1(p_i(t)) + \varkappa = t,$$

where $\varkappa$ does not depend on t. We mention in passing that the relation we have just written, yields the awaited explicit solution to the inversion problem for a point $t \in J(C)$, provided $a_1(C)$ does not lie entirely in $\Theta + t$. In particular, a generic point t produces a generic effective divisor $\sum p_i(t)$ of degree g. By construction, $a_1(p_g(t)) - t \in \Theta$. Hence, in view of the symmetry of the Riemann divisor (this is a decisive step!), we have:

$$a_{g-1}\left(\sum_{i=1}^{g-1} p_i(t)\right) + \varkappa = \sum_{i=1}^{g-1} a_1(p_i(t)) + \varkappa = t - a_1(p_g(t)) \in -\Theta = \Theta.$$

Thus $a_{g-1}(C_{g-1}) + \varkappa \subset \Theta$. The reverse inclusion can be derived from the general properties of principal polarization. For this step, too, there is a more direct proof (see Griffiths-Harris [1978]), which uses an algebraic variant of Jacobi's inversion theorem: $a_g(C_g) = J(C)$.

We return now to the general case of a curve C over a field k.

Example 1. Suppose $g = 2$. Then a_1 is an embedding and the polarization divisor $a_1(C)$ is isomorphic to the curve C itself. This is a case where C can be reconstructed uniquely up to isomorphism from its p.p. Jacobian $J(C)$.

Remark. This fact is a special case of Torelli's theorem: a curve is determined up to isomorphism by its p.p. Jacobian. (More about it can be found in the survey by V. S. Kulikov and P. V. Kurchanov on Hodge structures; see also Griffiths-Harris [1978] and Arbarello et al. [1985].)

Every polarization divisor $a_{g-1}(C_{g-1}) \subset J(C)$ can be obtained through (10) from one divisor

$$\Theta = \mu_{g-1}(C_{g-1}) \subset \operatorname{Pic}^{g-1} C,$$

which is called the *canonical polarization divisor.* Its points are the linear equivalence classes of divisors $D \in \operatorname{Div}^{g-1} C$ such that $l(D) \geq 1$. On applying the Riemann-Roch formula we obtain:

Lemma. *$K/\sim - \Theta = \Theta$, where $K/\sim \in \operatorname{Pic}^{2g-2} C$ is the canonical class.*

Corollary. *The linear equivalence class of a divisor $D \in \operatorname{Div}^{g-1} C$ is a point of symmetry, or a theta characteristic, of the canonical polarization divisor $\Theta \subset \operatorname{Pic}^{g-1} C$ if and only if $2D \sim K$.*

Definition. A linear equivalence class $D/\sim \in \operatorname{Pic}^{g-1} C$ such that $2D \sim K$, is called a *theta characteristic* of C.

Thus the theta characteristics of a curve C are those of the canonical polarization divisor $\Theta \in \operatorname{Pic}^{g-1} C$. Of course, the parity of a theta characteristic of C is defined according to its parity as a theta characteristic of Θ. In fact, there is a more direct way of defining it.

Proposition. *Each theta characteristic $D/\sim$ of the curve C has the same parity as $l(D)$.*

This results from a theorem of Riemann on the singularities of the theta divisor: the multiplicity of the canonical polarization divisor Θ at a point $D/\sim \in \operatorname{Pic}^{g-1} C$ is equal to $l(D)$ (see Arbarello et al. [1985] and a later chapter in this Encyclopaedia).

Further on we assume that $\operatorname{char} k \neq 2$. Then, by Theorem 3 of Sect. 1.6, there are $2^{g-1}(2^g + 1)$ even and $2^{g-1}(2^g - 1)$ odd theta characteristics on C. Via (10) for even theta characteristics $D'/\sim$, the canonical polarization divisor $\Theta \subset \operatorname{Pic}^{g-1} C$ yields all possible Riemann theta divisors on a p.p.

Jacobian $J(C)$. In other words, all such p.p. divisors are of the form $\Theta - D'/\sim$.

Example 2. If a theta characteristic $D/\sim$ does not lie on Θ, then it is even and $l(D) = 0$. Even theta characteristics $D/\sim$ such that $l(D) = 0$, are said to be *nondegenerate*. Similarly, an odd theta characteristic $D/\sim$ is said to be *nondegenerate* if $l(D) = 1$. Thus a nondegenerate odd theta characteristic contains a unique effective divisor. For a generic curve, all theta characteristics are nondegenerate.

Example 3. Let C be a hyperelliptic curve, let $p_1, \ldots, p_{2g+2}$ be the ramification points of its hyperelliptic projection, and let D be a divisor in the corresponding hyperelliptic linear system g_2^1. Then the theta characteristics on C are of the form

$$(mD + p_{i_1} + \ldots + p_{i_{g-1-2m}})/\sim,$$

where $-1 \leq m \leq (g-1)/2$ and the points p_{i_j} are distinct. These classes coincide only when $m = 1$ if

$$-D + p_{i_1} + \ldots + p_{i_{g+1}} \sim -D + p_{j_1} + \ldots + p_{j_{g+1}}$$

and $\{i_1, \ldots, i_{g+1}, j_1, \ldots, j_{g+1}\} = \{1, \ldots, 2g+2\}$. The parity of these characteristics is equal to that of $m+1$. They are nondegenerate for $m \leq 0$.

Example 4. A generic curve of genus 3 is isomorphic to its canonical curve, which is a generic plane quartic $C \subset \mathbb{P}^2$. On this curve are $2^2(2^3-1) = 28$ odd theta characteristics $D/\sim$, where D denotes the unique effective representative. By the geometric interpretation of the Riemann-Roch formula, there is a one-to-one correspondence $D \mapsto \overline{D}$ between the odd theta characteristics and the bitangents to C. Therefore a generic quartic has 28 bitangents (cf. Example 5 of Chap. 2, Sect. 3.15).

2.6. Jacobian of a Curve of Genus 1. Let C be a curve of genus 1. Then the Abel mapping $a_1 \colon C \to J(C)$, $q \mapsto (q-p)/\sim$, is an isomorphism, which makes C into an elliptic curve, with $0 = p$ (cf. the Example in Sect. 1.2). In particular, we have a canonical isomorphism $a_1 \colon C \to J(C)$, $q \mapsto (q-0)/\sim$, of the elliptic curve C with its Jacobian. What this means is that, for any divisor Σp_i of degree 0 on the elliptic curve, we have:

$$\sum p_i = \sum p_i/\sim.$$

In the left-hand side, $\sum$ is understood in the sense of the group law on C.

Example. Consider a nonsingular plane cubic $C \subset \mathbb{P}^2$. A hyperplane section divisor D determines a class $D/\sim \in \operatorname{Pic}^3 C$. Obviously, the mapping $\operatorname{Pic}^1 C \to \operatorname{Pic}^3 C$, $p \mapsto 3p$, is equivalent to the isogeny $\mathbf{3} \colon J(C) \to J(C)$, and

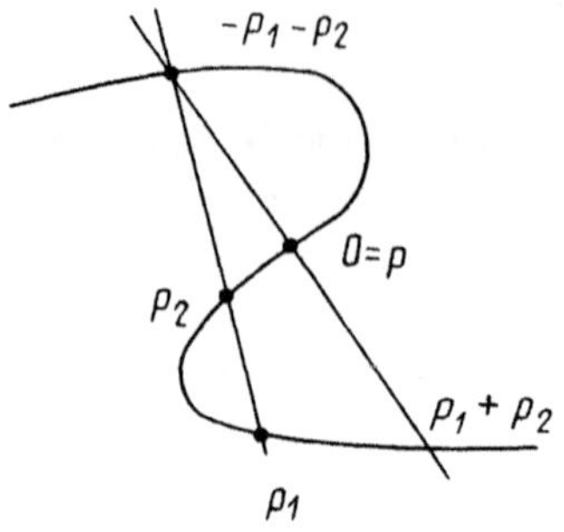

Fig. 28

hence surjective. Therefore $D \sim 3p$ for some $p \in C$. Hence, by linear normality, C is the image of the embedding $\varphi_{3p}\colon C \hookrightarrow \mathbb{P}^2$. Using this, it is a simple matter to prove that every nonsingular plane cubic in characteristic $\neq 2$ can be written, after a suitable choice of coordinates, in Weierstrass normal form (9) (see Theorem 2 of Sect. 1.7). The point p at infinity is an inflection point of C, and the tangent at that point is the line at infinity. We put on C an elliptic curve structure by setting $0 = p$. Then, for any hyperplane section divisor, we see that $p_1 + p_2 + p_3 = p_1 + p_2 + p_3 - 3 \cdot 0 = (p_1 + p_2 + p_3 - 3 \cdot 0)/\!\sim\, = 0$. So, $p_1 + p_2 + p_3 = 0$ and we obtain the well-known geometric interpretation of the group structure on C (see Fig. 28). Thus all inflection points p on C satisfy the relation $3p = 0$. By Theorem 1 of Sect. 1.6, if follows that, for $\operatorname{char} k \neq 3$, the curve C has nine inflection points. In the case where $\operatorname{char} k = 3$, there are three inflection points on a general cubic and only one on a supersingular one. Further, it is easy to express the geometric interpretation of the group law in coordinate form. If C is given in Weierstrass normal form (9) with 0 at infinity, and

$$p_1 = (x_1, y_1), \quad p_2 = (x_2, y_2)$$

are two distinct points on C, then we have:

$$p_1 + p_2 = \left(-x_1 - x_2 + \tfrac{1}{4}\left(\frac{y_1 - y_2}{x_1 - x_2}\right)^2, \right.$$

$$\left. (x_1 + x_2)\left(\frac{y_1 - y_2}{x_1 - x_2}\right) - \tfrac{1}{4}\left(\frac{y_1 - y_2}{x_1 - x_2}\right)^3 - \frac{x_1 y_2 - x_2 y_1}{x_1 - x_2} \right)$$

and $-p_1 = (x_1, -y_1)$. In this way it becomes evident that the operations are rational, and hence regular. Moreover, in the complex case the formula for $p_1 + p_2$ is directly connected with the addition theorem for the $\wp$-function (cf. Hurwitz-Courant [1922, 1964]).

This completes our presentation. Some further questions from algebraic curve theory are discussed in subsequent chapters of this Encyclopaedia.

References*

Ahlfors, L.V. [1966] Lectures on Quasiconformal Mappings. Van Nostrand, Princeton, New Jersey, 1966; 146 pp., Zbl. 138, 60.

Ahlfors, L.V., Sario, L. [1960] Riemann Surfaces. Princeton Univ. Press, Princeton, New Jersey, 1960; 382 pp., Zbl. 196, 338.

Alekseev, V.B. [1976] Abel's Theorem in Problems and Solutions. Nauka, Moscow, 1976; 207 pp. (Russian).

Arbarello, E., Cornalba, M., Griffiths, P., Harris, J. [1985] Geometry of Algebraic Curves; Vol. 1. Springer, New York, Berlin, Heidelberg, 1985; 386 pp., Zbl. 559.14017.

Arbarello, E., Sernesi, E. [1978] Petri's approach to the study of the ideal associated to a special divisor. Invent. Math. 49 (1978), 99–119, Zbl. 399.14019.

Artin, M. [1969] Algebraization of formal moduli, I; in: Global Analysis, Papers in honor of K. Kodaira, Univ. of Tokyo Press – Princeton Univ. Press, 1969; 21–71, Zbl. 205, 504.

Babbage, D.W. [1939] A note on the quadrics through a canonical curve. J. Lond. Math. Soc. 14 (1939), 310–315, Zbl. 26, 347.

Belyĭ, G.V. [1979] Galois extensions of a maximal cyclotomic field. Izv. Akad. Nauk SSSR, Ser. Mat. 43 (1979), 267–276. English transl.: Math. USSR, Izv. 14, 247–256 (1980), Zbl. 409.12012.

Bers, L. [1960] Quasiconformal mappings and Teichmüller's theorem; in: Analytic Functions. Princeton Math. Ser. 24, 1960; 89–119, Zbl. 100, 289.

Chebotarev, N.G. [1948] Theory of Algebraic Functions. Gostekhizdat, Moscow – Leningrad, 1948; 396 pp. (Russian).

Chevalley, C. [1951] Introduction to the Theory of Algebraic Functions of One Variable. Am. Math. Soc. (Math. Surveys No. 6), New York, 1951; 188 pp., Zbl. 45, 323.

Clemens, C.H. [1980] A Scrapbook of Complex Curve Theory. Plenum Press, New York, 1980; 186 pp., Zbl. 456.14016.

Deligne, P., Mumford, D. [1969] The irreducibility of the space of curves of given genus. Publ. Math. Inst. Hautes Etud. Sci. 36, 1969; 75–109, Zbl. 181, 488.

Dold, A. [1972] Lectures on Algebraic Topology. Springer, Berlin, Heidelberg, New York, 1972; 377 pp., Zbl. 234.55001.

Dubrovin, B.A., Novikov, S.P., Fomenko, A.T. [1980] Modern Geometry: Methods and Applications. Nauka, Moscow, 1980; 760 pp. English transl.: Springer, Berlin, Heidelberg, New York, 1984, Zbl. 433.53001.

Dubrovin, B.A., Novikov, S.P., Fomenko, A.T. [1984] Modern Geometry: Methods of Homology Theory. Nauka, Moscow, 1984; 344 pp. English transl.: Springer, Berlin, Heidelberg, New York, 1990, Zbl. 582.55001.

Eisenhart, L.P. [1947] An Introduction to Differential Geometry. Princeton Univ. Press, Princeton, New Jersey, 1947; 304 pp., Zbl. 33, 18.

Enriques, F. [1919] Sulle curve canoniche di genere p dello spazio a $p-1$ dimensioni. Rend. Accad. Sci. Ist. Bologna, 23 (1919), 80–82.

Forster, O. [1977] Riemannsche Flächen. Springer, Berlin, Heidelberg, New York, 1977; 223 pp., Zbl. 381.30021.

Fricke, R., Klein, F. [1897, 1912] Vorlesungen über die Theorie der Automorphen Funktionen (2 vol.). Teubner, Leipzig, 1897, 1912; 634 pp., 668 pp., Jbuch 28, 334.

Griffiths, P., Harris, J. [1978] Principles of Algebraic Geometry. Wiley, New York, 1978; 813 pp., Zbl. 408.14001.

* For the convenience of the reader, references to reviews in Zentralblatt für Mathematik (Zbl.), compiled using the MATH database, and Jahrbuch über die Fortschritte der Mathematik (Jbuch) have, as far as possible, been included in this bibliography.

Grothendieck, A. [1961–62] Technique de descente et théorèmes d'existence en géométrie algébrique, V. Les schémas de Picard : théorèmes d'existence. Sémin. Bourbaki 14, Exp. 232 (1961–62), 19 pp., Zbl. 238.14014.
Grothendieck, A. [1966] On the de Rham cohomology of algebraic varieties. Publ. Math. Inst. Hautes Etud. Sci. 29, 1966; 95–103, Zbl. 145, 176.
Gunning, R.C. [1972] Lectures on Riemann Surfaces: Jacobi Varieties. Princeton Univ. Press, Princeton, New Jersey, 1972; 189 pp., Zbl. 387.32008.
Gunning, R.C., Rossi, H. [1965] Analytic Functions of Several Complex Variables. Prentice-Hall, Englewood Cliffs, New Jersey, 1965; 317 pp., Zbl. 141, 86.
Halphen, G. [1882] Mémoire sur la classification des courbes gauches algébriques. J. Ecole Polytechnique 52 (1882), 1–200.
Harris, J. [1984] On the Kodaira dimension of the moduli space of curves, II. The even-genus case. Invent. Math. 75 (1984), 437–466, Zbl. 542.14014.
Harris, J. [1986] On the Severi problem. Invent. Math. 84 (1986), 445–461, Zbl. 596.14017.
Harris, J., Mumford, D. [1982] On the Kodaira dimension of the moduli space of curves. Invent. Math. 67 (1982), 23–86, Zbl. 506.14016.
Hartshorne, R. [1977] Algebraic Geometry. Springer, New York, Berlin, Heidelberg, 1977; 496 pp., Zbl. 367.14001.
Hartshorne, R. [1981–82] Genre des courbes algébriques dans l'espace projectif (d'après L. Gruson et C. Peskine). Sémin. Bourbaki, Exp. No. 592 (1981–82), Astérisque 92–93, 301–313, Zbl. 524.14037.
Hilbert's Problems [1969] Nauka, Moscow, 1969; 225 pp., Zbl. 213, 288. German transl.: Ostwalds Klassiker Bd. 252, Leipzig, Akad. Verlagsges. Geest & Portig (1971), 2nd ed. 1979.
Hirsch, M. [1976] Differential Topology. Springer, New York, Berlin, Heidelberg, 1976; 221 pp., Zbl. 356.57001.
Hörmander, L. [1966] An Introduction to Complex Analysis in Several Variables. Van Nostrand, Princeton, New Jersey, 1966; 208 pp., Zbl. 138, 62.
Hurwitz, A. [1891] Über Riemann'sche Flächen mit gegebenen Verzweigungspunkten. Math. Annalen 39 (1891), 1–61, Jbuch 23, 429.
Hurwitz, A. (ed. by Courant, R.) [1922, 1964] Vorlesungen über allgemeine Funktionentheorie und elliptische Funktionen (R. Courant, Hrsg. und ergänzt durch einen Abschnitt über geometrische Funktionentheorie). Springer (Grundlehren der Math. Wissenschaften 3), Berlin, Heidelberg, New York, 1922, 1964; 706 pp., Zbl. 135, 121.
Katsylo, P.I. [1984] The rationality of moduli spaces of hyperelliptic curves. Izv. Akad. Nauk SSSR 48 (1984), 705–710. English transl.: Math. USSR, Izv. 25, 45–50 (1985), Zbl. 593.14017.
Kleiman, S.L. [1977] The enumerative theory of singularities; in: Real and complex singularities, Proc. Nord. Summer Sch., Symp. Math., Oslo 1976, 297–396 (1977), Zbl. 385.14018.
Kostrikin, A.I., Manin, Yu.I. [1980] Linear Algebra and Geometry. Moscow State Univ., Moscow, 1980; 318 pp. (Russian), Zbl. 532.00002.
Kra, I. [1972] Automorphic Forms and Kleinian Groups. Benjamin, Reading, Mass.; 1972; 464 pp., Zbl. 253.30015.
Lang, S. [1982] Introduction to Algebraic and Abelian Functions (2nd ed.). Springer, New York, Berlin, Heidelberg, 1982; 169 pp., Zbl. 513.14024, Zbl. 255.14001.
Lewittes, J. [1964] Riemann surfaces and the theta function. Acta Math., 111 (1964), 37–61, Zbl. 125, 318.
Manin, Yu. I. [1972] Cubic Forms. Nauka, Moscow, 1972; 304 pp. English transl.: North Holland Publ. Co., Amsterdam etc. (1974), Zbl. 255.14002.
Massey, W. [1967, 1977] Algebraic Topology: An Introduction. Harcourt, Brace & World – Springer, New York, Berlin, Heidelberg, 1967, 1977; 261 pp., Zbl. 361.55002, Zbl. 153, 249.
Milne, J.S. [1980] Étale Cohomology. Princeton Univ. Press, Princeton, New Jersey, 1980; 323 pp., Zbl. 223.14012.

Mumford, D. [1965] Geometric Invariant Theory. Springer, Berlin, Heidelberg, New York, 1965; 146 pp., Zbl. 147, 393.
Mumford, D. [1966] Lectures on Curves on an Algebraic Surface. Princeton Univ. Press, Princeton, New Jersey, 1966; 200 pp., Zbl. 187, 427.
Mumford, D. [1970] Abelian Varieties. Oxford Univ. Press, London – Bombay, 1970; 242 pp., Zbl. 223.14022.
Mumford, D. [1975] Curves and Their Jacobians. The Univ. of Michigan Press, Ann Arbor, 1975; 104 pp., Zbl. 316.14010.
Mumford, D. [1976] Algebraic Geometry I: Complex Projective Varieties. Springer, New York, Berlin, Heidelberg, 1976; 186 pp., Zbl. 356.14002.
Narasimhan, R. [1968] Analysis on Real and Complex Manifolds. North-Holland, Amsterdam, 1968; 246 pp., Zbl. 188, 258.
Nœther M. [1882] Zur Grundlegung der Theorie der algebraischen Raumcurven. Verlag der königlichen Akademie der Wissenschaften, Berlin, 1882; 120 pp., Jbuch 14, 669.
Petri, K. [1923] Über die invariante Darstellung algebraischer Funktionen einer Veränderlichen. Math. Ann. 88 (1923), 242–289, Jbuch 49, 264.
Petrovskiĭ, I.G. [1961] Lectures on Partial Differential Equations. 3rd ed. Fizmatgiz, Moscow, 1961; 400 pp. (Russian), Zbl. 115, 81; Zbl. 38, 255.
Rado, T. [1925] Über den Begriff der Riemannschen Fläche. Acta Szeged 2 (1925), 101–121, Jbuch 51, 273.
Saint-Donat, B. [1973] On Petri's analysis of the linear system of quadrics through a canonical curve. Math. Ann., 206 (1973), 157–175, Zbl. 315.14010.
Sario, L., Nakai, M. [1970] Classification Theory of Riemann Surfaces. Springer, Berlin, Heidelberg, New York, 1970; 446 pp., Zbl. 199, 406.
Serre, J-P. [1956] Géométrie algébrique et géométrie analytique. Ann. Inst. Fourier, 6 (1956), 1–42, Zbl. 75, 304.
Serre, J-P. [1959] Groupes algébriques et corps de classes. Hermann, Paris, 1959; 202 pp., Zbl. 97, 356.
Shabat, B.V. [1969] Introduction to Complex Analysis. Nauka, Moscow, 1969; 576 pp. French transl.: Mir, Moscow, 1990, Zbl. 188, 379.
Shafarevich, I.R. [1972] Basic Algebraic Geometry. Nauka, Moscow, 1972, 567 pp. English transl.: Grundlehren der mathematischen Wissenschaften 213. Springer, New York, Berlin, Heidelberg, 1974, Zbl. 318.14001.
Shafarevich, I.R. [1986] Basic notions of algebra; in: *Itogi Nauki Tekh., Ser. Sovrem. Probl. Mat. Fundam. Napravleniya* 11. VINITI, Moscow, 1986, 290 pp. English transl.: Encycl. Math. Sci. 11: Algebra I, Springer, Berlin, Heidelberg, New York, 1990, Zbl. 655.00002, Zbl. 711.16001.
Shimura, G. [1971] Introduction to the Arithmetic Theory of Automorphic Functions. Princeton Univ. Press, Princeton, New Jersey, 1971; 267 pp., Zbl. 221.10029.
Shokurov, V.V. [1971] The Noether-Enriques theorem on canonical curves. Mat. Sb., Nov. Ser. 86 (1971), 367–408. English transl.: Math. USSR, Sb. 15, 361–401 (1972), Zbl. 225.14017.
Shokurov, V.V. [1979] Existence of a straight line on Fano varieties. Izv. Akad. Nauk SSSR 43 (1979), 922–964. English transl.: Math. USSR, Izv. 15, 173–209 (1980), Zbl. 422.14019.
Siegel, C.L. [1948–49] Analytic functions of several complex variables (Lectures delivered at the Institute for Advanced Study). Princeton, New Jersey, 1948–49; 200 pp., Zbl. 36, 50.
Spivak, M. [1965] Calculus on Manifolds. Benjamin, New York, 1965; 146 pp., Zbl. 141, 54.
Springer, G. [1957] Introduction to Riemann Surfaces. Addison-Wesley, Reading, Mass., 1957; 307 pp., Zbl. 78, 66.
Walker, R. [1950] Algebraic Curves. Princeton Univ. Press, Princeton, 1950; 201 pp., Zbl. 39, 377.
Weil, A. [1949] Number of solutions of equations over finite fields. Bull. Amer. Math. Soc. 55 (1949), 497–508, Zbl. 32, 394.

Wells, R.O. [1973] Differential Analysis on Complex Manifolds. Prentice-Hall, New York, 1973; 252 pp., Zbl. 262.32005.

Weyl, H. [1923] Die Idee der Riemannschen Flächen. Teubner, Leipzig, Berlin, 1923; 183 pp. 3rd ed. 1955, Zbl. 68, 60.

Whittaker, E.T., Watson, G.N. [1902, 1927, 1962] A Course of Modern Analysis. Cambridge Univ. Press, Cambridge, 1902, 1927, 1962; 608 pp., Zbl. 105, 269.

Zariski, O., Samuel, P. [1958] Commutative Algebra; Vol. 1. Van Nostrand, Princeton, 1958; 329 pp., Zbl. 81, 265; 2nd ed. Springer, Berlin, Heidelberg, New York, 1975, Zbl. 313.13001.

Zarkhin, Yu.G., Parshin, A.N. [1986] Finiteness problems in diophantine geometry (Appendix to the Russian translation of S. Lang's *Diophantine Geometry*); in: Lang, S., Foundations of Diophantine Geometry. Mir, Moscow, 1986; pp. 369–438. English transl.: Transl., II. Ser., Am. Math. Soc. 143, 35–102 (1989), Zbl. 644.14007.

II. Algebraic Varieties and Schemes

V. I. Danilov

Translated from the Russian
by D. Coray

Contents

Introduction

This survey deals with the foundations of algebraic geometry: we introduce its basic objects and present their main properties. Roughly speaking, algebraic geometry is concerned with systems of algebraic equations and their solutions. Its approach is to examine right away the whole set of solutions, which is regarded as forming a single geometric object, endowed with a topology, a sheaf of functions, etc. The mappings between such objects correspond to some algebraic transformations of the solutions.

Algebraic varieties, affine or projective, were initially considered over the field of real or complex numbers; transcendental methods were widely used (cf. the historical account in Shafarevich [1972]). The similarities noted between the theory of algebraic curves over $\mathbb{C}$ (Riemann surfaces) and the theory of algebraic numbers stimulated the search for a common algebraic foundation. After several preliminary attempts (the evolution of the concept of algebraic variety is described in Dieudonné [1969], Dolgachev [1972], and Shafarevich [1972]), the notion of scheme was elaborated, which enables one to speak the language of geometry when dealing with systems of algebraic equations over arbitrary commutative rings. Actually, schemes do not cover the whole range of objects studied by algebraic geometry (let us mention formal schemes, algebraic spaces, and the like), but they play a fundamental and central role in modern algebraic geometry.

Nevertheless, we focus our attention not on schemes, but on the notion of algebraic variety over an algebraically closed field, which is easier to understand and closer to intuition. The theory of these varieties, at the elementary level, is built much like that of differentiable or analytic manifolds. So, in Chapter 1 we stress these analogies: atlases, morphisms, vector bundles, sheaves, differential calculus. However, many of the concepts introduced in algebraic geometry are specific to algebraic varieties. Such are for instance the notions of irreducibility, completeness, rational map, dimension, and singular points. These are discussed in Chapter 2, together with some deeper properties of algebraic varieties. The greater part of Chapter 3 is devoted to projective geometry (degree, linear sections and projections, linear systems of divisors, Chow variety); there is also an account of the theory of intersection. Thus, only Chapter 4 is properly devoted to schemes. It essentially contains the basic definitions and the generalization to schemes of the notions and results which will be familiar from the previous three chapters. The cohomology theories on algebraic varieties, which also belong to the foundations, will be the subject of a separate article.

It is assumed that the reader is familiar with the general mathematical notions of set, topological space, field and algebra, vector space and polynomial, and – to a lesser degree – category and functor (cf. Shafarevich [1986]).

Some knowledge of differentiable or analytic manifolds, and of sheaves, is also desirable, but not absolutely necessary. We have endeavoured to provide most basic results with a sketch of proof, with the idea that this is indispensable for understanding the theory.

Chapter 1
Algebraic Varieties: Basic Notions

The aim of this chapter is to give a precise meaning to the following words: an algebraic variety is an object which is defined locally by some polynomial equations. The main distinction between algebraic varieties and differentiable or complex analytic manifolds (see Bourbaki [1967–1971], Chirka [1985], or Lang [1962]) lies in the choice of the local models. In the differentiable and complex analytic cases, these are the open subsets of $\mathbb{R}^n$ or $\mathbb{C}^n$. A local model of an algebraic variety is a subset of the coordinate space which is given by polynomial equations. This makes sense only if we fix a ground field K, over which both the polynomials and the solutions will be considered. In order to simplify the algebraic aspect of the question as much as possible and to concentrate on geometry, we shall assume in the first three chapters that the field K is algebraically closed. In the present chapter we shall also examine the simplest notions from algebraic geometry that have direct analogues in the differentiable and analytic cases.

§ 1. Affine Space

1.1. Base Field. The reader is free to assume that the base field K is the field $\mathbb{C}$ of complex numbers. However, even when dealing with the complex numbers, we shall proceed in a purely algebraic way; that is, we shall use the operations of addition and multiplication but no such thing as a limit. Hence our considerations will be valid for any field. At this point we may recall the subdivision made by A. Weil into classical methods (which rely on the properties of the field of reals, or of complex numbers, and stem from topology, analysis, differential equations or analytic function theory) and abstract methods, which are based on algebra and can be applied over an arbitrary base field.

There are also some more potent reasons for developing the theory over arbitrary fields, including fields of positive characteristic. In the first place, this is needed for applications to number theory. Secondly, even for the proof of assertions over $\mathbb{C}$, it is often convenient to use some properties of varieties over finite fields.

The only important property of the field $\mathbb{C}$ that we shall assume to hold for K, is that it is algebraically closed. This means that the 'Fundamental Theorem of Algebra' holds over K: every polynomial with coefficients in K splits into linear factors (see Sect. 1.5 for a higher dimensional generalization). Let us mention that an algebraically closed field is always infinite. By analogy with $\mathbb{C}$, the elements of K are also called numbers, or constants.

1.2. Affine Space. Let n be an integer. By n-dimensional (coordinate) *affine space* we mean the n-th Cartesian power K^n of K. An element of K^n is a sequence $(x_1, \dots, x_n)$ of n numbers $x_i \in K$. Such n-tuples can be added together coordinatewise, or multiplied by a constant; so K^n is a vector space over the field K. However, algebraic geometry supplies K^n with a weaker structure; in fact, among all mappings $K^n \to K$ (functions), one selects the so-called algebraic, or regular, functions.[5]

Now what functions on K^n is it natural to call algebraic? First of all the constants, which are identified with the numbers in K. Then the coordinate functions, that is, the projection maps $T_i : K^n \to K$, where $T_i(x_1, \dots, x_n) = x_i$. And, finally, the functions that are built from them through the elementary algebraic operations of addition and multiplication. These functions are called *regular* (so as to distinguish them from rational functions; these will also come up, but a little later). Thus, regular functions are expressed polynomially in terms of the coordinate functions T_i. Moreover, as K is infinite, we can identify the ring of regular functions on K^n with the polynomial ring $K[T_1, \dots, T_n]$ in the variables $T_1, \dots, T_n$ with coefficients in K.

One could also regard as regular all functions of the form $1/f$, where f is regular and nowhere zero. But such a function f is necessarily constant, so this does not lead to anything new. Here we use the fact that K is algebraically closed, since over $\mathbb{R}$ the function $1 + t^2$ does not vanish for any $t \in \mathbb{R}$.

1.3. Algebraic Subsets. The algebraic subsets of K^n are defined by systems of algebraic equations. An algebraic equation is an expression $f = 0$, where f is a polynomial in $T_1, \dots, T_n$. Given a family $F = (f_r,\ r \in R)$ of polynomials, the family of equations $(f_r = 0,\ r \in R)$ – or $F = 0$ for short – is called a *system of algebraic equations.* A solution (also called a zero, or a root) of this system is any point $x \in K^n$ such that $f_r(x) = 0$ for all $r \in R$. The set of all solutions to $F = 0$ is denoted by $V(F)$, or $[F = 0]$.

Definition. A subset of K^n is said to be *algebraic* if it is of the form $V(F)$ for some family F of polynomials in $T_1, \dots, T_n$.

For example, the empty set, and also K^n, are algebraic (take $F = \{1\}$, respectively, $F = \{0\}$). The intersection of any number of algebraic subsets is again algebraic, since $\bigcap V(F_j) = V(\bigcup F_j)$. The union of any *finite* number of algebraic subsets is also algebraic. Indeed, $V(F_1) \cup V(F_2) = V(F_1 \cdot F_2)$, where $F_1 \cdot F_2$ consists of all products of the form $f_1 f_2$ with $f_1 \in F_1$ and $f_2 \in F_2$. On the other hand, the complement of an algebraic subset $V \subset K^n$ is not algebraic (except for $V = \varnothing, K^n$).

[5] This way of defining a structure is fairly common in mathematics. For instance, on the set $\mathbb{C}$ of complex numbers one may consider the following increasingly general sets of functions: linear, affine, polynomial, analytic, differentiable, continuous, measurable, and – lastly – arbitrary. Then $\mathbb{C}$ will be, respectively, an object of linear algebra, of affine geometry, of algebraic or analytic geometry, a smooth manifold, a topological or a measurable space, and simply a continuum.

Here are some more tangible examples. Every point $x \in K^n$ is an algebraic subset. The zero set $V(f)$ of a single (nonconstant) function f is called an algebraic *hypersurface*. Hypersurfaces in K^2 are called affine plane curves. They are traditionally represented by pictures like those in Fig. 1.

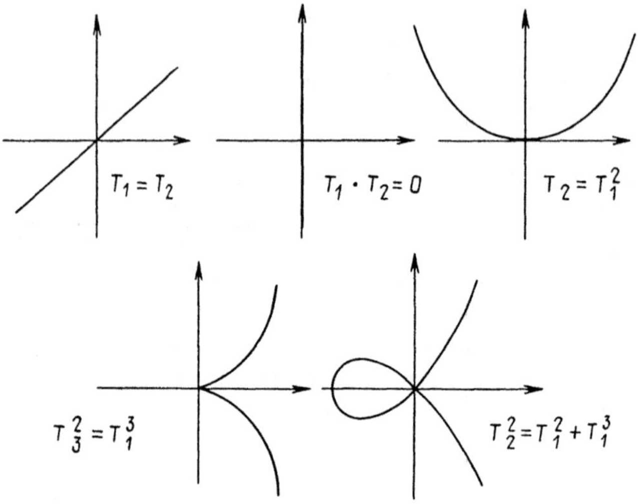

Fig. 1

1.4. Systems of Algebraic Equations; Ideals. Different systems of equations can have the same set of solutions. Indeed, if we adjoin to the system F the polynomial $\Sigma f_j g_j$, where $f_j \in F$ and $g_j \in K[T_1, \ldots, T_n]$, the set of solutions will remain unchanged. We shall say that $\Sigma f_j g_j$ can be expressed algebraically in terms of the family F. Two families, F and F', are said to be equivalent if every member of F can be expressed algebraically in terms of F', and conversely. Clearly, F and F' are equivalent if and only if they generate the same ideal in the ring $K[T_1, \ldots, T_n]$. Going over to ideals is useful because of *Hilbert's Basis Theorem*:

Theorem. *Every ideal in the polynomial ring $K[T_1, \ldots, T_n]$ is generated by a finite set of elements.*

In other words, the ring of polynomials over any field is noetherian. As a corollary we get that every system of algebraic equations is equivalent to a *finite* system of equations, or that every algebraic subset is the intersection of a finite number of hypersurfaces.

As we have already said, equivalent systems of equations have identical sets of solutions; however, two nonequivalent systems can also define the same subset. The reason for this is quite simple: the polynomials f, f^2, f^3, etc., have the same zeros. In other words, taking out a root does not modify the zeros. In view of this we shall say that two families, F and F', are weakly equivalent if every element $f \in F$ has a power f^r that can be expressed algebraically in terms of F', and conversely. Again, weakly equivalent systems of equations have identical sets of solutions. As we shall see now, the converse is also true. In any case, for each algebraic subset $V \subset K^n$ there is a largest ideal defining V, namely, the ideal $I(V)$ of all regular functions vanishing at all points of V.

1.5. Hilbert's Nullstellensatz. Let us start from the simplest situation. It is clear that the unit ideal $I = K[T_1, \ldots, T_n]$ defines the empty subset $V(I)$. Though this is much less obvious, the converse is also true; this assertion is called Hilbert's *Weak Nullstellensatz.*

Theorem. *If the ideal $I \subset K[T_1, \ldots, T_n]$ is not the unit ideal then $V(I)$ is nonempty.*

It is essential here that the field K is algebraically closed, for $1 + t^2 \neq 0$ for every $t \in \mathbb{R}$. As this theorem plays an important role, we sketch its proof briefly. We may assume that the ideal I is maximal (among non-unit ideals, as usual). Then the quotient algebra $K[T_1, \ldots, T_n]/I$ is a field, which contains K. We shall prove that these two fields coincide. Then, if we denote by t_i the image of T_i in K, we see that the point $t = (t_1, \ldots, t_n)$ belongs to $V(I)$. Hence it remains to show that the field $K[T_1, \ldots, T_n]/I$ is isomorphic to K (this assertion is analogous to the Gel'fand-Mazur theorem on maximal ideals in Banach algebras). Since K is algebraically closed, this assertion is a consequence of a purely algebraic lemma:

Lemma. *Let K be an arbitrary field, and L a K-algebra of finite type. Then, if L is a field, it is algebraic over K.*

The proof of this lemma exploits the very useful notion of integral dependence: for further details about it we refer to Atiyah-Macdonald [1969], Bourbaki [1961–1965] or Zariski-Samuel [1958,1960]. Let $A \subset B$ be two commutative rings. An element $b \in B$ is said to be *integral* over A if it satisfies an equation of integral dependence $b^m + a_1 b^{m-1} + \ldots + a_m = 0, \quad a_i \in A$. The main point here is that the leading coefficient is equal to 1. If every element of B satisfies such an equation, we say that the algebra B is integral over A. Sums and products of integral elements are again integral; so the set of integral elements in B is a subalgebra of B, called the *integral closure* of A in B.

We move on to the proof of the lemma. Suppose the algebra L is generated by the elements $x, x_1, \ldots, x_n$. Since L is a field, it contains the field $K(x)$. By induction, applied to $K(x) \subset L$, we may assume that the elements $x_1, \ldots, x_n$ are algebraic over $K(x)$. If x is algebraic over K, there is nothing to prove. We may therefore assume that x is transcendental over K, so that the ring $K[x]$ is isomorphic to the ring of polynomials in x. Since the x_i are algebraic over $K(x)$, there exist polynomials $f_i(T)$ with coefficients in $K[x]$ such that

$f_i(x_i) = 0$. If we denote by $g \in K[x]$ the product of the leading coefficients of the f_i, we find that the x_i are integral over the ring $A = K[x][g^{-1}]$. But then the algebra L is integral over A. It follows immediately that A, too, is a field. Indeed, let a be a nonzero element of A. Since $a^{-1} \in L$, a^{-1} is integral over A. This means that $a^{-m} + a_1 a^{-m+1} + \ldots + a_m = 0$, with $a_j \in A$, that is, $1 + a_1 a + \ldots + a_m a^m = 0$, whence $a^{-1} = -(a_1 + \ldots + a_m a^{m-1}) \in A$. On the other hand, A is obviously not a field; for instance, $1 + g$ is not invertible in A. This contradiction proves the lemma and the theorem.

Corollary (Hilbert's Nullstellensatz). *Let I be an ideal in $K[T_1, \ldots, T_n]$, and suppose the polynomial f vanishes at all points of the set $V(I) \subset K^n$. Then $f^r \in I$ for some integer $r \geq 0$.*

In the space K^{n+1} with coordinates $T_0, T_1, \ldots, T_n$ we consider the subset $V' = K \times V$ of all zeros of the polynomials in I. The function $1 - T_0 \cdot f$ is nonzero at all points of V'. By the preceding theorem, $1 - T_0 f$ and I generate the unit ideal in $K[T_0, T_1, \ldots, T_n]$. Writing this out and setting $1/f$ in place of T_0, we find $f^r \in I$, as required.

§ 2. Affine Algebraic Varieties

2.1. Affine Varieties. When we think of algebraic substitutions and transforms of the solutions of algebraic equations, we are led to the concept of a mapping between algebraic sets.

Let $V \subset K^n$ and $W \subset K^m$ be two algebraic subsets. A mapping $f\colon V \to W$ is said to be *regular* (or a morphism) if it is given by m regular functions $f_1, \ldots, f_m \in K[T_1, \ldots, T_n]$, that is, if it extends to a regular mapping of the ambient spaces $K^n \to K^m$. The composite of two regular maps is still regular. Hence algebraic sets, together with regular mappings, form a category. The objects of this category are called *affine algebraic varieties* (or, simply, affine varieties). Thus the coordinate space K^n, considered as an affine variety, is denoted by $\mathbb{A}^n$ and is called n-dimensional *affine space* (or *line* if $n = 1$, or *plane* if $n = 2$).

Different algebraic sets (which can even be embedded in different spaces K^n) may then turn out to be isomorphic, that is, in some sense identical. Thus the 'line' $T_2 = T_1$ and the 'parabola' $T_2 = T_1^2$, in K^2, are isomorphic to each other, and also isomorphic to the affine line $\mathbb{A}^1$.

A *regular function* on an algebraic set V is a regular mapping of V into K. Regular functions can be added and multiplied together, so that they form a ring (and even a K-algebra) $K[V]$. Given an algebraic subset $V \subset K^n$, the algebra $K[V]$ identifies with the quotient algebra $K[T_1, \ldots, T_n]/I(V)$. Also, the embedding $V \subset K^n$ can be recovered from the generators $t_i = T_i|_V$ of $K[V]$.

One can also think of a morphism in terms of regular functions. A mapping $f\colon V \to W$ is regular if and only if, for every regular function $g \in K[W]$, the function $f^*(g) = g \circ f$ is regular on V. In this case the map $f^*\colon K[W] \to K[V]$ is a K-algebra homomorphism. Conversely, every such homomorphism $K[W] \to K[V]$ is induced by a morphism $V \to W$.

This way of viewing an affine variety as a set V with an additional structure – namely the algebra $K[V]$ of regular functions on V – is conceptually very useful. In the style of H. Weyl one might say that the formula $F = 0$ looks so enticingly like an equation that it could lure us into making some mechanical calculations, so we replace it by the ring $K[V]$. Thus we turn away from non-essential characteristics and take equally into consideration all equations that can be derived from the original one by means of a rational transformation of the variables.

2.2. Abstract Affine Varieties. The K-algebra $K[V]$ of regular functions on an algebraic set V has two specific properties. First of all, it is of finite type, that is, it is generated by finitely many elements. Secondly, as an algebra of functions with values in a field K, it is reduced, that is, it has no nilpotent elements (other than 0). Finally, it follows from Hilbert's Nullstellensatz that, by associating with a point $x \in V$ the maximal ideal $I(x) = \{f \in K[V],\ f(x) = 0\}$, we get a bijection between V and the set $\operatorname{Specm} K[V]$ of all maximal ideals of the ring $K[V]$.

These properties enable us to give an abstract definition of an *affine variety* over K as a triple $(X, K[X], \varphi)$, where X is a set, $K[X]$ a reduced K-algebra of finite type, and φ a bijection of X onto $\operatorname{Specm} K[X]$.

The elements of X are the points of this variety, while those of $K[X]$ are called its regular functions. In fact, given $x \in X$ and $f \in K[X]$, it makes sense to talk about the value $f(x)$ of f at the point x. By definition, it is the image of f under the composite map

$$K[X] \xrightarrow{\alpha} K[X]/\varphi(x) \underset{\approx}{\xleftarrow{\beta}} K,$$

where α is projection onto the quotient algebra, and β the structure K-homomorphism, which is one-to-one by virtue of Hilbert's Nullstellensatz. The fact that φ is bijective means that both the points and the functions are in good supply: there are enough functions to distinguish the points, and enough points to realize all K-algebra homomorphisms $K[X] \to K$. In what follows, we shall no longer write φ, and the values at the points will be understood.

With this terminology, a *morphism* of $(X, K[X])$ into $(Y, K[Y])$ is a pair (f, f^*), consisting of a mapping $f\colon X \to Y$ and a K-algebra homomorphism $f^*\colon K[Y] \to K[X]$, such that $f^*(g)(x) = g(f(x))$ for every $g \in K[Y]$ and $x \in X$. In point of fact, each of f and f^* is determined by the other.

Every abstract affine variety $(X, K[X])$ is isomorphic to an algebraic subset of some suitable affine space K^n. For it suffices to take some generators $t_1, \ldots, t_n$ of the algebra $K[X]$ and to use them to embed X in K^n.

2.3. Affine Schemes. Suppose that, in the definition of an abstract affine variety, we forget about the requirement that the ring $K[X]$ should be reduced. Then the object obtained will be called an *affine algebraic K-scheme* (more briefly: an affine scheme). An element of $K[X]$ defines, as in Sect. 2.2, a mapping $X \to K$; but in general these elements cannot be identified with functions. Indeed, some nonzero elements of $K[X]$ can give rise to functions which are identically zero on X. Besides, it follows from Hilbert's Nullstellensatz that this can happen only with nilpotent elements of $K[X]$.

Morphisms of affine schemes are defined in exactly the same words as in Sect. 2.2 for affine varieties. But now the map f^* is no longer determined by f. Let us give a few examples of schemes.

Example 1. Every affine variety is an affine scheme. And conversely, given any affine scheme X, there is a canonically associated affine variety $X_{\text{red}} = (X, K[X]/I(X))$. Here $I(X)$ is the ideal consisting of all the elements of $K[X]$ that vanish at all points of X, that is, the ideal of its nilpotent elements.

Example 2. Let A be any commutative K-algebra of finite type. The associated affine scheme $(\operatorname{Specm} A, A)$ is also denoted by $\operatorname{Specm} A$. Each homomorphism of K-algebras gives rise to a morphism of affine schemes (in the opposite direction). Thus, the category of affine schemes is antiequivalent to the category of K-algebras of finite type.

Example 3. Let $(X, K[X])$ be an affine scheme, and I an ideal of $K[X]$. Then the subset $V(I)$ of zeros of I, together with the ring $K[X]/I$, is an affine scheme. We say it is the *subscheme* of X defined by the ideal I. For instance, the subscheme of X defined by the ideal $I(X)$ is the associated variety X_{red}.

Let us consider, for example, the subscheme of the line $\mathbb{A}^1$ defined by the equation $T^2 = 0$ (or by the ideal (T^2)). Its ring $K[T]/(T^2)$ consists of all expressions of the form $a + bT$, with $a, b \in K$. Such an expression 'remembers', so to speak, not only the value of the function at the point 0 (that is, a), but also its derivative (that is, b). That is why, also in the general case, the subscheme defined by an ideal I is thought of as an 'infinitesimal neighbourhood' of the set $V(I) \subset \mathbb{A}^n$.

Just like affine varieties, affine schemes can be realized as subschemes of some suitable affine spaces $\mathbb{A}^n$. We shall not make any systematic use of the language of schemes until Chapter 4, but we must know about it. Many natural constructions lead up directly to schemes, even though we may start from varieties. It is of course always possible to go over to the associated variety, but in so doing we may lose some important geometric information. There must be some good motivation for schemes to appear in books like Arnol'd, Varchenko & Guseĭn-Zade [1982] and Griffiths-Harris [1978], which are so far-away from them.

We introduce a few simple methods for building some new affine varieties from old ones.

2.4. Products of Affine Varieties. Let X and Y be affine varieties; then the Cartesian product $X \times Y$ also has a natural structure of affine variety. More precisely, the product appears naturally as the affine scheme $(X \times Y, K[X] \otimes_K K[Y])$. Now it is a theorem that this scheme is a variety. To this effect one must check that a nonzero element $\Sigma f_j \otimes g_j$ of the tensor product $K[X] \otimes_K K[Y]$ yields a nonzero function $\Sigma f_j(x)g_j(y)$ on $X \times Y$. We may assume that the f_j are linearly independent. Then, if $y \in Y$ is any point where some $g_j(y) \neq 0$, the function $\Sigma f_j g_j(y)$ on X is different from zero.

For instance, $\mathbb{A}^n \times \mathbb{A}^m$ is isomorphic to $\mathbb{A}^{n+m}$.

The product $X \times Y$ has two canonical morphisms, namely, the projections onto X and Y, which make it into the direct product in the category of affine varieties. The *graph of a morphism* $f: X \to Y$ is the subvariety Γ_f of $X \times Y$ defined by the equations $1 \otimes g = f^*(g) \otimes 1$, with $g \in K[Y]$. As a set, Γ_f consists of those pairs $(x, y) \in X \times Y$ for which $f(x) = y$. As a special case, the *diagonal* $\Delta_X \subset X \times X$ (the graph of the identity morphism $X \to X$) is a subvariety of $X \times X$.

2.5. Intersection of Subvarieties. Let Y and Z be subvarieties of an affine variety X. Then the intersection $Y \cap Z$ is also a subvariety of X. However, it is more natural and more correct to think of this intersection as the *subscheme* of X defined by the ideal $I(Y) + I(Z)$. Then the occurrence of nilpotent elements in the scheme $Y \cap Z$ bears witness to the nontransversality of Y and Z.

Example 1. Let $X = \mathbb{C}^2$ with coordinates T and S. Let Y be the 'parabola' $[S = T^2]$, and Z the 'horizontal' line $[S = 0]$. The set-theoretic intersection of Y and Z consists of one point. As for the scheme-theoretic intersection, which is defined by the ideal $(S, S - T^2) = (S, T^2)$, it has a more interesting structure: in particular, it is isomorphic to the scheme of Example 3 above. This relates to the fact that the line Z is tangent to the curve Y (cf. Fig. 2).

If, instead of Z, we consider another 'horizontal' line $Z_y = [S = y]$, with $y \in \mathbb{C} - \{0\}$, then the intersection $Y \cap Z_y$ consists of two points $(\pm\sqrt{y}, y)$, even in the scheme-theoretic sense.

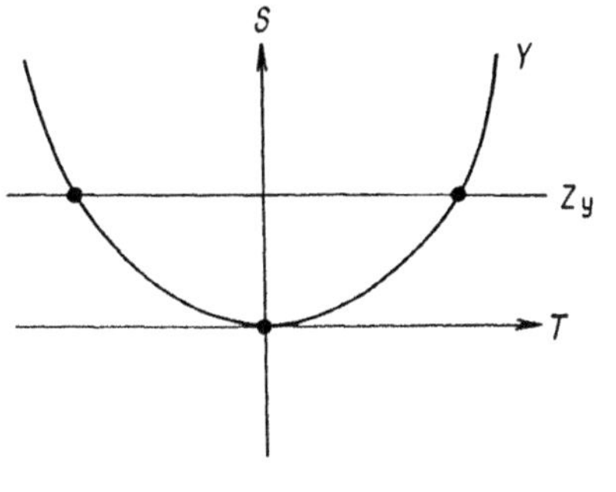

Fig. 2

Example 2. Let Y be a subvariety of X, and Y' a subvariety of X'. The intersection of $Y \times X'$ and $X \times Y'$ in $X \times X'$ is the variety $Y \times Y'$. This agrees with our intuitive feeling that the varieties $Y \times X'$ and $X \times Y'$ are situated transversally in $X \times X'$.

2.6. Fibres of a Morphism. Let $f\colon X \to Y$ be a morphism of affine varieties. In general the image $f(X)$ is not a subvariety of Y. Consider, for example, the 'hyperbola' $X = [T_1T_2 = 1]$ in K^2, and let f be projection onto the T_1-axis. Then $f(X) = K - \{0\}$ is not an algebraic subset of K.

With inverse images the situation looks better. For a point $y \in Y$ the subset $f^{-1}(y) = \{x \in X,\ f(x) = y\}$ is an algebraic subvariety of X. However, it is again better to view it as the *subscheme* of X defined by the ideal $f^*(\mathfrak{m}_y)K[X]$, where $\mathfrak{m}_y$ is the maximal ideal of the point $y \in Y$. The scheme $f^{-1}(y)$ is called the *fibre* of f over the point y. It is isomorphic to the intersection of the graph Γ_f with the 'horizontal' $X \times \{y\}$ in $X \times Y$. The terminology evokes the fact that the variety X is, so to speak, fibred into the varieties (or the schemes) $f^{-1}(y)$, where y runs through the points of Y.

Once again, the presence of nilpotent elements in the ring of the scheme $f^{-1}(y)$ testifies to some special property of the morphism, like ramification, multiple fibres, etc. Thus, Fig. 2 represents the graph of the map $f\colon \mathbb{C} \to \mathbb{C}$, where $f(x) = x^2$. For $y \neq 0$ the fibre $f^{-1}(y)$ is reduced and consists of two distinct points $\pm\sqrt{y}$; for $y = 0$ these two points merge into one 'double' point 0, which is also the reason why nilpotent elements occur. Let us examine two more examples.

Example 1. Let $f\colon K \to K^2$ be the map taking t to the point (t^2, t^3). Its graph Γ_f is the curve in K^3 which is given parametrically as (t, t^2, t^3), $t \in K$, or by the equations $T_1 = T^2$, $T_2 = T^3$. The image of f is the curve $C \subset K^2$ with equation $T_1^3 = T_2^2$. Although f does yield a bijection between K and C, it is not an isomorphism of affine varieties. Indeed the fibre $f^{-1}(P)$ contains some nilpotent elements (the graph Γ_f is tangent to the T-axis). This is revealed geometrically by the fact that the point P is singular on C (cf. Fig. 3).

Example 2. An even more striking situation occurs when the field K has positive characteristic $p > 0$. Let $F\colon K \to K$ be the map given by the formula $F(x) = x^p$ (or $S = T^p$); this is called the *Frobenius morphism*. Set-theoretically it is one-to-one (if $x^p = x'^{\,p}$ then $(x - x')^p = x^p - x'^{\,p} = 0$ and $x = x'$), but again it is not an isomorphism. Moreover, for every point $y \in K$, the fibre $F^{-1}(y)$ is given by the ideal $(T^p - y) = (T - \sqrt[p]{y}\,)^p$, and therefore it is non-reduced. The map F is critical at all points (this can be seen also by computing the derivative: $dT^p/dT = pT^{p-1} \equiv 0$). Its graph is tangent to the horizontal line at every point; nevertheless, F is nonconstant!

More generally, let $f\colon X \to Y$ be a morphism of affine varieties, and $Z \subset Y$ a subscheme, defined by an ideal $J \subset K[Y]$. The inverse scheme of Z under

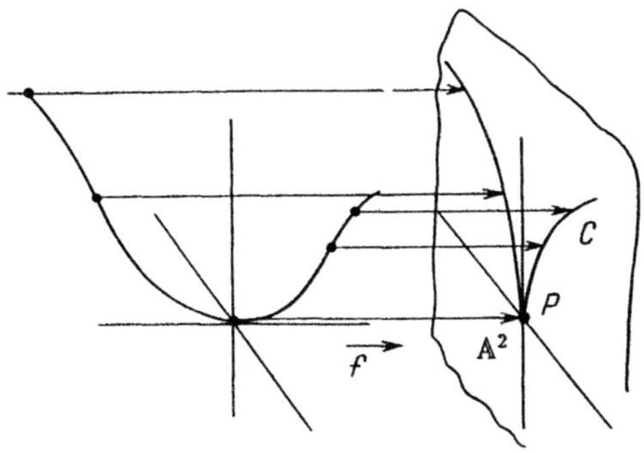

Fig. 3

f is the subscheme $f^{-1}(Z)$ in X given by the ideal $f^*(J)K[X]$. For instance, the intersection of two subvarieties, Y and Z, in X can be looked upon as the inverse image of the diagonal $\Delta_X \subset X \times X$ under the embedding $Y \times Z \subset X \times X$, or as the intersection of $Y \times Z$ with Δ_X in $X \times X$. This procedure of *reduction to the diagonal* is used quite often. The graph Γ_f is the inverse image of the diagonal under the morphism $id \times f \colon X \times Y \to X \times X$. It is not a casual fact that the operations of product, intersection, and inverse image are interrelated. Indeed all three of them are special cases of the more general fibre product operation; see Sect. 4.2.

2.7. The Zariski Topology. As in Sect. 1.3, we call algebraic subset (or subvariety) of an affine variety X the set $V(I)$ of common zeros of the functions in some ideal $I \subset K[X]$. As before, the algebraic subsets of X are closed with respect to intersections and finite unions. Hence we may declare that they are the closed sets of some topology on X, which is called the *Zariski topology*.

Let $f \colon X \to Y$ be a morphism of affine varieties. As we have seen, the inverse image $f^{-1}(V)$ of any algebraic subset $V \subset Y$ is algebraic in X. This means that f is a continuous map in the Zariski topology. In particular, every regular function is continuous. Conversely, the Zariski topology is the weakest topology in which the points are closed, and the regular functions continuous. If Y is a subvariety of X then the topology on Y coincides with that induced from X.

It is worth mentioning that the image of an affine variety is not necessarily open or closed. Let us consider for instance the morphism $f \colon \mathbb{A}^2 \to \mathbb{A}^2$ defined by $f(x,y) = (x, xy)$ (cf. Fig. 4). Its image consists of the point $(0,0)$ and the open set $U = \{(x,y),\ x \neq 0\}$. This set $U \cup \{(0,0)\}$ is not closed, even locally. However, it is made up of (two) locally closed pieces: the point $(0,0)$ and U. Later on we shall see that this reflects a general fact: the image of an algebraic variety is always the union of finitely many locally closed pieces.

The Zariski topology is a very natural one; in the abstract case it is hard to think of anything better. Still, in many respects it looks rather unusual,

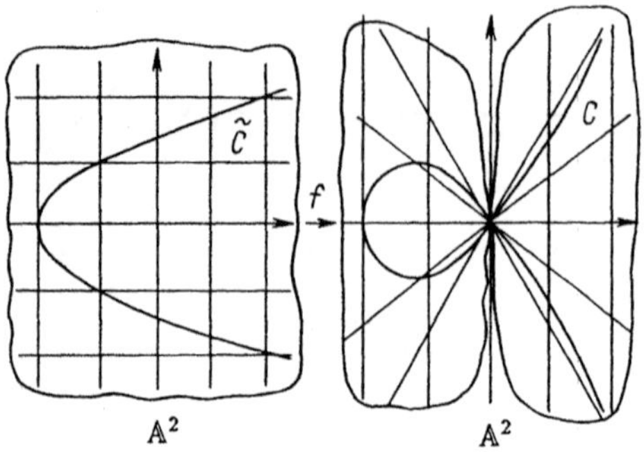

Fig. 4

compared with the ordinary metric topology on $\mathbb{C}^n$. Of course the polynomial functions on $\mathbb{C}^n$ are continuous in the topology given by the Euclidean metric. It follows that the classical topology on $\mathbb{C}^n$ is stronger than the Zariski topology. In other words, a Zariski open (respectively, closed) subset is also open (respectively, closed) in the classical topology. The converse is not true; for instance, in $\mathbb{C}$ the algebraic sets are exactly $\mathbb{C}$ and its finite subsets. Thus the Zariski open subsets are 'very big'; in particular, the Zariski topology is highly non-Hausdorff.

A further difference with the classical topology is that the Zariski topology on the product $X \times Y$ of two affine varieties is stronger than the product of the Zariski topologies on X and Y. So, in $\mathbb{A}^2 = \mathbb{A}^1 \times \mathbb{A}^1$ there are many infinite algebraic subsets which are not made up of vertical and horizontal lines (for example, the diagonal).

Although there is quite a distance between the Zariski topology and the classical one, they are not divided by an impassable chasm. Here is the easiest footbridge joining them: if an open subset $U \subset X$ is Zariski dense, it is dense also in the classical topology. More subtle is the connectedness theorem: a set that is connected in the Zariski topology is also connected in the classical one. Results of this kind are explained in more detail in the article on the cohomology of algebraic varieties. They enable us to apply to complex algebraic varieties the methods of algebraic topology (homotopy, cohomology, etc.) and analysis (periods of integrals, Hodge theory); these methods are presented in Griffiths-Harris [1978]. Transcendental methods act as a powerful incentive to search for algebraic analogues and thus contribute to the subsequent development of abstract algebraic geometry.

2.8. Localization. The Zariski topology makes it possible to define regular functions in a more local fashion. Let $U \subset X$ be an open subset of an affine variety X, and $f \in K[X]$ a function that does not vanish at any point of U. Then the function $1/f$ is defined at every point of U and can be considered a 'regular' function on U in view of its algebraic origin (cf. Sect. 1.2). We must then also regard as regular the functions of the form g/f, where $g \in K[X]$.

More generally, we say that a function $h\colon U \to K$ is *regular at a point* $x \in U$ if there exist two functions $f, g \in K[X]$ such that $f(x) \neq 0$ and $h = g/f$ in some neighbourhood of x. More precisely we can say that h coincides with g/f on the set $U \cap \mathcal{D}(f)$, where $\mathcal{D}(f) = X - V(f) = \{x' \in X,\ f(x') \neq 0\}$. The sets of the form $\mathcal{D}(f)$ are called the *basic open subsets* of X. Clearly, they form a basis for the Zariski topology on X.

The functions on U that are regular at every point of U form a ring, which is denoted by $\mathcal{O}_X(U)$. If $U' \subset U$ then the restriction of functions from U to U' yields a homomorphism $\mathcal{O}_X(U) \to \mathcal{O}_X(U')$ of rings (or of K-algebras). This object $\mathcal{O}_X$ – which will play an important role later on – is called the structure sheaf of rings on X. Clearly, $K[X] \subset \mathcal{O}_X(X)$; as a matter of fact, equality holds.

Proposition. *If X is an affine variety then $K[X] = \mathcal{O}_X(X)$.*

Indeed, suppose the function $h\colon X \to K$ is regular at every point $x \in X$. Then $h = g_x/f_x$ in $\mathcal{D}(f_x)$ and $f_x(x) \neq 0$. By Hilbert's Nullstellensatz, the functions f_x, $x \in X$, generate the unit ideal in $K[X]$. Hence there exists a decomposition $1 = \Sigma\, a_x f_x$, with $a_x \in K[X]$. It follows that $h = h \cdot 1 = \Sigma\, a_x h f_x = \Sigma\, a_x g_x \in K[X]$.

This proposition allows us to talk about regular functions with no risk of ambiguity. The decomposition $1 = \Sigma\, a_x f_x$ plays a role similar to that of a partition of unity in the theory of differentiable manifolds.

2.9. Quasi-affine Varieties. Let again U be an open subset of an affine variety X. In general the pair $(U, \mathcal{O}_X(U))$ is not an affine variety. First of all, the K-algebra $\mathcal{O}_X(U)$ may not be finitely generated. Secondly, there may be 'few' points in U, that is, the mapping $U \to \operatorname{Specm} \mathcal{O}_X(U)$ (see Sect. 2.2) may not be surjective.

Example. We want to show that $U = \mathbb{A}^n - \{0\}$ is not affine for $n \geq 2$. To this effect we shall verify that $\mathcal{O}_{\mathbb{A}^n}(U)$ coincides with $K[\mathbb{A}^n]$. In other words, every regular function on $\mathbb{A}^n - \{0\}$ extends to a regular function on $\mathbb{A}^n$. This property is reminiscent of Hartogs's theorem in the theory of analytic functions, and departs sharply from the situation that prevails in the differentiable case.

Indeed, let f be a function regular on U. We cover U by the sets $\mathcal{D}(T_i)$, the T_i being coordinates on $\mathbb{A}^n$. Then the restriction of f to $\mathcal{D}(T_i)$ is of the form $g_i/T_i^{r_i}$, with $g_i \in K[T_1, \ldots, T_n]$ and $r_i \geq 0$; we may further assume that g_i is not divisible by T_i. Since the restrictions coincide on $\mathcal{D}(T_1) \cap \mathcal{D}(T_2)$, we see that $T_1^{r_1} g_2 = T_2^{r_2} g_1$. Now, from the uniqueness of the decomposition into prime factors in the polynomial ring $K[T_1, \ldots, T_n]$, we conclude that $r_1 = r_2 = 0$ and $g_1 = g_2 = f$.

On the other hand, the basic open sets $\mathcal{D}(f) \subset X$ are affine varieties. Related to this we have the following two facts: The ring $\mathcal{O}_X(\mathcal{D}(f))$ of regular functions on $\mathcal{D}(f)$ coincides with the ring $K[X][f^{-1}]$ of fractions of the form g/f^r, with $g \in K[X]$ and $r \geq 0$. Further, the Zariski topology on $\mathcal{D}(f)$ is induced by the Zariski topology on X.

In any case, the open subsets of affine varieties look locally like affine varieties. They are called *quasi-affine algebraic varieties.*

2.10. Affine Algebraic Geometry. Though algebraic geometry deals chiefly with projective varieties, it is worth mentioning that affine algebraic geometry also has its own, often unexpectedly hard, problems. Difficulties arise already for the simplest affine varieties, namely, affine space $\mathbb{A}^n$. Serre's problem on vector bundles over $\mathbb{A}^n$ was solved only comparatively recently (Suslin [1976], Quillen [1976]). Here is another famous question: suppose the variety $X \times \mathbb{A}^m$ is isomorphic to $\mathbb{A}^{n+m}$; is it true that X is isomorphic to $\mathbb{A}^n$? An affirmative answer (which is obvious for $n = 1$) was obtained only recently for $n = 2$ (Miyanishi [1981]); for $n > 2$ the question is open.

Perhaps the reason for the difficulties lies in the fact that the space $\mathbb{A}^n$ (at least for $n > 1$) is very 'flexible'. The automorphisms of $\mathbb{A}^1$ are easily seen to be of the form $T' = aT + b$, with $a, b \in K$ and $a \neq 0$. That $\mathbb{A}^n$ (for $n > 1$) has quite a few more automorphisms is made clear by the example of the triangular transformation:

$$\begin{aligned} T_1' &= T_1 + f_0, \\ T_2' &= T_2 + f_1(T_1), \\ &\dots\dots\dots\dots\dots \\ T_n' &= T_n + f_{n-1}(T_1, \dots, T_{n-1}), \end{aligned}$$

where $f_i \in K[T_1, \dots, T_i]$. In particular, every finite subset of $\mathbb{A}^n$, where $n > 1$, can be carried by an automorphism into any other finite subset with the same cardinality. For $n = 2$ every automorphism of $\mathbb{A}^n$ is generated by triangular and linear automorphisms. This is not known, and almost certainly false, for $n > 2$. These questions are closely related to the problem of linearizing the action of algebraic groups on $\mathbb{A}^n$.

Finally, one should mention the so-called Jacobian problem. Consider a map $f\colon \mathbb{C}^n \to \mathbb{C}^n$ defined by polynomials $f_1, \dots, f_n$ in $\mathbb{C}[T_1, \dots, T_n]$, and suppose the Jacobian $\det(\partial f_j/\partial T_i)$ does not vanish anywhere on $\mathbb{C}^n$. (We may assume it is identically equal to 1.) The Jacobian conjecture says that f must then be an isomorphism. For a discussion of this problem see Bass, Connell & Wright [1982].

§ 3. Algebraic Varieties

It was already perceived quite a long time ago that, by considering only affine varieties, one gets an incomplete picture of what goes on, as if one could see only part of the actual variety. This is connected with the fact that affine space is non-compact: we do not control the behaviour 'at infinity'. For instance, any two lines in the affine plane meet, unless they are parallel. It is convenient to postulate that even parallel lines meet, albeit in an 'infinitely distant point'. Adjoining these points to affine space $\mathbb{A}^n$ makes it into projective space $\mathbb{P}^n$. Another nice feature of the projective viewpoint is that affinely different curves, such as the ellipse, the parabola, and the hyperbola, turn out to be simply different affine parts of the projective conic. That is why algebraic geometry has always been preeminently a projective geometry. So we have to proceed now from affine varieties to the more general algebraic varieties.

3.1. Projective Space. The easiest way to define n-dimensional *projective space* $\mathbb{P}^n$ is to say it is the set of lines in the vector space K^{n+1}. Every line, that is, every one-dimensional vector subspace $L \subset K^{n+1}$, is given by a nonzero vector $(x_0, \ldots, x_n) \in K^{n+1}$, which is determined up to multiplication by a nonzero constant $\lambda \in K^* = K - \{0\}$. Therefore we may regard $\mathbb{P}^n$ as the quotient space $K^{n+1} - \{0\} / K^*$.

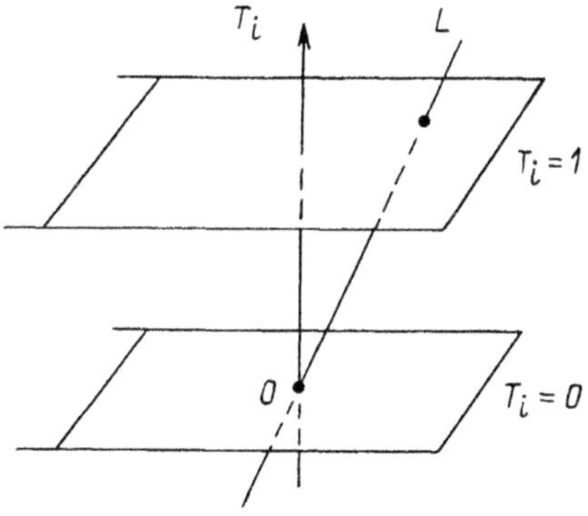

Fig. 5

The coordinate functions $T_0, \ldots, T_n$ on K^{n+1} are called the *homogeneous coordinates* on $\mathbb{P}^n$. However, one must be careful that the T_i, like any nonconstant polynomial in the T_i, are not functions on $\mathbb{P}^n$. Such expressions as T_j/T_i can be viewed as functions, but not on the whole of $\mathbb{P}^n$: only on the subset $U_i = \mathbb{P}^n - H_i$, where H_i consists of the points $(x_0, \ldots, x_n)$ with $x_i = 0$. In other words, U_i consists of those lines $L \subset K^{n+1}$ which project isomorphically onto the i-th coordinate axis. For fixed i, the functions $\xi_j^{(i)} = T_j T_i^{-1}$,

$j = 0, 1, \ldots, n$, define a one-to-one correspondence between U_i and the affine subspace $T_i = 1$ in K^{n+1}. Under this correspondence, H_i consists of the lines L lying in the hyperplane $T_i = 0$ and can be identified with $\mathbb{P}^{n-1}$. In this sense, $\mathbb{P}^n$ is obtained from the affine space $U_i \simeq K^n$ by adjunction of the hyperplane at infinity $H_i \simeq \mathbb{P}^{n-1}$.

The sets U_i form a covering of $\mathbb{P}^n$, and each of them has a natural structure of affine variety $\mathbb{A}^n$. Moreover, these structures agree on the intersections $U_i \cap U_j$. Indeed, we can regard $U_i \cap U_j$ as being the basic open set $\mathcal{D}(\xi_j^{(i)})$ in U_i, and also as the basic open subset $\mathcal{D}(\xi_i^{(j)})$ of U_j. In the former case, the ring of regular functions is generated by $\xi_0^{(i)}, \ldots, \xi_n^{(i)}, {\xi_j^{(i)}}^{-1}$; in the latter, by $\xi_0^{(j)}, \ldots, \xi_n^{(j)}, {\xi_i^{(j)}}^{-1}$. Now, these rings coincide. For example,

$$\xi_k^{(i)} = T_k/T_i = (T_k/T_j)(T_i/T_j)^{-1} = \xi_k^{(j)} \cdot {\xi_i^{(j)}}^{-1}$$

and

$${\xi_j^{(i)}}^{-1} = (T_j/T_i)^{-1} = T_i/T_j = \xi_i^{(j)}.$$

Conversely, the $\xi^{(j)}$ can be expressed by means of the $\xi^{(i)}$.

Thus we see that $\mathbb{P}^n$ looks locally like an affine variety. We may therefore talk about regular functions on $\mathbb{P}^n$ (admittedly, they are quite scarce: constants only), or about the algebraic subvarieties of $\mathbb{P}^n$ (which are quite numerous), its Zariski topology, etc. Some comparable ideas can be used not only for $\mathbb{P}^n$, but also for any geometric object that looks locally like an affine variety. The resulting theory of algebraic varieties has much resemblance to that of differentiable or analytic manifolds.

3.2. Atlases and Varieties. Let X be a topological space. An *affine chart* (or *coordinate neighbourhood*) in X is an open subset $U \subset X$ equipped with a structure of affine variety, with the requirement that the induced topology on U should coincide with the Zariski topology. We say that two charts, U and U', are *compatible* if, for every open subset $V \subset U \cap U'$, one has $\mathcal{O}_U(V) = \mathcal{O}_{U'}(V)$.

An *atlas* on X is a collection $\mathcal{A} = (U_i)_{i \in I}$ of mutually compatible affine charts covering X. Two atlases, $\mathcal{A}$ and $\mathcal{A}'$, are *equivalent* if their union is also an atlas, that is, if the charts of $\mathcal{A}$ are compatible with those of $\mathcal{A}'$.

By a structure of *algebraic variety* on X we mean an equivalence class of atlases. In what follows we shall restrict attention to the algebraic varieties that have a *finite* atlas. By a chart on X we mean an affine chart that belongs to some atlas defining the variety structure of X. Every point lies in an arbitrarily small chart.

Every affine variety is an algebraic variety. Every closed subset $Y \subset X$ of an algebraic variety comes equipped with a canonical structure of algebraic variety; Y is also called a *subvariety* (or a closed subvariety) of X. An open subset $U \subset X$ also has an obvious structure of algebraic variety.

The covering of projective space $\mathbb{P}^n$ by the U_i is an atlas and converts $\mathbb{P}^n$ into an algebraic variety. More generally, if V is any finite-dimensional vector space over K, we denote by $\mathbb{P}(V)$ the set of lines of V through the origin. If $l: V \to K$ is a nonzero linear map, we define $H_l \subset \mathbb{P}(V)$ to be the set of lines $L \subset \ker l$. Then $U_l = \mathbb{P}(V) - H_l$ consists of those lines L for which $l(L) = K$ and can be identified with the affine subspace $l^{-1}(1) \subset V$. The structures on the various U_l are compatible and make $\mathbb{P}(V)$ into an algebraic variety. Of course $\mathbb{P}^n = \mathbb{P}(K^{n+1})$.

3.3. Gluing. This operation yields some new varieties out of old ones. Let (X_i) be a finite covering of some set X, where each X_i has a structure of algebraic variety. We make two assumptions:

a) for every pair i, j the set $X_i \cap X_j$ is open in X_i and in X_j;
b) the algebraic variety structures induced on $X_i \cap X_j$ from X_i and from X_j coincide.

Then there exists on X a unique structure of algebraic variety such that the X_i are open subvarieties. We say that X is obtained by *gluing* (or pasting together) the varieties X_i.

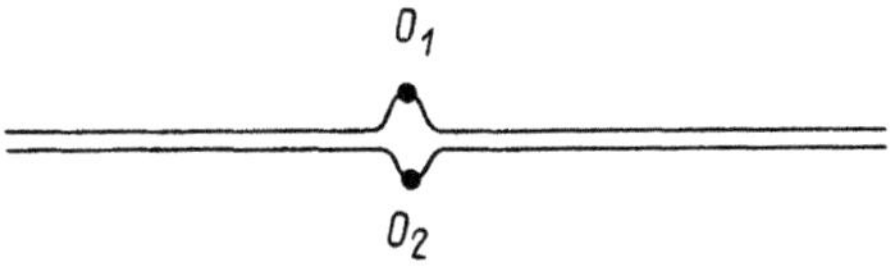

Fig. 6

One may, for instance, think of projective space $\mathbb{P}^n$ as the result of gluing the affine spaces U_i, $i = 0, 1, \ldots, n$. Here is another example. Suppose X_1 and X_2 are isomorphic to the affine line $\mathbb{A}^1$, and let T_1 and T_2 be coordinates on X_1, respectively, X_2. Let us identify $X_1 - \{0\}$ and $X_2 - \{0\}$ by setting $T_1 = T_2$. What we get is an *affine line with the point* 0 *doubled* (see Fig. 6). Such a variety occurs naturally as the set of orbits for the action $\lambda(x, y) = (\lambda x, \lambda^{-1} y)$ of the group K^* on the plane K^2.

Example. A good exercise on the theme of gluing is the construction of *torus embeddings*. We fix a lattice M, that is, a free abelian group of finite type (which is therefore isomorphic to $\mathbb{Z}^n$, but the basis is irrelevant). Let $S \subset M$ be a submonoid, that is, S contains 0 and is closed under addition. Then we can form the semigroup K-algebra $K[S]$. It is generated additively by all elements of the form x^m, with $m \in S$, multiplication being defined by the rule $x^m \cdot x^{m'} = x^{m+m'}$. If S is finitely generated as a monoid then the K-algebra $K[S]$ is of finite type and defines an affine variety, namely, $\operatorname{Specm} K[S]$. For instance, if $S = M$ we get the n-dimensional torus $\mathbb{T} = \operatorname{Specm} K[M] = \operatorname{Specm}[T_1, \ldots, T_n, T_1^{-1}, \ldots, T_n^{-1}]$.

We consider now in the dual lattice $M^* = \mathrm{Hom}(M, \mathbb{Z})$ a subset B which is contained in a $\mathbb{Z}$-basis of the group M^*. We can attach to it the following monoid in M:

$$B^\perp = \{m \in M, \quad b(m) \geq 0 \quad \forall b \in B\}.$$

The corresponding affine variety $\mathrm{Specm}\, K[B^\perp]$ will be denoted by X_B. (This variety is called a torus embedding because the torus $\mathbb{T}$ acts on it in a natural way.) If $B' \subset B$ then $B'^\perp \supset B^\perp$, which gives rise to a natural homomorphism of K-algebras $K[B^\perp] \to K[B'^\perp]$ and to the opposite morphism of varieties $X_{B'} \to X_B$. It is not difficult to check that the latter is an open immersion.

Now, given a collection Σ of such subbases B of M^*, it is possible to glue together the varieties X_B and $X_{B'}$ $(B, B' \in \Sigma)$ along the open pieces $X_{B \cap B'}$, so as to obtain a torus embedding X_Σ. For instance, $\mathbb{P}^n$ is obtained from $\Sigma = \{B_0, \ldots, B_n\}$, where $B_0 = \{e_1, \ldots, e_n\}$, and

$$B_i = \{e_1, \ldots, \widehat{e_i}, \ldots, e_n, -e_1 - \ldots - e_n\} \quad \text{for } i = 1, \ldots, n.$$

What makes the interest of torus embeddings, is that various objects on X_Σ (like invertible sheaves and their cohomology, differential forms, etc.) can be described in combinatorial terms depending on Σ. For instance, invertible sheaves are represented by polyhedra in $M \otimes \mathbb{R}$, and their sections by the integer points on these polyhedra. For further details, see Danilov [1978].

3.4. The Grassmann Variety. Let again V be a vector space over K. We denote by $G(k, V)$ (or $G(k, n)$ if $n = \dim V$) the set of k-dimensional subspaces $W \subset V$; for $k = 1$ we get $\mathbb{P}(V)$. Generalizing the construction of projective space, we shall give $G(k, V)$ the structure of an algebraic variety, called the *Grassmann variety.*

Let $V = V' \oplus V''$ be a direct decomposition, with $\dim V' = k$. To each such decomposition we shall attach the set $U(V', V'')$, consisting of the subspaces $W \subset V$ which project isomorphically onto V'. These subspaces can be identified with the graphs of linear maps from V' to V''. Hence $U(V', V'') \simeq \mathrm{Hom}_k(V', V'') \simeq V'' \otimes V'^*$ is naturally identified with a vector space of dimension $k(n-k)$ and is endowed with the structure of an affine variety. It is an immediate verification that all these charts $U(V', V'')$ are compatible and give $G(k, V)$ an algebraic variety structure. For further details on the Grassmannian, see Griffiths-Harris [1978] and Grothendieck-Dieudonné [1971].

3.5. Projective Varieties. A closed subset of projective space is said to be a *projective variety.* We exhibit a general method for producing such varieties.

Let V be a vector space over K. We define a *cone* in V to be an algebraic subvariety $C \subset V$ which is invariant under scalar multiplication, that is, multiplication by a constant. To every cone C we associate the subset $\mathbb{P}(C) \subset \mathbb{P}(V)$ consisting of the lines $L \subset C$. The set $\mathbb{P}(C)$ is closed in $\mathbb{P}(V)$. Indeed, provided we identify a chart U_l (where $l \colon V \to K$ is a linear map) with the affine subspace $l^{-1}(1) \subset V$, the set $\mathbb{P}(C) \subset U_l$ is seen to be identical with the intersection $C \cap l^{-1}(1)$, which is obviously closed in $l^{-1}(1)$.

In the coordinates $T_0, \dots, T_n$ on V, the cone C is given by homogeneous equations $f_j(T_0, \dots, T_n) = 0, \quad j \in J$. Then $\mathbb{P}(C) \cap U_i$ is given by the equations $f_j(T_0/T_i, \dots, T_n/T_i) = 0$. The equations $f_j = 0$ are called the *homogeneous equations* of $\mathbb{P}(C)$.

Conversely, every projective variety $X \subset \mathbb{P}(V)$ is of the form $\mathbb{P}(C)$ for some cone $C \subset V$. Indeed, let (U_i) be the standard atlas of $\mathbb{P}^n$, and suppose $X \cap U_i$ is given by equations $f_j^{(i)}(T_0/T_i, \dots, T_n/T_i) = 0, \quad j \in J_i$. Then, for large m, $T_i^m f_j^{(i)}(T_0/T_i, \dots, T_n/T_i) = g_j^{(i)}(T_0, \dots, T_n)$ is a homogeneous form in $T_0, \dots, T_n$, and the equations $g_j^{(i)} = 0, \quad j \in J_i, \quad i = 0, 1, \dots, n$, define X in $\mathbb{P}^n$.

The simplest projective varieties are the linear ones. If $W \subset V$ is a vector subspace, the subvariety $\mathbb{P}(W) \subset \mathbb{P}(V)$ is said to be *linear*. If W is a hyperplane in V then $\mathbb{P}(W)$ is called a *hyperplane* in $\mathbb{P}(V)$. We define the *linear hull* of a set to be the intersection of all the linear varieties that contain it. For two distinct points, x and y, it is nothing but the projective line $\overline{xy}$, and so forth. To give a hyperplane $W \subset V$ is the same as giving a line $W^\perp$ in the dual space V^*, and conversely. Hence the set of all hyperplanes in $\mathbb{P}(V)$ is also a projective space, namely, $\mathbb{P}(V^*)$.

Every vector space V can be regarded as an affine part of projective space $\mathbb{P}(V \oplus K)$, more precisely as the complementary set to the hyperplane $\mathbb{P}(V) \subset \mathbb{P}(V \oplus K)$. If $X \subset V$ is an algebraic variety then the closure of X in $\mathbb{P}(V \oplus K)$ is a projective variety. This is a standard way to proceed from affine to projective varieties (which, by the way, depends on the embedding $X \subset V$). If $\xi_1, \dots, \xi_n$ are coordinates on V, projectivization looks as follows. Let $f(\xi_1, \dots, \xi_n)$ be a polynomial of degree d; its homogenization is the homogeneous, degree d polynomial $\tilde{f}(T_0, \dots, T_n) = T_0^d f(T_1/T_0, \dots, T_n/T_0)$. Now if X is given by equations $f_j = 0$ then its projectivization $\overline{X}$ is defined by the equations $\tilde{f}_j = 0$.

§ 4. Morphisms of Algebraic Varieties

4.1. Definitions. Let X be an algebraic variety described by an atlas (X_i), and Y an affine variety. We say that a map $f\colon X \to Y$ is *regular* if the restriction of f to every chart X_i has this property. In particular, we have the notion of a regular function. For any open set $U \subset X$, we denote by $\mathcal{O}_X(U)$ the K-algebra of functions regular on U. If $U' \subset U$, there is a restriction homomorphism $\mathcal{O}_X(U) \to \mathcal{O}_X(U')$.

Suppose now Y is an arbitrary algebraic variety. A continuous mapping $f\colon X \to Y$ is called a *morphism* (or a regular map) *of algebraic varieties* if, for every chart $V \subset Y$, the induced mapping $f^{-1}(V) \to V$ is regular. In other words, for every regular function g on the open subset $V \subset Y$, the function $f^*(g) = g \circ f$ must be regular on $f^{-1}(V)$. This means that f^* yields an algebra homomorphism $\mathcal{O}_Y(V) \to \mathcal{O}_X(f^{-1}(V))$.

The composite of two morphisms is again a morphism, so that algebraic varieties form a category. The canonical injection of a closed subvariety is a morphism, and we say that a morphism $Y \to X$ is a *closed immersion* if it yields an isomorphism of Y onto a closed subvariety of X. If $f\colon X \to Y$ is a morphism, and $Y' \subset Y$ a closed subvariety, then $f^{-1}(Y')$ is a closed subvariety of X (cf. Sect. 2.4). In particular, for a point $y \in Y$ the variety $f^{-1}(y) \subset X$ is called the *fibre of the morphism* f over y.

A variety X provided with a morphism $f\colon X \to Y$ is sometimes called a variety over Y, or a Y-variety. X is thereby viewed as the family of algebraic varieties $X_y = f^{-1}(y)$, parametrized by the points $y \in Y$. Given two Y-varieties, say, $f\colon X \to Y$ and $f'\colon X' \to Y$, a morphism from f to f' is a morphism $\varphi\colon X \to X'$ such that $f = f' \circ \varphi$. Each fibre $f^{-1}(y)$ is mapped into the corresponding fibre $f'^{-1}(y)$, so we get a family of morphisms $\varphi_y\colon X_y \to X'_y$.

4.2. Products of Varieties. Let X and Y be two algebraic varieties, with defining atlases (X_i) and (Y_j). Then $(X_i \times Y_j)$ is an atlas for the product $X \times Y$, so $X \times Y$ is also an algebraic variety. An easy verification shows that $X \times Y$ is the direct product of X and Y in the category of varieties.

In particular, for any variety X, the diagonal mapping $\Delta\colon X \to X \times X$ $(\Delta(x) = (x, x))$ is a morphism, though in general it is not a closed immersion. In other words, the diagonal in $X \times X$ may fail to be closed. An example is furnished by the 'affine line with a point doubled' from Sect. 3.3. If, in spite of that, the diagonal in $X \times X$ is closed then we say that the variety X is *separated.* (One should not confuse this notion with the question whether X is Hausdorff as a topological space!) Any affine variety, for instance, is separated (cf. Sect. 2.4). The class of separated varieties is closed under taking direct products or going over to subvarieties. We will check below that projective space – and hence any projective variety – is separated. In what follows we shall therefore deal exclusively with separated varieties.

That a variety is non-separated has to do with the fact that, when we obtain it by gluing some of its affine pieces, these are glued imperfectly. To be precise, one has the following *separatedness criterion:* a variety X, described by an atlas (X_i), is separated if and only if the image of $X_i \cap X_j$ under the canonical injection into $X_i \times X_j$ is closed. In fact, the image of $X_i \cap X_j$ in $X_i \times X_j$ is just the intersection of $X_i \times X_j$ with the diagonal in $X \times X$.

Let us apply this criterion to the standard atlas (U_i), $i = 0, 1, \ldots, n$, of projective space $\mathbb{P}^n$ (cf. Sect 3.1). It is easy to check that the image of $U_i \cap U_j$ in $U_i \times U_j$ is given by the equations

$$\xi_k^{(i)} = \xi_j^{(i)} \xi_k^{(j)}, \quad \xi_k^{(j)} = \xi_k^{(i)} \xi_i^{(j)}, \quad k = 0, \ldots, n,$$

whence we see that $\mathbb{P}^n$ is separated.

The category of algebraic varieties possesses not only direct products, but also fibre products. This gives it a distinct advantage over the category of differentiable manifolds. We demonstrate the existence, limiting ourselves to separated varieties. Let $f\colon X \to Z$ and $g\colon Y \to Z$ be two varieties over Z. We define the *fibre product* of X and Y over Z to be the following subvariety of $X \times Y$:

$$X \times_Z Y = \{(x, y) \in X \times Y, \quad f(x) = g(y)\}.$$

More correctly, it is the inverse image of the diagonal Δ_Z under the morphism $f \times g\colon X \times Y \to Z \times Z$; this immediately makes $X \times_Z Y$ into a scheme. The fibre product contains as special cases the direct product (where Z is a point), a fibre (where $Y \to Z$ is the inclusion of a point), and the intersection of subvarieties.

A commutative diagram

$$\begin{array}{ccc} X \times_Z Y & \xrightarrow{g'} & X \\ {\scriptstyle f'}\downarrow & & \downarrow{\scriptstyle f} \\ Y & \xrightarrow{g} & Z \end{array}$$

is called a *Cartesian square.* Looking at it in a slightly nonsymmetric way, we may say that the fibre product is an operation which turns a Z-variety X into a Y-variety $X \times_Z Y$. This operation is called a *base extension.* The fibre of f' above a point $y \in Y$ is isomorphic to the fibre of f over the point $g(y)$. Base extension is the direct analogue of the notion of induced fibre bundle in topology.

When, in particular, $g\colon Y \to Z = Y$ is the identity morphism, the fibre product consists of the pairs $(x, y) \in X \times Y$ such that $f(x) = y$. For understandable reasons it is called the *graph* Γ_f of the morphism $f\colon X \to Y$. If Y is separated then Γ_f is a closed subset of $X \times Y$. The projection $\Gamma_f \to X$ is an isomorphism, and every morphism $f\colon X \to Y$ factorizes into the closed immersion $X \xrightarrow{\sim} \Gamma_f \subset X \times Y$ and the projection map $X \times Y \to Y$.

4.3. Equivalence Relations. The dual notion to that of fibre product is the amalgamated sum, which is the universal completion of the diagram

$$\begin{array}{ccc} R & \longrightarrow & Y \\ \downarrow & & \vdots \\ X & \dashrightarrow & X \amalg_R Y\,. \end{array}$$

It exists rather rarely. We examine briefly the special case of equivalence relations. An *equivalence relation* on a variety X is a closed subvariety R of $X \times X$ that is set-theoretically an equivalence relation. (We leave it to the reader to formulate the scheme-theoretic variant of this definition.) The question of existence of the quotient-variety X/R is rather delicate and far from being solved; see the discussion in Artin [1971], Grothendieck-Demazure [1977], and Mumford [1965]. In one simple case the answer is affirmative: it is when both projections $R \rightrightarrows X$ are local isomorphisms. Basically, it is a gluing (cf. Sect. 3.3).

Any variety X can be regarded as a quotient-variety U/R, where U is an affine variety, and the projections $R \rightrightarrows U$ are local isomorphisms. For this one must take an atlas (U_i) and set $U = \coprod_i U_i$, and $R = U \times_X U$. More generally, we could define any variety as a pair (U, R), where U is an affine variety, $R \subset U \times U$ an equivalence relation on U, and the projections $R \rightrightarrows U$ are local isomorphisms. Then it is natural to consider that a morphism of a pair (U, R) into a pair (U', R') is a morphism $f\colon U \to U'$ such that $(f \times f)(R) \subset R'$. However, such a 'simple' solution to the problem of defining a variety would be incorrect. Actually, giving a pair (U, R) is essentially like giving one atlas, but one still needs to identify equivalent atlases (see Sect. 3.1 and 4.1). The proper formulation of this identification is left to the interested reader.

If, in the definition of a variety as a pair (U, R), we replace the condition "*the projections $R \rightrightarrows U$ are local isomorphisms*" by the weaker requirement that "*the projections $R \rightrightarrows U$ be étale morphisms*" (cf. § 5 of Chapter 2), then we obtain the very interesting notion of an *algebraic space*, which generalizes that of an algebraic variety. Algebraic spaces are interesting in that they occur as the result of many algebro-geometric constructions (quotient varieties, blowdowns, schemes of moduli); cf. Artin [1971].

In broad outline the genesis of the concept of variety is as follows. In the beginning we have the point and the affine line $\mathbb{A}^1$. Fibre products lead to the affine spaces $\mathbb{A}^n$ and their subvarieties, the affine varieties. Quotients of affine varieties by (locally isomorphic) equivalence relations produce all algebraic varieties. Finally, algebraic spaces come from étale equivalence relations.

4.4. Projection. Some important classes of morphisms are furnished by the operations of linear and multilinear algebra. We begin with the linear ones. Consider the map $\pi\colon \mathbb{A}^{n+1} - \{0\} \to \mathbb{P}^n$ that sends a nonzero point $x \in K^{n+1}$ into the line $Kx \subset K^{n+1}$, regarded as a point of $\mathbb{P}^n$. This map is clearly regular. If f is a regular function on $\mathbb{P}^n$ then $\pi^*(f)$ is regular on $\mathbb{A}^{n+1} - \{0\}$. As we saw in Sect. 2.9, such functions identify with the polynomials in $T_0, \dots, T_n$. Moreover, $\pi^*(f)$ is invariant under scalar multiplication, so the corresponding polynomial must have degree zero; hence it is a constant. Thus we have found that *every regular function on $\mathbb{P}^n$ is constant.* This simple result is the prototype of numerous finiteness theorems for projective varieties.

We shall now look at the projective variant of projection. Let $H \simeq \mathbb{P}^n$ be a hyperplane in $\mathbb{P}^{n+1}$, and p a point not lying on H. For every point $x \in \mathbb{P}^{n+1} - \{p\}$, the line $\overline{px}$ in $\mathbb{P}^{n+1}$ meets H in a single point $\pi_p(x)$. This gives rise to a mapping

$$\pi_p : \mathbb{P}^{n+1} - \{p\} \to H \simeq \mathbb{P}^n,$$

which is obviously regular. This is called the *linear projection from p* (or: *with center p*). If we identify $\mathbb{P}^{n+1} - H$ with $\mathbb{A}^{n+1}$, we recover the preceding example.

If a subvariety $X \subset \mathbb{P}^{n+1}$ does not pass through the point p then the restriction to X of the projection map defines a morphism $X \to \mathbb{P}^n$. Later on we shall also discuss the case when $p \in X$. Projection from a point can be iterated; alternatively, one can project at once from a linear subvariety. More generally, let $f : V \to W$ be any homomorphism of vector spaces over K. Then f gives rise to a natural morphism $\mathbb{P}(V) - \mathbb{P}(\ker f) \to \mathbb{P}(W)$, which is called a *collineation*. In particular, an automorphism of V yields an automorphism of $\mathbb{P}(V)$. Later on we shall see that every automorphism of $\mathbb{P}^n$ is a collineation.

The operations of multilinear algebra lead to three famous morphisms: the embeddings of Veronese, Segre, and Plücker.

4.5. The Veronese Embedding. Given two vector spaces $W \subset V$, going over to the k-th symmetric power yields an inclusion $\mathrm{Sym}^k W \subset \mathrm{Sym}^k V$. Hence we obtain a mapping of the corresponding Grassmannians. The most interesting case is when W is one-dimensional; then $\mathrm{Sym}^k W$ also has dimension 1, and we get the *Veronese mapping*

$$v_k : \mathbb{P}(V) \to \mathbb{P}(\mathrm{Sym}^k V).$$

In order to convince ourselves that it is regular, we shall use a coordinate system $T_0, \dots, T_n$ on V, and on $\mathrm{Sym}^k V$ the coordinates $T^a = T_0^{a_0} \cdot \ldots \cdot T_n^{a_n}$, where $a = (a_0, \dots, a_n)$ is a vector with nonnegative integer coordinates, and $\sum a_i = k$. Then v_k is the map sending a point $x = (x_0, \dots, x_n)$ to the point $v_k(x)$ with coordinates $(x^a = x_0^{a_0} \cdot \ldots \cdot x_n^{a_n})$.

It is easy to check that v_k is a closed immersion. Its image is given by the quadratic equations $T^a T^b = T^{a'} T^{b'}$, with $a + b = a' + b'$. The image of $\mathbb{P}^1$ under the Veronese mapping $v_k : \mathbb{P}^1 \to \mathbb{P}^k$ is the *rational normal curve* of degree k. For $k = 2$ it is the conic with equation $T_0 T_2 = T_1^2$. For $k = 3$ it is the curve in $\mathbb{P}^3$ with equations

$$T_0 T_3 = T_1 T_2, \quad T_0 T_2 = T_1^2, \quad T_1 T_3 = T_2^2.$$

If we project this curve from the point $p = (0, 1, 0, 0)$ into the plane $\mathbb{P}^2$, what we get is essentially Example 1 of Sect. 2.6.

The image of the embedding $v_2 : \mathbb{P}^2 \to \mathbb{P}^5$ is called the *Veronese surface.*

4.6. The Segre Embedding. If $W \subset V$ and $W' \subset V'$ are two inclusions of vector spaces then we have an inclusion $W \otimes W' \subset V \otimes V'$. Hence again we obtain a mapping of the corresponding Grassmannians. In particular, if W and W' are one-dimensional then $W \otimes W'$ also has dimension 1, and we get the *Segre mapping*

$$s\colon \mathbb{P}(V) \times \mathbb{P}(V') \to \mathbb{P}(V \otimes V').$$

In coordinates, to the points $x = (x_0, \ldots, x_n)$ and $y = (y_0, \ldots, y_m)$ one attaches the point $s(x, y)$ with coordinates $(x_i y_j)$, $i = 0, \ldots, n$, $j = 0, \ldots, m$. The image of s is given in the homogeneous coordinates R_{ij} of $\mathbb{P}(V \otimes V')$ by the quadratic equations $R_{ij}R_{kl} = R_{il}R_{kj}$. One checks easily that the Segre mapping is a closed immersion. In particular, *the product of two projective varieties is projective.*

The simplest case is the embedding of $\mathbb{P}^1 \times \mathbb{P}^1$ in $\mathbb{P}^3$; its image is the quadric with equation $XY = ZT$. As the Segre embedding maps each fibre $\mathbb{P}^1 \times \{y\}$ into a line of $\mathbb{P}^3$, the quadric $s(\mathbb{P}^1 \times \mathbb{P}^1)$ is covered by two families of lines. This is quite visible on the hyperboloid of one sheet.

4.7. The Plücker Embedding. We use here the exterior power. If W is a k-dimensional subspace of a vector space V then $\Lambda^k W$ is a line in $\Lambda^k V$. Hence we get the *Plücker mapping*

$$p\colon G(k, V) \to \mathbb{P}(\Lambda^k V).$$

Once again we can convince ourselves that p is a closed immersion, whose image is given by quadratic equations (cf. Griffiths-Harris [1978] and Hodge-Pedoe [1952]). We obtain as a corollary that *the Grassmann varieties are projective.*

For $k = 1$ we recover the isomorphism $G(1, V) \simeq \mathbb{P}(V)$. For $k = n - 1$, owing to the isomorphism $\Lambda^{n-1} V \simeq V^* \otimes \Lambda^n V \simeq V^*$, we find again the isomorphism $G(n-1, V) \simeq \mathbb{P}(V^*)$. Therefore the simplest nontrivial example is $G(2, 4)$, the variety of lines in $\mathbb{P}^3$. The morphism p embeds $G(2, 4)$ in $\mathbb{P}(\Lambda^2 K^4) \simeq \mathbb{P}^5$ as the hypersurface with equation

$$T_{12}T_{34} - T_{13}T_{24} + T_{14}T_{23} = 0.$$

Chapter 6 of Griffiths-Harris [1978] is devoted to a study of this variety and of its sections.

§ 5. Vector Bundles

An algebraic variety may have some additional structure, compatible with its structure as a variety. We shall examine briefly the notion of algebraic group and dwell in more detail on vector bundles.

5.1. Algebraic Groups. Suppose we have a set G, on which we are given a structure of algebraic variety together with a group structure. We say that these two structures are compatible (and define on G a structure of *algebraic group*) if the multiplication map $\mu\colon G \times G \to G$ and the inverse mapping $\iota\colon G \to G$ are regular.

For instance, the set of nondegenerate matrices $\mathrm{GL}(n, K)$ is an algebraic group, which is even affine as a variety. In particular, $\mathrm{GL}(1, K) = K^*$ is called the *multiplicative group* and is sometimes denoted by G_m. Another example of an affine group is the *additive group* K, whose group law is addition; it is sometimes denoted by G_a. An entirely different example is provided by the group law on a plane curve of the third degree, which will be discussed in Chapter 3. This is a special case of what is known as an *abelian variety* (cf. Mumford [1970]).

We define a homomorphism of algebraic groups to be a group homomorphism $f: G \to H$ which is at the same time a morphism of algebraic varieties. Its kernel, $\ker f = f^{-1}(e)$, is again an algebraic group. For instance, multiplication by a constant defines a homomorphism of G_a into itself. If the field K has positive characteristic p then the Frobenius morphism $x \mapsto x^p$ is also a homomorphism of algebraic groups, and is injective. A further example is the Artin-Schreier map, $x \mapsto x - x^p$, whose kernel identifies with the prime subfield $\mathbb{F}_p \subset K$. Raising to the n-th power, $x \mapsto x^n$, is a surjective homomorphism $G_m \to G_m$. Its kernel, which is denoted by μ_n, is isomorphic to the group of n-th roots of unity in the field K.

It is not part of our plan to go into a detailed exposition of the beautiful and very developed theory of algebraic groups (cf. Borel [1969], Humphreys [1975] or Serre [1959]). Two notions, however, must be mentioned. First, the action of an algebraic group G on an algebraic variety X. This is defined by a morphism of varieties, $\rho: G \times X \to X$, which satisfies two axioms: $\rho(e, x) = x$ and $\rho(g, \rho(g', x)) = \rho(gg', x)$. We may observe that these axioms express the commutativity of certain diagrams; for instance, the second axiom reflects the commutativity of the diagram

$$\begin{array}{ccc} G \times G \times X & \xrightarrow{\mu \times id} & G \times X \\ \big\downarrow{\scriptstyle id \times \rho} & & \big\downarrow{\scriptstyle \rho} \\ G \times X & \xrightarrow{\ \rho\ } & X \end{array}$$

The other notion is that of an algebraic family of groups (G_x), parametrized by the points of a variety X, that is, a 'group' in the category of X-varieties. We shall be mainly interested in the case of a family of vector spaces.

5.2. Vector Bundles. A *vector bundle* on a variety X is an X-variety $p: E \to X$, equipped with a 'zero-section' $0: X \to E$, an 'addition' operation, that is, an X-morphism $+: E \times_X E \to E$, and an operation of 'multiplication by constants', $K \times E \to E$, which is also an X-morphism. Addition is required to be commutative and associative, while multiplication must be distributive, etc., as in the definition of a vector space. For each point $x \in X$, the fibre $E_x = p^{-1}(x)$ has the structure of a vector space over K. Hence a vector bundle E can be thought of as a family of vector spaces E_x, each one growing above its own point $x \in X$.

We define a homomorphism of vector bundles to be a morphism of X-varieties which commutes with the operations 'zero', 'addition', and 'multiplication'. In other words, a fibre is carried into a fibre, by a homomorphism of vector spaces. Hence the vector bundles on X form a category Vect_X. Further, a base extension $f\colon X \to Y$ yields a functor $f^*\colon \mathrm{Vect}_Y \to \mathrm{Vect}_X$.

Here are some examples. Every (finite-dimensional) vector space V can be viewed as a vector bundle over a point. For any variety X, the bundle $X \times V \to X$ is called the *trivial* vector bundle (of type V, or: of rank $\dim V$). A vector bundle $p\colon E \to X$ is said to be *locally trivial* if there exists an atlas (X_i) such that the induced bundles $p^{-1}(X_i) \to X_i$ are trivial. By a known procedure, these bundles are given by cocycles $g_{ij}\colon X_i \cap X_j \to \mathrm{Aut}\, V$. For a locally trivial bundle, the dimension of the fibres E_x is a locally constant function of x. Arbitrary vector bundles do not share this property, since the dimension of some individual fibres can jump.

It can be shown that any vector bundle can be seen – locally over the base – as a sub-vector bundle of some trivial vector bundle. Given a homomorphism $\varphi\colon E \to F$ of vector bundles, there exists a kernel bundle $\ker \varphi = \varphi^{-1}(0_F)$, where 0_F is the zero-section of F. Unfortunately, this is no longer true for the cokernels of φ, and it is not always possible to define a quotient bundle F/E. This is one reason why coherent sheaves are widely preferred to vector bundles. Nevertheless, if E is a locally trivial vector subbundle of F then the quotient bundle F/E always exists.

5.3. Tautological Bundles. Let $G(k,V)$ be the Grassmann variety of k-dimensional vector subspaces of the vector space V. We consider in $G(k,V) \times V$ the subset S of all pairs (W,v) such that $v \in W$. It is easy to conceive that this is a vector subbundle of the trivial vector bundle $V_{G(k,V)} = G(k,V) \times V$ over $G(k,V)$. It is called the *tautological* bundle, or the *universal subbundle*, on $G(k,V)$. It is locally trivial of rank k. The quotient bundle $Q = V_{G(k,V)}/S$ is called the *universal quotient bundle* on $G(k,V)$.

In particular, we have on projective space $\mathbb{P}(V)$ a tautological *line* bundle S.

Later on we shall associate with every variety its tangent bundle, which plays an important role in the study of the variety.

5.4. Constructions with Bundles. If E and F are vector bundles on X, their fibre product $E \times_X F$, regarded as an X-variety, is also a vector bundle. It is denoted by $E \oplus F$ and is called the *direct sum* of the two bundles. In fact, on each fibre it is the direct sum of the fibres $E_x \oplus F_x$. I know of no other general construction for vector bundles. But, for locally trivial bundles, all natural vectorial constructions carry over: the tensor product $E \otimes F$, the symmetric and exterior powers, $\mathrm{Sym}^k E$ and $\Lambda^p E$, the dual bundle $E^\vee$, etc. For further details, see Bourbaki [1967–1971] or Lang [1962].

By mimicking the construction of projective space, for a vector bundle $E \to X$, one can build the corresponding projective bundle $\mathbb{P}_X(E) \to X$, whose fibres are the projective spaces $\mathbb{P}(E_x)$. One can also talk about conical subbundles of E, that is, subvarieties $C \subset E$ which are invariant under the action of K on E. Such cones $C \subset E$ lead to subvarieties $\mathbb{P}_X(C) \subset \mathbb{P}_X(E)$, as in Sect. 3.5.

§ 6. Coherent Sheaves

An alternative, more convenient, way of giving linear objects is provided by coherent sheaves of modules. In view of their great importance we shall say a few words about them, even though this will distract us from geometry. For further details about sheaves we refer to Godement [1958], Hartshorne [1977], and Serre [1955].

6.1. Presheaves. Let X be an arbitrary topological space, and $\mathrm{Op}(X)$ the category of the open subsets of X. A *presheaf* of sets (respectively, of groups, of modules or of rings) is a contravariant functor F from the category $\mathrm{Op}(X)$ to the category of sets (respectively, of groups, of modules or of rings). In other words, for each open subset $U \subset X$ we are given a set $F(U)$, and for each inclusion of open subsets $V \subset U$ we are given a map $\rho_{U,V}\colon F(U) \to F(V)$. Furthermore, $\rho_{U,U}$ must be the identity map, and if $W \subset V \subset U$ then $\rho_{U,W} = \rho_{V,W} \circ \rho_{U,V}$.

The elements of $F(U)$ are also referred to as the sections of F over U, and the mappings ρ as the restriction maps. One also writes $\rho_{U,V}(s) = s|_V$. This terminology is explained by the following example. Let $f\colon Y \to X$ be some variety over X. For an open subset $U \subset X$, a section of f over U is defined to be a morphism $g\colon U \to Y$ such that the composition map $f \circ g$ is the canonical injection of U in X. Let us denote by $\tilde{Y}(U)$ the set of all sections of f over U; this gives us a presheaf. A special case is the presheaf $\mathcal{O}_X$ constructed in Sect. 4.1.

A *morphism of presheaves* $\varphi\colon F \to G$ is a collection of mappings $\varphi_U\colon F(U) \to G(U)$, where U runs through $\mathrm{Op}(X)$, which are compatible with the restriction maps. In particular, if $\varphi\colon Y \to Z$ is a morphism of X-varieties then we have a morphism of presheaves $\tilde{\varphi}\colon \tilde{Y} \to \tilde{Z}$. This way of regarding varieties as being presheaves happens to be useful when looking for generalizations of the concept of algebraic variety.

6.2. Sheaves. A presheaf F on X is said to be a *sheaf* if it satisfies the following axiom:

Suppose we are given a family (U_i) of open subsets of X, together with sections $s_i \in F(U_i)$ which agree on the intersections (that is, $s_i|_{U_i \cap U_j} = s_j|_{U_i \cap U_j}$ for any i and j). Then there exists a section $s \in F(\bigcup U_i)$ such that $s_i = s|_{U_i}$ for every i, and it is unique with this property.

For instance, the presheaf $\widetilde{Y}$ we have constructed in Sect. 6.1, starting from the X-variety Y, is a sheaf. Other examples of sheaves include the presheaf $\mathcal{O}_X$ of regular functions on an algebraic variety X, the presheaf of smooth functions on a differentiable manifold, the presheaf of continuous functions on a topological space, etc. It may be said that sheaves arise whenever the sections can be given by local conditions.

Of course, not every presheaf is a sheaf. Thus, the constant presheaf, which to each set U associates a fixed set A, is very rarely a sheaf. It is, however, possible to attach to every presheaf F what is, in some sense, its closest sheaf, F^+. It is constructed as follows. For any open subset $U \subset X$, we denote by $\mathrm{Cov}(U)$ the set of all open coverings of U. For any covering $\mathcal{U} = (U_i) \in \mathrm{Cov}(U)$, we define $F(\mathcal{U})$ to be the set of all families of sections $s_i \in F(U_i)$ that agree on the intersections $U_i \cap U_j$. Further, if a covering $\mathcal{U}'$ is a refinement of $\mathcal{U}$ then there is a canonical map $F(\mathcal{U}) \to F(\mathcal{U}')$. Hence we get a direct system $F(\mathcal{U})$, $\mathcal{U} \in \mathrm{Cov}(U)$. We now define $F^+(U)$ as the direct limit of this system. It is easy to verify that:

a) F^+ is a sheaf;
b) there is a natural morphism of presheaves from F to F^+;
c) any morphism of F into a sheaf G factorizes through F^+.

This operation allows us to extend to sheaves all operations on presheaves: to begin with, one performs the operation in the category of presheaves, and then one goes over to the associated sheaf. In particular, one can do with sheaves of sets all that one can do with sets. Thus, sheaves of sets serve as a good model for set theory, thereby formalizing the notion of a 'variable set'. For our part, we shall be more interested in sheaves of modules.

6.3. Sheaves of Modules. We recall that the structure sheaf $\mathcal{O}_X$ on an algebraic variety X is a sheaf of rings, and even of K-algebras. Hence we can consider sheaves of modules over $\mathcal{O}_X$. For some time, that X is an algebraic variety will be relatively unimportant to us; so we can think we are talking about any topological space X, equipped with a sheaf of commutative rings $\mathcal{A}$. Such an object is named a *ringed space.* Before moving on to sheaves of modules, let us stop a while to look at the sheaf-theoretic definition of an algebraic variety. In Sect. 2.8, an affine variety was equipped with a sheaf of K-algebras. We can now define an algebraic variety as a ringed space which is locally isomorphic to an affine variety. Differentiable and analytic manifolds, supervarieties, etc., can be defined in a very similar way.

So, let $(X, \mathcal{A})$ be a ringed space. A sheaf F on X is said to be a *sheaf of $\mathcal{A}$-modules* if, for every open subset $U \subset X$, the set $F(U)$ is endowed with a structure of module over the ring $\mathcal{A}(U)$, compatible with the restriction maps. All notions from the theory of modules (homomorphisms, kernels, cokernels, exact sequences, direct sums, tensor products, etc.) carry over to sheaves of modules. For instance, $F \otimes_{\mathcal{A}} G$ is the sheaf associated with the presheaf $U \mapsto F(U) \otimes_{\mathcal{A}(U)} G(U)$. A subtle point, which is well known, has to do with the cokernel of a morphism of sheaves of modules $\varphi\colon F \to G$. Again, this is the sheaf associated with the presheaf $U \mapsto G(U)/\varphi(F(U))$. Hence, saying that a sequence of sheaves

$$0 \to F \to G \to H \to 0$$

is exact, indicates only that the following sequence is exact:

$$0 \to F(X) \to G(X) \to H(X),$$

while in general $G(X) \to H(X)$ is not an epimorphism. The deviation from surjectivity is controlled by the cohomology of the sheaf F.

The sheaf of $\mathcal{A}$-modules $\mathcal{A}^{(I)}$, where I is any set, is said to be *free* of rank $\operatorname{Card}(I)$. A sheaf that is locally isomorphic to a free sheaf of modules is said to be *locally free*. An important role is played by the locally free sheaves of rank one, that is, by the sheaves that are locally isomorphic to $\mathcal{A}$: they are called *invertible*. The tensor product $F \otimes_{\mathcal{A}} G$ of two invertible sheaves is also invertible, and $F \otimes_{\mathcal{A}} F^* \simeq \mathcal{A}$, where $F^* = \operatorname{Hom}_{\mathcal{A}}(F, \mathcal{A})$ is the dual sheaf. Hence the set $\operatorname{Pic}(X)$ of classes of invertible sheaves of $\mathcal{A}$-modules up to isomorphism is an abelian group, which is named the *Picard group* of X.

Suppose now $f\colon X \to Y$ is a morphism of varieties, and F a sheaf of modules on X. The *direct image* of F under f is the sheaf of modules f_*F on Y which to an open subset $V \subset Y$ associates $F(f^{-1}(V))$. It is somewhat more complicated to define the *inverse image* sheaf f^*G for a sheaf G of $\mathcal{O}_Y$-modules. As a preliminary step we define a presheaf $f^{\cdot}G$ on X. For any open set $U \subset X$ we set $(f^{\cdot}G)(U) = \varinjlim G(V)$, where the direct limit extends over the open subsets $V \subset Y$ containing $f(U)$. This is a presheaf of modules over $f^{\cdot}\mathcal{O}_Y$. Then f^*G is the sheaf of modules associated with the presheaf $f^{\cdot}G \otimes_{f^{\cdot}\mathcal{O}_Y} \mathcal{O}_X$. Further, f^* preserves local freeness, rank, and invertibility. In particular, f^* yields a homomorphism of the Picard groups $\operatorname{Pic} Y \to \operatorname{Pic} X$.

6.4. Coherent Sheaves of Modules. We revert now to algebraic varieties. A sheaf of modules on a variety is said to be *quasi-coherent* (respectively, *coherent*) if it is locally isomorphic to the cokernel of a morphism of free sheaves (respectively, of free sheaves of finite rank).

Example 1. Let X be an affine variety, and M a module over the ring $K[X]$. For any open set $U \subset X$, we write $\widetilde{M}(U) = M \otimes_{K[X]} \mathcal{O}_X(U)$. This defines a sheaf of modules $\widetilde{M}$ on X. The correspondence $M \mapsto \widetilde{M}$ preserves tensor products, exactness, and so on. In particular, $\widetilde{M}$ is quasi-coherent (coherent if M is of finite type). Besides, any quasi-coherent (respectively, coherent) sheaf of modules is of this form.

Example 2. Let $p\colon E \to X$ be a vector bundle. We may attach to it the sheaf $\mathcal{L}_X(E)$ of linear forms on E. To be precise, for any open subset $U \subset X$, the sections of this sheaf over U are those regular functions on $p^{-1}(U)$ which are linear on the fibres of p. What we get is a coherent sheaf of modules, which is free (locally free, respectively, invertible) if the vector bundle E is trivial (locally trivial, respectively, of rank 1).

In particular, for the tautological line bundle S on projective space $\mathbb{P} = \mathbb{P}(V)$, the invertible sheaf $\mathcal{L}_{\mathbb{P}}(S)$ is denoted by $\mathcal{O}_{\mathbb{P}}(1)$ and is called the *tautological,* fundamental, or twisting invertible sheaf on $\mathbb{P}$. Its global sections form a vector space $H^0(\mathbb{P}, \mathcal{O}_{\mathbb{P}}(1))$, which is canonically isomorphic to the space V^* of linear forms on V. The m-th tensor power of $\mathcal{O}_{\mathbb{P}}(1)$ is also invertible and is denoted by $\mathcal{O}_{\mathbb{P}}(m)$.

The properties of being coherent or quasi-coherent are preserved when going over to the kernel and the cokernel of a morphism of sheaves of modules, and also under tensor product and inverse image. Further, the direct image of a quasi-coherent sheaf is quasi-coherent. However, the direct image of a coherent sheaf is not coherent in general. Here are two fairly typical examples.

Example 3. Let f be the morphism mapping $\mathbb{A}^1$ into a point. Then $f_*\mathcal{O}_{\mathbb{A}^1}$ is essentially the space $H^0(\mathbb{A}^1, \mathcal{O}_{\mathbb{A}^1}) = K[T]$, which is infinite-dimensional over K. Hence $f_*\mathcal{O}_X$ is not coherent in this case.

Example 4. We consider the embedding of $\mathbb{A}^1 - \{0\}$ in $\mathbb{A}^1$. The direct image of the structure sheaf leads to the $K[T]$-module $K[T, T^{-1}]$, which is not of finite type, either.

In Chapter 2 we shall get to know the important class of proper morphisms, which are such that the direct image of a coherent sheaf is coherent. In the forthcoming § 7 we shall see yet another family of coherent sheaves: the sheaves of differential forms.

6.5. Ideal Sheaves. A further important source of examples of coherent sheaves is provided by ideal sheaves. Let Y be a subvariety of X, and $\mathcal{I}(Y)$ the subsheaf of $\mathcal{O}_X$ consisting of the sections that vanish on Y. This is a coherent sheaf of ideals of $\mathcal{O}_X$, and the quotient sheaf $\mathcal{O}_X/\mathcal{I}(Y)$ is isomorphic to $\mathcal{O}_Y$.

Conversely, let $\mathcal{I} \subset \mathcal{O}_X$ be a coherent sheaf of ideals. The support of the quotient sheaf $\mathcal{O}_X/\mathcal{I}$ (that is to say, the set of points $x \in X$ where $\mathcal{I}_x \neq \mathcal{O}_{X,x}$) is a closed subset of X. (Here and further on, F_x denotes – as usual – the *stalk* of the sheaf F at the point x, that is, $\varinjlim F(U)$, where U runs through all open neighbourhoods of x.) In the spirit of § 2, the ringed space $(\operatorname{Supp}(\mathcal{O}_X/\mathcal{I}), \mathcal{O}_X/\mathcal{I})$ may be called the *subscheme* of X defined by the sheaf of ideals $\mathcal{I} \subset \mathcal{O}_X$.

More generally, we may define an *algebraic scheme* to be a ringed space which is locally isomorphic to a subscheme of affine space. The systematic study of schemes is postponed until Chapter 4, so we shall merely say a few words about them now. Algebraic schemes differ from varieties only in that they may have nilpotent elements in their structure sheaf. Hence we can think of them as 'infinitesimal fattenings' of algebraic varieties. By killing the nilpotents, that is, by reducing the scheme X, we obtain an algebraic variety X_{red} with the same set of points. As we have only begun to see, many natural constructions lead up precisely to schemes and go over for arbitrary schemes.

6.6. Constructions of Varieties. One application of coherent sheaves is connected with globalizing the local constructions examined earlier. Suppose given an algebraic variety X, together with a quasi-coherent sheaf $\mathcal{A}$ of $\mathcal{O}_X$-algebras of finite type. The last condition means that $\mathcal{A}$ is generated locally by finitely many sections. Globalizing the construction of the affine variety Specm of § 2, we shall build a scheme $S = \mathrm{Specm}_X(\mathcal{A})$, together with a morphism $\pi\colon \mathrm{Specm}_X(\mathcal{A}) \to X$. This is done as follows. Suppose first that X is affine. Then $\mathrm{Specm}_X(\mathcal{A}) = \mathrm{Specm}\,\mathcal{A}(X)$, and the morphism π is dual to the structure K-algebra homomorphism $K[X] \to \mathcal{A}(X)$. If $\mathcal{D}(f)$ is a basic open subset of X then, $\mathcal{A}$ being quasi-coherent, we have

$$\mathcal{A}(\mathcal{D}(f)) = \mathcal{A}(X) \otimes_{K[X]} K[\mathcal{D}(f)] = \mathcal{A}(X)(f^{-1}).$$

Therefore $\mathrm{Specm}\,\mathcal{A}(\mathcal{D}(f))$ can be identified with the open subset $\pi^{-1}(\mathcal{D}(f))$. In the general case, one must start from an atlas (X_i) of the variety X and obtain S by gluing together the schemes $\mathrm{Specm}\,\mathcal{A}(X_i)$. Note that $\pi_*(\mathcal{O}_S) = \mathcal{A}$, so that the sections of $\mathcal{A}$ materialize as the regular functions on S. Morphisms of the form $\mathrm{Specm}_X(\mathcal{A}) \to X$ are said to be *affine*.

Two special cases of this construction are particularly important. The first one is the construction of the vector bundle $\mathbb{V}(F)$ associated with a coherent sheaf of modules F. In this case, one takes for $\mathcal{A}$ the sheaf of symmetric algebras $\mathrm{Sym}_{\mathcal{O}_X}(F)$ of F. This construction commutes with base extension. In particular, for any point $x \in X$, the fibre of $\mathbb{V}(F)$ over x identifies with the vector space $F(x)^*$, dual to $F(x) = F_x/\mathfrak{m}_x F_x$, where $\mathfrak{m}_x$ is the maximal ideal of $\mathcal{O}_{X,x}$. Furthermore, the sections of F correspond to the functions on $\mathbb{V}(F)$ that are linear on each fibre. Moreover, F is isomorphic to the sheaf $\mathcal{L}_X(\mathbb{V}(F))$ of linear forms on $\mathbb{V}(F)$ (cf. § 5).

The other special case is useful for constructing projective bundles. Suppose the sheaf of algebras $\mathcal{A}$ is graded, $\mathcal{A} = \bigoplus_{k\geq 0} \mathcal{A}_k$, and assume:

a) $\mathcal{A}_0 = \mathcal{O}_X$;
b) the sheaf $\mathcal{A}_1$ is coherent;
c) $\mathcal{A}_1$ generates $\mathcal{A}$ over $\mathcal{A}_0$.

Then the canonical morphism of sheaves of $\mathcal{O}_X$-algebras $\mathrm{Sym}(\mathcal{A}_1) \to \mathcal{A}$ is surjective, which induces a closed immersion of the X-schemes

$$C = \mathrm{Specm}_X(\mathcal{A}) \hookrightarrow \mathbb{V}(\mathcal{A}_1).$$

As $\mathcal{A}$ is graded, the subscheme C is invariant under the action of K by scalar multiplication on the vector bundle $\mathbb{V}(\mathcal{A}_1)$, so that C represents a cone bundle. The corresponding projective bundle $\mathbb{P}_X(C) \to X$ (cf. § 5) is called the *projective spectrum* of the graded sheaf of algebras $\mathcal{A}$. We denote it by $\mathrm{Proj}(\mathcal{A})$. The construction of the projective spectrum also commutes with base extension. Morphisms of the form $\mathrm{Proj}(\mathcal{A}) \to X$ are said to be *projective*. For further details, see Dieudonné [1969], Grothendieck-Dieudonné [1961], Hartshorne [1977] or Mumford [1966].

§ 7. Differential Calculus on Algebraic Varieties

The differential of a map and tangent space are two key notions in any theory of varieties. They serve as tools for linearizing various problems.

7.1. Differential of a Regular Function. Let $f(T_1, \dots, T_n)$ be a polynomial (or a regular function on $\mathbb{A}^n$). Using the habitual differentiation rules for sum and product, one can define the partial derivatives $\partial f/\partial T_i$ in a purely formal manner, without any limiting process. They are again polynomials. The *differential* of f at a point $x \in \mathbb{A}^n$ is the linear mapping

$$d_x f \colon K^n \to K,$$

which sends a point $\xi = (\xi_1, \dots, \xi_n) \in K^n$ to the number

$$(d_x f)(\xi) = \sum_{i=1}^{n} (\partial f/\partial T_i)(x)\, \xi_i.$$

We may also say that $d_x f$ is the linear part of f at x, for

$$f(x + \xi) = f(x) + \sum_{i=1}^{n} (\partial f/\partial T_i)(x)\, \xi_i + \dots,$$

where ... stands for terms of order ≥ 2 in ξ. As the $\partial f/\partial T_i$ are polynomials, the maps $d_x f$ paste together into one regular mapping $df \colon K^n \times \mathbb{A}^n \to K$, such that $df(\xi, x) = (d_x f)(\xi)$.

The structure of the fibre $f^{-1}(f(x))$ in a neighbourhood of x depends whether or not the differential $d_x f$ vanishes. If $d_x f \neq 0$ then the variety $f^{-1}(f(x))$ looks 'much like' the variety of zeros of the differential $d_x f$; near x it is 'smooth'. It is natural to regard the linear space of zeros of $d_x f$ as 'tangent' to $f^{-1}(f(x))$ at x. If, on the other hand, $d_x f = 0$ then the picture gets more complicated. We have drawn below the graph of the function $f = T_1^2 + T_1^3 - T_2^2$ and its zero-level section $f^{-1}(0)$. The differential vanishes only at the origin, where the curve $f^{-1}(0) = [T_2^2 = T_1^2 + T_1^3]$ has a 'singularity'. The remaining curves $f^{-1}(c)$ of a fixed level are all 'smooth'.

Note that the differential may be equal to zero at all points, even if f is nonconstant (cf. the Frobenius morphism in § 2). However, this occurs only in very special situations:

Proposition. *Let f be a nonconstant polynomial, and suppose the differentials $d_x f$ are zero at all points $x \in f^{-1}(0)$. Then f is the p-th power of some polynomial, where p is the characteristic of the field K. In particular, f is reducible.*

Indeed, assume f is irreducible and suppose $\partial f/\partial T_i$ vanishes at the points of $f^{-1}(0)$. It follows from Hilbert's Nullstellensatz that $\partial f/\partial T_i$ is divisible by f. Now, the degree of $\partial f/\partial T_i$ is less than the degree of f, whence $\partial f/\partial T_i$ is the zero polynomial. In characteristic zero, this implies that f is constant. In characteristic $p > 0$, we can only assert that f is a function of $T_1^p, \ldots, T_n^p$. But then $f = g^p$ for some polynomial g.

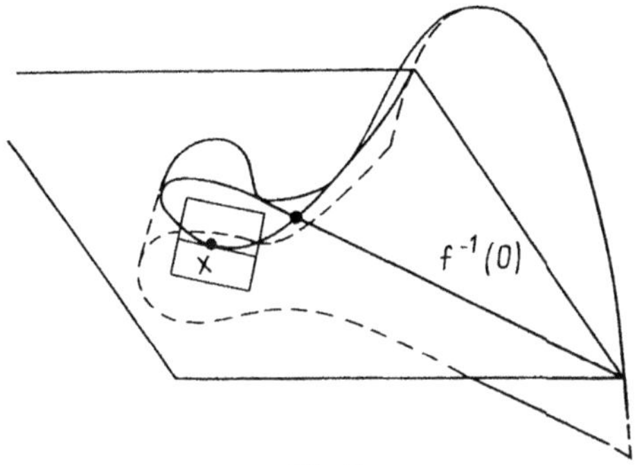

Fig. 7

7.2. Tangent Space. Let now X be the subvariety (or, better, the subscheme) of $\mathbb{A}^n$ defined by an ideal $I \subset K[T_1, \ldots, T_n]$, and let $x \in X$. We shall say that a vector $\xi \in K^n$ is tangent to X at the point x if $(d_x g)(\xi) = 0$ for every function g in I (or only for a set of generators of the ideal I). The set of all such vectors, which forms a vector subspace of K^n, is denoted by $T_x X$ and is called the *tangent space* to X at x.

Clearly, every regular function f on X has a differential $d_x f \colon T_x X \to K$. More generally, if $f \colon X \to Y$ is a morphism of affine varieties then there is a well-defined differential $d_x f \colon T_x X \to T_{f(x)} Y$, which is K-linear.

This definition of tangent space is fairly evident, but it uses the embedding $X \subset \mathbb{A}^n$. It is easy to see how we can get rid of that. Let $\mathfrak{m}_x \subset K[X]$ be the maximal ideal of all functions that vanish at the point $x \in X$. By associating with every function its differential, we obtain a K-linear homomorphism $\delta \colon \mathfrak{m}_x \to (T_x X)^*$. It is easy to figure out that it is surjective and that its kernel coincides with $\mathfrak{m}_x^2$. This allows us to give an invariant definition of $T_x X$ as the space dual to $\mathfrak{m}_x/\mathfrak{m}_x^2$ (Zariski), and also to 'paste together' all tangent spaces $T_x X$ into one tangent bundle TX. However, it is worth while spending some time on another useful notion.

7.3. Tangent Cone. Given a point x on a variety $X \subset \mathbb{A}^n$, one can attach to it what is called the tangent cone C_xX, which lies in the tangent space T_xX. Intuitively one may regard it as the set of limiting positions of secants $\overline{xx'}$, as $x' \in X$ tends to x.

Again we start from the case of a single function f and write its decomposition into homogeneous forms, say, $f = \sum_d f_d$. The nonzero form of least degree, f_{d_0}, is called the *initial,* or leading form of f. It is clear that the behaviour of f in the vicinity of the origin is determined principally by its initial form. The subvariety (more precisely, the subscheme) of zeros of f_{d_0} is called the tangent cone to the hypersurface $[f = 0]$.

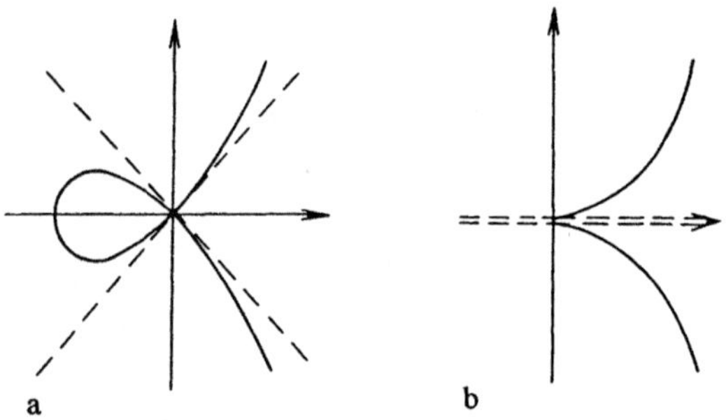

Fig. 8

Examples. Let X, as in Fig. 8a), be the zero-level contour line of the function $f = T_1^2 + T_1^3 - T_2^2$ (cf. Fig. 7). The tangent cone at the origin is given by the equation $T_1^2 = T_2^2$; it is the union of the two lines $T_1 = T_2$ and $T_1 = -T_2$. Similarly, the tangent cone to the curve $T_2^2 = T_1^3$ is given by the equation $T_2^2 = 0$, which represents the line $T_2 = 0$ doubled; see Fig. 8b).

We now give the general definition. Suppose we have a subscheme $X \subset \mathbb{A}^n$, with defining ideal I, and let x be a point of X. Without loss of generality we may assume that x is 0. The *tangent cone* C_xX is the subscheme of K^n given by the zeros of the initial forms of all $g \in I$. (In general it is not sufficient to take only a set of generators of I.) It is clear that C_xX lies in the tangent space T_xX, and spans it as a vector space. (To be more precise, the tangent space to C_xX at the origin coincides with T_xX.) In a certain sense, the variety X 'looks much like' C_xX in a neighbourhood of x, but this must not be understood literally, as though it were a local isomorphism (cf. Fig. 8b)).

To give a more invariant definition of the tangent cone, we shall need the following algebraic concept. Let $\mathfrak{a}$ be an ideal in a ring A. By the associated graded ring of the pair $(A, \mathfrak{a})$, we mean the graded ring

$$\operatorname{gr}(A, \mathfrak{a}) = \bigoplus_{k \geq 0} (\mathfrak{a}^k / \mathfrak{a}^{k+1}).$$

For further details on this topic, see Atiyah-Macdonald [1969], Bourbaki [1961–1965], Serre [1965] or Zariski-Samuel [1958,1960]. With the preceding notation, it can be shown that the quotient ring of $K[T_1, \ldots, T_n]$ by the ideal of leading forms of I is isomorphic to $\operatorname{gr}(K[X], \mathfrak{m}_x)$. For any variety X, the cone $C_x X$ can therefore be defined as $\operatorname{Specm}(\operatorname{gr}(\mathcal{O}_{X,x}, \mathfrak{m}_x))$.

The tangent cone, and the tangent space alike, are functorial objects. Given a morphism $f: X \to Y$ and a point $x \in X$, the linear mapping $d_x f: T_x X \to T_{f(x)} Y$ carries the cone $C_x X$ into the cone $C_{f(x)} Y$.

7.4. Smooth Varieties and Morphisms. The points where a variety appears simplest are those where it looks like its tangent space.

Definition. A point x on a variety X is said to be *smooth* (or nonsingular, or regular; one also says that X is smooth at x) if $C_x X = T_x X$. In the contrary case, x is said to be *singular.*

As we saw in Sect. 7.1, almost every point of a hypersurface in $\mathbb{A}^n$ is smooth. We shall see later that this holds for any variety X. Thus the set $\operatorname{Sing} X$ of singular points on X is closed and nowhere dense. If $\operatorname{Sing} X$ is empty then we say that X is *smooth.* Smooth varieties are those which look most like differentiable or analytic manifolds, which are smooth by definition. Of course, $\mathbb{A}^n$ is smooth, as well as $\mathbb{P}^n$ and the Grassmann varieties. If X is a smooth variety, and $f: X \to K$ is a regular function, then the variety $f^{-1}(f(x))$ is smooth at the points where $d_x f \neq 0$. Plainly, this extends to morphisms of smooth varieties. We rather introduce the relative variant of smoothness for a family of varieties.

Definition. A morphism $f: X \to Y$ is said to be *smooth at a point* $x \in X$ if the differential map $d_x f: C_x X \to C_{f(x)} Y$ can be viewed like the projection of $V \times C_{f(x)} Y$ onto its second factor, where V is a vector space. In the contrary case, we say that f *degenerates* at x. A morphism which is nowhere degenerate is said to be *smooth.*

We mention a few simple properties of smooth morphisms:

a) If X and Y are smooth then f is smooth if and only if the induced maps on the tangent spaces are surjective.
b) A variety X is smooth if and only if the morphism that maps X into a point is smooth.
c) If Y is a smooth variety, and $f: X \to Y$ is a smooth morphism, then the variety X is also smooth.
d) If $f: X \to Y$ is smooth then each fibre $f^{-1}(y)$ is smooth.

Assertions c) and d) are special cases of a more general statement:

e) Smoothness is preserved under composition and base extension.

7.5. Normal Bundle. Let $Y \subset X$ be two varieties, both being smooth and affine. It is natural to define the normal space to Y in X at a point $y \in Y$ to be the quotient space $N_{Y|X,y} = T_yX/T_yY$. The differentials of functions in the ideal $I(Y)$ yield some linear maps on $N_{Y|X,y}$, which are zero for functions in $I(Y)^2$. Hence $N_{Y|X,y}$ is dual to the space $I(Y)/(I(Y)^2 + \mathfrak{m}_y)$. This suggests a way of gluing together these normal spaces into one normal bundle.

Let Y be now any subvariety (or subscheme) of X, defined by an ideal sheaf $\mathcal{I} \subset \mathcal{O}_X$. The quotient $\mathcal{I}/\mathcal{I}^2 = \mathcal{I} \otimes_{\mathcal{O}_X} \mathcal{O}_Y$, being concentrated on Y, can be viewed as a coherent sheaf on Y. We call it the *conormal* sheaf. The vector bundle $\mathbb{V}(\mathcal{I}/\mathcal{I}^2)$ on Y associated with this sheaf (cf. § 6), is called the *normal* bundle to Y in X and is denoted by $N_{Y|X}$.

In the spirit of Sect. 7.3, one can define the *normal-cone bundle* $C_{Y|X}$ as $\mathrm{Specm}_Y(\mathrm{gr}(\mathcal{O}_X, \mathcal{I})) = \mathrm{Specm}_Y(\bigoplus_{k \geq 0} \mathcal{I}^k/\mathcal{I}^{k+1})$. This is a closed subscheme of $N_{Y|X}$.

7.6. Tangent Bundle. The easiest way to define the *tangent bundle* $TX \to X$ of a variety X is to say it is the normal bundle to the diagonal in $X \times X$. The fibre of this bundle TX over a point x is isomorphic to $N_{x|X}$, that is, T_xX. If $f: X \to Y$ is a morphism then $f \times f: X \times X \to Y \times Y$ carries the diagonal into the diagonal. By functoriality, this yields a homomorphism $Tf: TX \to TY$ of the tangent bundles, which is compatible with f and coincides on each fibre with $d_xf: T_xX \to T_{f(x)}Y$.

Examples. a) $T(X \times Y) \simeq TX \times TY$.

b) If $Y \subset X$ then there is a canonical exact sequence of vector bundles on Y:

$$0 \to TY \to TX|_Y \to N_{Y|X}.$$

c) The tangent bundle $TG(k, V)$ of the Grassmann variety $G(k, V)$ can be identified with the bundle $S^* \otimes Q$, where S and Q are, respectively, the universal subbundle and the universal quotient bundle on $G(k, V)$.

In particular, on tensoring by S^* the sequence of vector bundles

$$0 \to S \to V_{\mathbb{P}(V)} \to Q \to 0$$

on projective space $\mathbb{P}(V)$, we obtain the exact sequence

$$0 \to K_{\mathbb{P}(V)} \to V_{\mathbb{P}(V)} \otimes S^* \to T\mathbb{P}(V) \to 0.$$

7.7. Sheaves of Differentials. The conormal sheaf to the diagonal in $X \times X$, regarded as an $\mathcal{O}_X$-module, is called the *cotangent* sheaf, or *sheaf of differential 1-forms* on X. We denote it by Ω^1_X. Its sections, which correspond to linear functions on (each fibre of) TX, are called *differential 1-forms.* Examples of 1-forms are the differentials df of functions on X. For $x \in X$ the differential map d yields an isomorphism $\mathfrak{m}_x/\mathfrak{m}_x^2 \simeq \Omega^1_X(x)$. Differentials are functorial in the sense that a morphism of varieties $f: X \to Y$ yields a sheaf homomorphism $f^*: \Omega^1_Y \to \Omega^1_X$. Further details on sheaves and modules of differentials will be found in Grothendieck-Dieudonné [1964–1967], Hartshorne [1977], and Manin [1970], [1971].

Example 1. The sheaf $\Omega^1_{\mathbb{A}^n}$ is free on the generators $dT_1, \ldots, dT_n$.

Example 2. On $\mathbb{P}^n$ there is an exact sequence of sheaves

$$0 \to \Omega^1_{\mathbb{P}^n} \to K^{n+1} \otimes_K \mathcal{O}_{\mathbb{P}^n}(-1) \to \mathcal{O}_{\mathbb{P}^n} \to 0,$$

which is dual to the sequence of bundles from Example 7.6 *c*).

Example 3. If $Y \subset X$ is a subscheme, with defining ideal sheaf $\mathcal{I} \subset \mathcal{O}_X$, there is an *adjunction* exact sequence, dual to that of Example *b*) in Sect. 7.6:

$$\mathcal{I}/\mathcal{I}^2 \xrightarrow{\delta} \Omega^1_X \otimes_{\mathcal{O}_X} \mathcal{O}_Y \to \Omega^1_Y \to 0.$$

The homomorphism δ, which is induced by the differentiation map d, is often injective (for instance, if X and Y are smooth).

Example 4. Let $X \subset \mathbb{P}^2$ be a smooth curve of degree 3. Then the ideal sheaf $\mathcal{I}$ of X in $\mathbb{P}^2$ is isomorphic to $\mathcal{O}_{\mathbb{P}^2}(-3)$, and $\mathcal{I}/\mathcal{I}^2 \simeq \mathcal{O}_X(-3)$. Hence $\Omega^1_X \otimes \mathcal{O}_X(-3)$ is isomorphic to the second exterior power of $\Omega^1_{\mathbb{P}^2} \otimes \mathcal{O}_X$, that is, to the restriction of $\Lambda^2\Omega^1_{\mathbb{P}^2}$ to X. Now, one can see from Example 2 that $\Lambda^2\Omega^1_{\mathbb{P}^2} \simeq \mathcal{O}_{\mathbb{P}^2}(-3)$. It follows that $\Omega^1_X \simeq \mathcal{O}_X$; so there is on X a nontrivial regular differential 1-form.

This form can, of course, be written down explicitly. To that effect, we consider in $\mathbb{A}^3$ the following three (rational) 1-forms:

$$\omega_0 = \frac{T_1 dT_2 - T_2 dT_1}{\partial F/\partial T_0}, \quad \omega_1 = \frac{T_2 dT_0 - T_0 dT_2}{\partial F/\partial T_1}, \quad \omega_2 = \frac{T_0 dT_1 - T_1 dT_0}{\partial F/\partial T_2},$$

where $F(T_0, T_1, T_2)$ is the homogeneous third-degree polynomial which defines X. As the degrees of the numerators and denominators of the ω_i are equal to 2, each form ω_i is homogeneous and defines a 1-form on $\mathbb{P}^2$, which is regular off the curve $\mathcal{D}_i = [\partial F/\partial T_i = 0]$. Now comes the main point: all three forms coincide when restricted to X. This is because

$$3F = \sum \frac{\partial F}{\partial T_i} T_i \quad \text{and} \quad dF = \sum \frac{\partial F}{\partial T_i} dT_i$$

both vanish over X. Therefore the $\omega_i|_X$ produce a well-defined form ω on X, which is regular. Indeed, X being smooth, the $\mathcal{D}_i$ do not contain any point of X in their intersection.

We observe further that, if F is a polynomial of degree $d \geq 3$ then one can multiply the forms ω_i by any homogeneous polynomial of degree $d-3$, so as to obtain – by the same argument – a regular 1-form on the curve

$[F=0] \subset \mathbb{P}^2$. Hence, on a smooth curve of degree d in $\mathbb{P}^2$, there are $\frac{1}{2}(d-1)(d-2)$ (linearly independent) regular differential forms.

The exterior powers $\Lambda^p(\Omega^1_X)$ of the sheaf of $\mathcal{O}_X$-modules Ω^1_X are denoted by Ω^p_X. We call them *sheaves of p-differentials* on X. As in the case of 1-forms, their behaviour is contravariant. They allow the standard operations, namely (Bourbaki [1967–1971], Cartan [1967]):

a) exterior multiplication;
b) convolution with vector fields (that is, with sections of the tangent bundle $TX \to X$);
c) exterior differentiation, $d\colon \Omega^p_X \to \Omega^{p+1}_X$.

Besides, for algebraic varieties, one can investigate vector fields, differential operators, differential equations, connectedness, and the other notions from differential geometry (cf. Bourbaki [1967–1971], Griffiths-Harris [1978], and Wells [1973]).

Chapter 2
Algebraic Varieties: Fundamental Properties

In Chapter 1 the theory of algebraic varieties was developed along the lines of the theory of differentiable manifolds. In the present chapter, we shall analyse some notions and properties that are specific to algebraic varieties and either have no differentiable analogue or require an essentially different approach. The former comprise the notions of irreducibility and normality, rational maps, and blowings-up. The latter, those of completeness, dimension, and smoothness. This is also where the fundamental properties of algebraic varieties are established, namely, the theorem on the dimension of fibres, the theorem on the constructibility of the image, the connectedness theorem, the Zariski Main Theorem, the completeness of projective varieties, the finiteness theorem, etc. For this we shall have recourse to somewhat more algebra than we did in Chapter 1.

§ 1. Rational Maps

1.1. Irreducible Varieties. Let us start from the plane curve $[T_1T_2=0] \subset \mathbb{A}^2$. It is plainly made up of two pieces, namely, the lines $[T_1=0]$ and $[T_2=0]$ meeting in the point 0 (cf. Fig. 1). Now, these lines cannot be further decomposed into simpler closed subsets. It turns out that any algebraic variety can be written out as the union of finitely many irreducible components.

We begin with a general definition. We say that a topological space X is *irreducible* if it is not the union of two proper closed subsets. Equivalently, every nonempty open subset is dense in X. In particular, an irreducible space is connected, though the converse is not true, as can be seen from the above example. The closure of an irreducible subspace is irreducible. So is the image of an irreducible set under a continuous mapping.

An algebraic variety is said to be *irreducible* if it is irreducible in the Zariski topology.

Example 1. Affine space $\mathbb{A}^n$ (whence also $\mathbb{P}^n$) is irreducible. Indeed, suppose $\mathbb{A}^n = V(f) \cup V(g)$; then $f \cdot g = 0$. Now, the polynomial ring $K[T_1, \ldots, T_n]$ is an integral domain; hence either f or g is equal to 0. This means that either $V(f)$ or $V(g)$ coincides with $\mathbb{A}^n$. More generally, an affine variety X is irreducible if and only if the ring $K[X]$ is an integral domain.

Example 2. If two varieties, X and Y, are irreducible then so is their product $X \times Y$. To this end we remark that every open set $U \subset X \times Y$ projects onto an open subset of Y. Indeed, the projection of U coincides with the union of the open sets $V_x \subset Y$, $x \in X$, such that $U \cap (\{x\} \times Y) = \{x\} \times V_x$.

Example 3. Let $f \in K[T_1, \ldots, T_n]$ be an irreducible polynomial. It is an easy consequence of Gauss' theorem on unique factorization in the ring $K[T_1, \ldots, T_n]$ that the hypersurface $V(f) \subset \mathbb{A}^n$ is irreducible. More generally, if $f = f_1^{m_1} \cdot \ldots \cdot f_r^{m_r}$ is a decomposition into irreducible factors then $V(f) = V(f_1) \cup \ldots \cup V(f_r)$ is the decomposition of $V(f)$ into irreducible components.

Definition. By an *irreducible component* of a variety X, we mean a maximal irreducible subset of X. Of course, it is closed.

Every algebraic variety can be expressed as a finite union of irreducible components. Indeed, if X is reducible, we decompose it into two smaller subvarieties, and so forth. That this process will stop after a finite number of steps is guaranteed by a specific property of the Zariski topology, namely that it is noetherian.

1.2. Noetherian Spaces. A topological space X is said to be *noetherian* if every descending sequence $Y_1 \supset Y_2 \supset \ldots$ of closed subsets of X is stationary, that is, there is an integer r such that $Y_r = Y_{r+1} = \ldots$. The following simple facts hold (see, for example, Bourbaki [1961–1965]):

a) Every subspace of a noetherian space is noetherian.

b) A space X is noetherian if and only if every open subset $U \subset X$ is quasi-compact (that is, from every open covering of U one can choose a finite subcovering).

c) X is noetherian if it is covered by finitely many noetherian subspaces.

Proposition. *An algebraic variety is noetherian.*

We recall that we have restricted our attention to varieties with a finite atlas. Hence, in view of *a*) and *c*), it is enough to check that affine space $\mathbb{A}^n$ is noetherian. Let $Y_1 \supset Y_2 \supset \dots$ be a descending chain of subvarieties of $\mathbb{A}^n$, and $I(Y_1) \subset I(Y_2) \subset \dots$ the corresponding ascending chain of ideals of $K[T_1, \dots, T_n]$. By Hilbert's Basis Theorem, the ideal $\bigcup_i I(Y_i)$ is generated by finitely many elements, which lie in some $I(Y_r)$. Then $I(Y_r) = I(Y_{r+1}) = \dots$, and $Y_r = Y_{r+1} = \dots$.

1.3. Rational Functions. A rational function in the variables $T_1, \dots, T_n$ is defined as the ratio f/g of two polynomials, f and g, in $T_1, \dots, T_n$, with $g \neq 0$. Note that it is not a function on the whole of $\mathbb{A}^n$, but only on the open subset $\mathcal{D}(g) \subset \mathbb{A}^n$ where g is different from zero. It is thereby uniquely determined by its restriction to any nonempty open subset $U \subset \mathcal{D}(g)$. Conversely, any regular function on an open set $U \subset \mathbb{A}^n$ can be represented by a rational function.

This suggests the following generalization to any algebraic variety X. A *rational function* on X is an equivalence class of regular mappings $f: U \to K$, where U is an open dense subset of X. Two such maps, say, $f: U \to K$ and $f': U' \to K$, are regarded as equivalent if they agree on $U \cap U'$. This really is an equivalence relation, because $U \cap U'$ is also dense in X. (The naïve definition of a rational function as the ratio of two regular functions is of little interest, since on $\mathbb{P}^n$ there are few regular functions.)

Rational functions can be added and multiplied together, so that the set $K(X)$ of all rational functions on the variety X is a ring. It is clear that $K(X)$ is the direct limit $\varinjlim \mathcal{O}_X(U)$ of the rings $\mathcal{O}_X(U)$, as U runs through the open dense subsets of X. If X is irreducible, $K(X)$ is even a field. Indeed, if $f: U \to K$ is a nonzero function, it is invertible on the nonempty (and therefore dense) open subset $U - f^{-1}(0)$. Further, for X irreducible, the field $K(X)$ coincides with the quotient field of the integral domain $K[U]$, where U is any affine chart of X. For arbitrary X, the ring $K(X)$ is the direct sum of the fields $K(X_i)$, where the X_i denote the irreducible components of X.

1.4. Rational Maps. In much the same way one defines a *rational map* of a variety X into a *separated* variety Y, as an equivalence class of morphisms $U \to Y$, where U is open and dense in X. Among all such U there exists a biggest one, which is called the *domain of definition* of the rational map f.

Example. Consider the transformation of $\mathbb{P}^n$ into itself, which carries a point with homogeneous coordinates $(x_0, \dots, x_n)$ into the point $(x_0^{-1}, \dots, x_n^{-1})$. It is defined if all $x_i \neq 0$. However, its domain of definition is more extended, as it includes the points having only one of the x_i equal to zero. Such a point, say, $(0, x_1, \dots, x_n)$ is mapped into $(1, 0, \dots, 0)$. In particular, for $n = 1$ the mapping is defined everywhere (and is nothing else than $(x, y) \mapsto (y, x)$). For $n = 2$, it is undefined only at the three points $(1,0,0)$, $(0,1,0)$, and $(0,0,1)$.

Rational maps have the notorious defect that their composition is not always defined. In fact, the image of the preceding map may happen to lie entirely outside the domain of definition of the next one. This will not happen if the image of any component is dense; a mapping with this property is said to be *dominant*. If $f\colon X \dashrightarrow Y$ is a dominant rational map of irreducible varieties, then there is an injection of the field $K(Y)$ into the field $K(X)$.

A rational map $f\colon X \dashrightarrow Y$ of irreducible varieties is said to be *birational* if it has a rational inverse $f^{-1}\colon Y \dashrightarrow X$. Equivalently, f^* establishes an isomorphism of the fields $K(Y)$ and $K(X)$. For instance, the mapping $x \mapsto (x^2, x^3)$ sets up a birational equivalence of the line $\mathbb{A}^1$ with the plane curve $C = \left[T_1^3 = T_2^2\right]$. We give a further example.

Example. Let $C \subset \mathbb{P}^2$ be an irreducible conic. Then stereographic projection (Fig. 9), that is, linear projection from a point $p \in C$, yields a birational equivalence of the conic C with the line $\mathbb{P}^1$. More generally, if $X \subset \mathbb{P}^n$ is an irreducible hypersurface of degree 2 then linear projection from a smooth point $p \in X$ yields a birational equivalence $X \dashrightarrow \mathbb{P}^{n-1}$. As we shall see later, a smooth cubic curve in $\mathbb{P}^2$ is no longer birationally equivalent to $\mathbb{P}^1$.

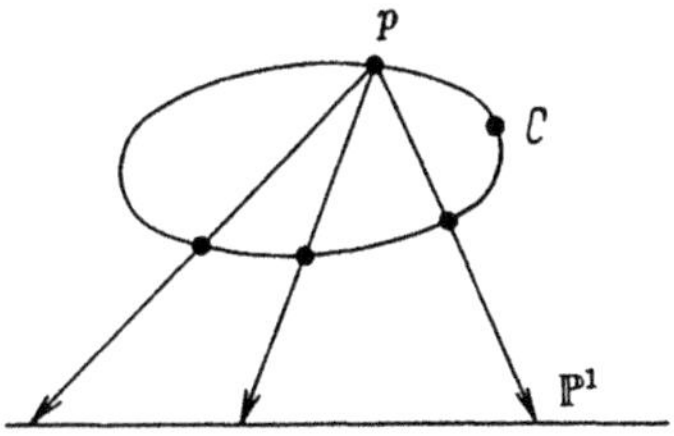

Fig. 9

Birational equivalence furnishes a weaker notion of equivalence than isomorphism. This, too, is a specific feature of algebraic geometry. The study of algebraic varieties up to birational transformations and the discovery of birational invariants constitute the object of birational geometry.

1.5. Graph of a Rational Map. There is yet another way to represent a rational map $f\colon X \dashrightarrow Y$ by a morphism. Suppose f is defined on an open dense subset $U \subset X$, and let $\Gamma \subset U \times Y$ be its graph (cf. Chap. 1, Sect. 4.2). The closure of Γ in $X \times Y$ is called the *graph of the rational mapping* f. We denote it by Γ_f. The projection map $p\colon \Gamma_f \to X$ is a morphism; and it is birational, since above U we know that p is an isomorphism. If q denotes the second projection $\Gamma_f \to Y$, then f appears as the composite of p^{-1} and the morphism q. A rational map f can be thought of as a multivalued mapping, which to a point x associates the set $f(x) = q(p^{-1}(x))$ (and is almost everywhere single-valued).

More generally, one defines an algebraic *correspondence* between two varieties X and Y as a closed subset $T \subset X \times Y$. Further the *image* of a point

$x \in X$ under T is defined as the subset $T(x) = q(p^{-1}(x))$ in Y, where p and q are the projections of T into X and Y.

Example. We consider the graph of the projection mapping $\pi\colon \mathbb{A}^{n+1} \dashrightarrow \mathbb{P}^n$ of Chap. 1, Sect. 4.4. This map is defined outside the origin $0 \in \mathbb{A}^{n+1}$. The graph of its restriction to $\mathbb{A}^{n+1} - \{0\}$ consists of all pairs (x, l), where x is a nonzero point of K^{n+1}, and l is the line in K^{n+1} passing through x, that is, $l = Kx$. Its closure in $\mathbb{A}^{n+1} \times \mathbb{P}^n$ is made up of the same pairs (x, l), with $x \in l$, except that now x may also be 0. So it is given by the equations

$$T_i T_j' = T_j T_i', \quad i, j = 0, 1, \ldots, n, \tag{$*$}$$

where the T_i are coordinates on $\mathbb{A}^{n+1}$, and the T_i' are the corresponding homogeneous coordinates on $\mathbb{P}^n$.

The variety thus obtained (that is, the graph of π) will be denoted by $\widetilde{\mathbb{A}}^{n+1}$. We shall have a look at its projection σ onto $\mathbb{A}^{n+1}$. To do this we cover $\widetilde{\mathbb{A}}^{n+1}$ by means of charts $\widetilde{U}_i$, $i = 0, 1, \ldots, n$, which are defined by the conditions $T_i' \neq 0$ (in other words, we take the inverse images of the charts U_i in $\mathbb{P}^n$; cf. Chap. 1, Sect. 3.1). Using $(*)$, we may express T_j as $T_i \xi_j^{(i)}$, where $\xi_j^{(i)} = T_j'/T_i'$. Hence, as coordinates for $\widetilde{U}_i$ we can choose T_i, together with the $\xi_j^{(i)}$, $j \neq i$, and they exhibit an isomorphism of $\widetilde{U}_i$ with $\mathbb{A}^{n+1}$. In these coordinates, the map σ takes the following shape:

$$\sigma^*(T_i) = T_i, \quad \sigma^*(T_j) = T_i \xi_j^{(i)} \quad \text{for } j \neq i.$$

From this we see that σ induces an isomorphism of $\widetilde{\mathbb{A}}^{n+1} - \sigma^{-1}(0)$ with $\mathbb{A}^{n+1} - \{0\}$. (This would also follow from the definition of the graph.) More importantly, $\sigma^{-1}(0)$ is given in the chart $\widetilde{U}_i$ by the *single* equation $T_i = 0$.

As a general fact, the fibre $\sigma^{-1}(0)$ consists of all pairs $(0, l)$, where l is any line $l \subset K^{n+1}$. Hence $\sigma^{-1}(0)$ is isomorphic to $\mathbb{P}^n$. It is called the *exceptional subvariety* of $\widetilde{\mathbb{A}}^{n+1}$. The point $0 \in \mathbb{A}^{n+1}$ is, as it were, blown up under the effect of σ^{-1}; it bursts in all directions, and each tangent direction has its own associated point on the exceptional variety $\sigma^{-1}(0)$ (cf. Fig. 10). For that reason the morphism σ is called the *blowing-up* of $\mathbb{A}^{n+1}$ at the point 0, or the *σ-process* with centre 0. And the variety $\widetilde{\mathbb{A}}^{n+1}$ is said to be the *blow-up* of $\mathbb{A}^{n+1}$ at the point 0.

Let us turn now to the second projection map $\tilde{\pi}\colon \widetilde{\mathbb{A}}^{n+1} \to \mathbb{P}^n$. The fibre of $\tilde{\pi}$ over a point $l \in \mathbb{P}^n$, that is, over a line $l \subset K^{n+1}$, consists of those points x which lie on l. It therefore identifies with the line l itself. Hence $\tilde{\pi}$ can be identified with the universal line bundle $S \to \mathbb{P}^n$ of Chap. 1, Sect. 5.3, and the exceptional variety E with the zero-section of S.

1.6. Blowing up a Point. A generalization of the foregoing construction is the blowing-up of a point on an arbitrary variety. Suppose first that X is embedded in $\mathbb{A}^{n+1}$ and passes through 0. We may then restrict the projection

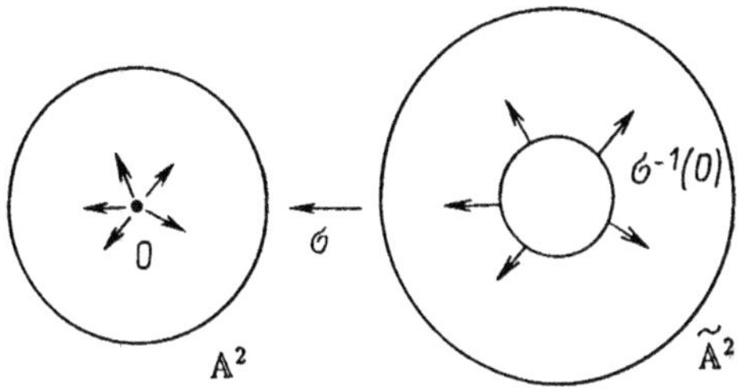

Fig. 10

map $\pi\colon \mathbb{A}^{n+1} \dashrightarrow \mathbb{P}^n$ to X and consider its graph, which will again be called the *blow-up* of X at the point 0. We denote it by $\widetilde{X}$. We can think of it as being the closure in $\widetilde{\mathbb{A}}^{n+1}$ of the set $\sigma^{-1}(X-\{0\})$. Hence the projection $\sigma_X\colon \widetilde{X} \to X$ yields an isomorphism between $\widetilde{X} - \sigma_X^{-1}(0)$ and $X - \{0\}$.

We look now at what happens above the point 0. We assert that the fibre $\sigma_X^{-1}(0)$ can be identified with the projectivization $\mathbb{P}(C_0X)$ of the tangent cone C_0X to X at 0. This agrees with our intuitive feeling, that blowing-up spreads out the tangent directions. We shall limit our attention to the case when X is given in $\mathbb{A}^{n+1}$ by a single equation $f=0$, with $f \in K[T_0,\dots,T_n]$. Suppose the decomposition $f = \sum\limits_d f_d$ into homogeneous forms of degree d starts from f_{d_0}. Then the variety $\sigma^{-1}(X)$ is given, in the chart $\widetilde{U}_i$ of $\widetilde{\mathbb{A}}^{n+1}$, by the zeros of the polynomial

$$\sigma^*(f) = f(T_i\xi_0^{(i)},\dots,T_i\xi_n^{(i)}) = T_i^{d_0}(\tilde{f}_{d_0} + T_i\tilde{f}_{d_0+1} + \dots).$$

Hence the total inverse image $\sigma^{-1}(X)$ consists of the exceptional variety $E = \sigma^{-1}(0)$ (with multiplicity d_0) plus $\widetilde{X}$, which is defined in the chart $\widetilde{U}_i$ by the equation

$$\tilde{f}_{d_0} + T_i\tilde{f}_{d_0+1} + \dots = 0.$$

We see also that the fibre $\sigma_X^{-1}(0) = \widetilde{X} \cap E$ is given in the chart $\widetilde{U}_i$ by the equations $T_i = 0$ and $\tilde{f}_{d_0} = 0$, and in $E \simeq \mathbb{P}^n$ by the initial form $f_{d_0} = 0$.

Let us examine, for example, the blow-up of the plane curve $C \subset \mathbb{A}^2$ with equation $Y^2 = X^2 + X^3$ (cf. Fig. 8 a)). On replacing Y by the expression $X\xi$, we get the equation $X^2\xi^2 = X^2 + X^3$. If now we remove the factor X^2, we obtain the equation $\xi^2 = 1 + X$, which defines $\widetilde{C}$ in one of the charts (cf. Fig. 4). In a similar fashion, the blow-up of the curve $Y^2 = X^3$ is given by the equation $\xi^2 = X$.

If now we want to blow up a point x on an arbitrary variety X, we have to embed some neighbourhood U of x in $\mathbb{A}^{n+1}$ in such a way that x goes to 0, and glue $X - \{x\}$ with $\widetilde{U}$ along the open subsets $U - \{x\}$ and $\widetilde{U} - \sigma_U^{-1}(0)$. This yields an X-variety $\sigma\colon \widetilde{X}_x \to X$. But the question arises whether this construction depends on the choice of U and of the embedding of U in $\mathbb{A}^{n+1}$.

That is why we shall present another construction, which is invariant and more general.

1.7. Blowing up a Subscheme. Let $Y \subset X$ be a subscheme, with defining ideal sheaf $\mathcal{I} \subset \mathcal{O}_X$. We form the graded sheaf of $\mathcal{O}_X$-algebras $\mathcal{A} = \bigoplus_{k \geq 0} \mathcal{I}^k$. The projective spectrum σ: $\mathrm{Proj}(\mathcal{A}) \to X$ of this algebra (cf. Chap. 1, Sect. 6.6) is called the *blow-up* of the variety X along the subscheme Y. We give three arguments to justify this definition.

a) Let $x \in X$ be a point that does not lie on Y. Then, in a neighbourhood of x, the ideal $\mathcal{I}$ coincides with $\mathcal{O}_X$ and $\mathcal{A} = \mathcal{O}_X[T]$, so its Proj is isomorphic to X. Hence σ is an isomorphism over $X - Y$.

b) Let us see what happens over Y. In view of the functoriality of Proj, the subscheme $\sigma^{-1}(Y)$ looks like Proj of the algebra

$$\mathcal{A} \otimes_{\mathcal{O}_X} \mathcal{O}_Y = \bigoplus_{k \geq 0} (\mathcal{I}^k / \mathcal{I}^{k+1}) = \mathrm{gr}(\mathcal{O}_X, \mathcal{I}).$$

Thus $\sigma^{-1}(Y) \to Y$ is built like the projectivization of the normal cone $C_{Y|X}$. It is important to notice that the subscheme $\sigma^{-1}(Y) \subset \widetilde{X}$ is given locally by one equation, as in Sect. 1.6. As a matter of fact, this property characterizes the blowing-up (cf. Hartshorne [1977]).

c) In the special case where $X = \mathbb{A}^{n+1}$ and $Y = \{0\}$, the present construction agrees with the one considered in Sect. 1.6. Indeed, let $T_0, \ldots, T_n$ be a coordinate system for $\mathbb{A}^{n+1}$, and $T'_0, \ldots, T'_n$ a set of generators for the ideal $\mathfrak{m}_0$, which are to be regarded as elements of degree 1 in the graded ring $A = \bigoplus_k \mathfrak{m}_0^k$. These elements satisfy the relations $T_i T'_j = T_j T'_i$, which is just $(*)$.

Blowing up serves as a means of studying the local structure of a variety X in a neighbourhood of a subvariety Y, and it enables one to contemplate the singularities as through a magnifying glass. Blowing up is also used for the resolution of singularities and for removing the points of indeterminacy of rational maps. For instance, the rational map of $\mathbb{P}^2$ into itself, given by the formula $(x, y, z) \mapsto (x^{-1}, y^{-1}, z^{-1})$, becomes regular after blowing up the three points $(1,0,0)$, $(0,1,0)$, and $(0,0,1)$.

§ 2. Finite Morphisms

2.1. Quasi-finite Morphisms. The simplest varieties are the finite ones. Before going over to the deeper study of infinite varieties, it is useful to look into families of finite varieties, that is, into morphisms with finite fibres. Contrary to expectation, these morphisms are said to be *quasi-finite,* the term 'finite morphism' being reserved for morphisms verifying some extra closedness property.

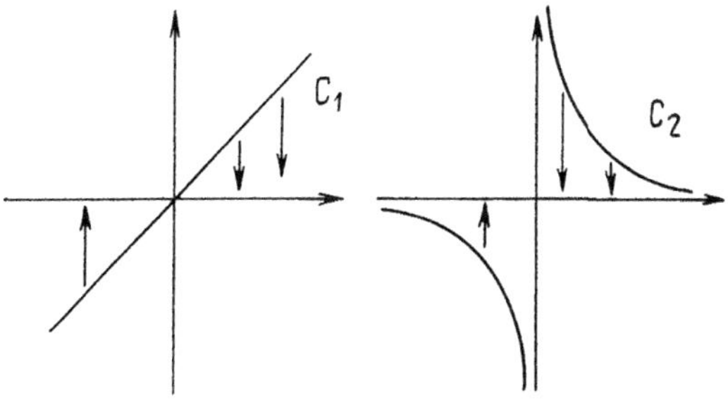

Fig. 11

Let $f\colon X \to Y$ be a quasi-finite morphism. The number of elements in a fibre $f^{-1}(y)$ may depend on the point $y \in Y$. By way of example, we may consider the two plane curves C_1 and C_2 in $\mathbb{A}^2$ defined by the equations $Y^2 - XY = 0$ and $XY^2 - Y = 0$, and look at their projection onto the x-axis (Fig. 11). For fixed x we have a quadratic equation with respect to y, and hence each fibre contains no more than two points. And there are indeed two points, providing $x \neq 0$. For $x = 0$, in both cases the fibre consists of one point, albeit for completely different reasons. In the former case, as $x \to 0$, the two points in the fibre merge into one. In the latter case, one of them 'goes to infinity'. In the former case, the morphism is considered to be finite; but not in the latter. In order to give a general definition, we must resort to some algebraic terminology.

2.2. Finite Morphisms. A morphism of affine varieties $f\colon X \to Y$ is said to be *finite* if the ring $K[X]$ is finitely generated not only as an algebra over $K[Y]$, but also as a $K[Y]$-module. (When this condition holds, $K[X]$ is called a finite $K[Y]$-algebra.) Another way of saying is that the algebra $K[X]$ is integral over $K[Y]$ (cf. § 1 of Chap. 1). Indeed, let $g \in K[X]$ and define M_n to be the $K[Y]$-submodule of $K[X]$ generated by $1, g, \dots, g^{n-1}$. If we assume $K[X]$ to be finite over $K[Y]$, the ascending chain of modules $M_1 \subset M_2 \subset \dots$ is stationary. Thus $M_r = M_{r+1}$ for some r, and g^r can be expressed in terms of $1, \dots, g^{r-1}$ with coefficients in $K[Y]$. Hence g is integral over $K[Y]$. The converse is obvious.

A morphism $f\colon X \to Y$ of arbitrary varieties is said to be *finite* if, for every chart $V \subset Y$, the variety $f^{-1}(V)$ is affine and the morphism $f^{-1}(V) \to V$ is finite. In this situation, the sheaf of $\mathcal{O}_Y$-algebras $\mathcal{A} = f_*(\mathcal{O}_X)$ is coherent, and the Y-variety X is the same as $\mathrm{Specm}_Y(\mathcal{A})$. Conversely, for any coherent sheaf of $\mathcal{O}_Y$-algebras $\mathcal{A}$, the morphism $\mathrm{Specm}_Y(\mathcal{A}) \to Y$ is finite.

Being a finite morphism is a property which is preserved under composition and base extension. Further, a finite morphism is quasi-finite; in fact, it is easy to verify that a finite K-algebra has only finitely many maximal ideals.

2.3. Finite Morphisms Are Closed. A morphism $f\colon X \to Y$ is said to be *closed* if the image $f(Z)$ of any closed subset $Z \subset X$ is closed in Y.

Theorem. *Every finite morphism is closed.*

For the proof we may assume that Y is affine, and $Z = X$. We shall prove that, if $y \notin f(X)$ then there is a function $g \in K[Y]$ such that $g(y) = 1$ and $f(X)$ is contained in the zero-set of g, that is, $K[X]$ is annihilated by $f^*(g)$. Let $A = K[Y]$, $B = K[X]$, and let $\mathfrak{m}$ be the maximal ideal of A corresponding to the point y. By virtue of Hilbert's Nullstellensatz, $y \notin f(X)$ if and only if $f^*(\mathfrak{m})B = B$. Now, since B is a finite A-module, the required assertion follows from what is known as *Nakayama's lemma*:

Lemma. *Let M be a finitely generated A-module, and $\mathfrak{a} \subset A$ an ideal such that $\mathfrak{a}M = M$. Then M is annihilated by some element of $1 + \mathfrak{a}$.*

By induction on the number of generators of M, the proof of the lemma is reduced to the case when M is generated by one element, say m. Then we have $m = am$ for some $a \in \mathfrak{a}$, and hence $(1 - a)m = 0$.

Usually this lemma is applied to the situation in which A is a local ring and $\mathfrak{a} \subset A$ is its maximal ideal. Then $\mathfrak{a}M = M$ (or $M \otimes_A (A/\mathfrak{a}) = 0$) implies $M = 0$.

2.4. Application to Linear Projections. Let $X \subset \mathbb{P}^n$ be a projective variety, and $p \in \mathbb{P}^n$ a point which does not lie on X. Then the linear projection $\pi\colon X \to \mathbb{P}^{n-1}$ with centre p is a finite morphism.

To see that, we blow up $\mathbb{P}^n$ at p and consider the corresponding morphism $\tilde{\pi}\colon \widetilde{\mathbb{P}^n} \to \mathbb{P}^{n-1}$. We may regard X as a subvariety of $\widetilde{\mathbb{P}^n}$, which does not meet the exceptional variety $E \subset \widetilde{\mathbb{P}^n}$. Now, we have already seen in § 1 that $\tilde{\pi}$ is a locally trivial bundle with fibre $\mathbb{P}^1$. We show – more generally – that if $Z \to Y$ is a locally trivial $\mathbb{P}^1$-bundle, E any section of this bundle, and $X \subset Z$ a closed subvariety which does not meet E, then X is finite over Y.

As this assertion is local in Y, we may assume that Y is affine, $Z = \mathbb{P}^1 \times Y$, and $E = \{\infty\} \times Y$. Our variety X lies in $\mathbb{A}^1 \times Y = (\mathbb{P}^1 \times Y) - E$, where it is defined by some ideal $I \subset A[T]$, where $A = K[Y]$. Given a polynomial $a_0T^n + \ldots + a_n$ in $A[T]$, we say that a_0 is its leading coefficient. We now form the ideal $I_0 \subset A$ of all leading coefficients of the polynomials in I. It is easy to see that the variety $V(I_0)$ is $X \cap (\{\infty\} \times Y)$, which is empty, so that $I_0 = A$. This means that I contains a polynomial of the form $T^n + \ldots + a_n$. Therefore the A-module $K[X] = A[T]/I$ is generated by the elements $1, T, \ldots, T^{n-1}$, so it is finite over A. This proves that X is finite over Y.

2.5. Normalization Theorems. By the theorem of Sect. 2.3, the set $\pi(X)$ is closed in $\mathbb{P}^{n-1}$. If it differs from $\mathbb{P}^{n-1}$, we can project it once more into $\mathbb{P}^{n-2}$, and so on. In the end what we obtain is a finite, surjective mapping $X \to \mathbb{P}^m$ for some m. In other words, every projective variety maps in a finite manner onto some projective space. For instance, any projective curve can be looked upon as a finite covering of $\mathbb{P}^1$. More important for us is the affine variant:

Proposition. *For every affine variety, there exists a finite, surjective morphism onto some affine space.*

Indeed, suppose $X \subset \mathbb{A}^n$ is an affine variety. We view $\mathbb{A}^n$ as embedded in $\mathbb{P}^n$ in the standard way and define $\overline{X}$ to be the closure of X in $\mathbb{P}^n$. If $X \neq \mathbb{A}^n$ then $\overline{X}$ does not contain the hyperplane at infinity, $H = \mathbb{P}^n - \mathbb{A}^n$. We choose now the centre of projection in H, but off $\overline{X}$. Then the projection map $\pi\colon X \to \mathbb{P}^{n-1}$ is a finite morphism, and $\pi^{-1}(\mathbb{P}^{n-1} - H) = X$. Hence $X \to \mathbb{P}^{n-1} - H = \mathbb{A}^{n-1}$ is also finite. Its image is closed in $\mathbb{A}^{n-1}$; so the process can be repeated until the image of X coincides with $\mathbb{A}^m$.

This proposition works as a fundamental tool for the local study of algebraic varieties. There is also a useful relative version: given a closed subvariety $X \subset \mathbb{A}^n \times Y$ which dominates Y, there exists a morphism of Y-varieties, $X \to \mathbb{A}^m \times Y$, which is finite and surjective over some nonempty open subset $V \subset Y$. To prove this, we again look at the closure of X in $\mathbb{P}^n \times Y \supset \mathbb{A}^n \times Y$. Moreover, as a centre of projection we select a point $p \in H$ such that $\{p\} \times Y$ is not contained entirely in $\overline{X}$. The resulting projection is finite, though not everywhere, but only over the open set $V = \{y \in Y,\ (p,y) \notin \overline{X}\}$.

2.6. The Constructibility Theorem. From this observation it can be seen in particular that the image $f(X)$ of a dominant morphism $f\colon X \to Y$ contains a nonempty open subset $V \subset Y$. By induction it is easy to derive the *Constructibility Theorem of Chevalley:*

Theorem. *If $f\colon X \to Y$ is a morphism of algebraic varieties then the image $f(X)$ is constructible in Y.*

A set is said to be *constructible* if it can be obtained from the open or closed subsets by means of a finite sequence of operations, each consisting of taking out an intersection, a union or a complement; cf. Fig. 4.

2.7. Normal Varieties. Affine space $\mathbb{A}^n$ has the following important qualitative property. If $X \to \mathbb{A}^n$ is a finite birational morphism then it is an isomorphism. In fact, we will show that $K[X] = K[\mathbb{A}^n]$. In view of the birationality, we can represent every element $r \in K[X]$ by an irreducible fraction f/g, with $f, g \in K[T_1, \ldots, T_n]$. Now, by the finiteness, r satisfies an equation $r^m + a_1 r^{m-1} + \ldots + a_m = 0$, with $a_i \in K[T_1, \ldots, T_n]$, that is, $f^m + a_1 f^{m-1} g + \ldots + a_m g^m = 0$. If h is an irreducible factor of g, it must therefore divide f^m, and hence also f (since the polynomial ring is a unique factorization domain). As this would contradict the irreducibility of f/g, it follows that g is a constant, and $r \in K[\mathbb{A}^n]$.

More generally, an algebraic variety X is said to be *normal* if every finite birational morphism $X' \to X$ is an isomorphism. Normality is a local property. As we shall see later, normal varieties enjoy a number of rather nice properties. For an affine variety X, normality means precisely that the ring $K[X]$ is integrally closed in its field of fractions. (One also says that the ring $K[X]$ is *normal.*) The above argument shows in fact that every unique factorization domain is normal.

The curve C of Example 1 in Chap. 1, Sect. 6 is not normal; and indeed the parametrization $\mathbb{A}^1 \to C$ given there is finite and birational, but it is not an isomorphism. Since the line $\mathbb{A}^1$ is normal, the morphism $\mathbb{A}^1 \to C$ is a normalization of C. A finite birational morphism $X^\nu \to X$ is called a *normalization* if the variety X^ν is normal. Normalization is clearly defined only up to isomorphism; what is more important is that it always exists. The construction of a normalization is based on two facts from commutative algebra. The first one is easy: integral closure commutes with localization. One can therefore normalize the affine charts of X and afterwards glue together the pieces obtained. The second fact is more subtle (cf. Bourbaki [1961–1965], Chap. 5, § 3): let A be an integral K-algebra of finite type with quotient field L, and $L \subset L'$ a finite field extension; then the integral closure of A in L' is finite over A.

2.8. Finite Morphisms Are Open. A morphism $f\colon X \to Y$ is said to be *open* if it carries the open subsets of X into open subsets of Y.

Theorem. *Let $f\colon X \to Y$ be a finite, dominant morphism, and assume Y to be normal. Then f is open.*

By shrinking Y, we may assume that Y and X are affine. Let $\mathcal{D}(g) \subset X$ be a basic neighbourhood of a point $x_0 \in X$, where $g \in K[X]$. We have to show that $f(\mathcal{D}(g))$ contains a neighbourhood of $y_0 = f(x_0)$. Let $P(T) = T^m + a_1T^{m-1} + \ldots + a_m = 0$ be the minimal equation of g over the field $K(Y)$. We assert that the $a_i \in K[Y]$. This depends on the following lemma:

Lemma. *Suppose A is an integral domain and g is integral over A, with minimal polynomial $T^m + a_1T^{m-1} + \ldots + a_m = 0$ over the quotient field of A. Then the coefficients $a_1, \ldots, a_m$ are integral over A.*

(In fact, if $g_1, \ldots, g_m$ are the roots of the minimal polynomial then, like g, they are integral over A. Now, the a_i can be written as symmetric polynomials in $g_1, \ldots, g_m$. Hence they, too, are integral over A.)

Let now Z be the subvariety of $Y \times \mathbb{A}^1$ that is given by the zeros of $P(T)$. The finite morphism $(f, g)\colon X \to Y \times \mathbb{A}^1$ factorizes through Z. In view of the minimality of $P(T)$, the variety X dominates Z, and so – according to Sect. 2.3 – X maps onto Z. Since $g(x_0) \neq 0$, one of the coefficients, a_i (say), is nonzero at the point y_0. Then, for every point $y \in Y$ such that $a_i(y) \neq 0$, the equation $T^m + a_1(y)T^{m-1} + \ldots + a_m(y) = 0$ has a nonzero root. This means that there exists $x \in X$ with $g(x) \neq 0$ and $y = f(x)$. This completes the proof of the theorem.

It is essential that Y should be normal, as can be seen from the normalization of the 'cross' $[T_1T_2 = 0]$. Besides, instead of assuming Y to be normal, we can assume that it is unibranch. A variety Y is said to be *unibranch* if the normalization morphism $Y^{\nu} \to Y$ is a bijection. Then, by the theorem of Sect. 2.3, it is a homeomorphism. For instance, the curve of Fig. 8 b) is unibranch, while that of Fig. 8 a) has two branches at the origin. The theorems of Sections 2.3 and 2.8 hold in a very general algebraic setting and are known as the theorems of Cohen-Seidenberg (the 'going-up' and 'going-down' theorems on prime ideals); cf. Bourbaki [1961–1965], Atiyah-Macdonald [1969] or Zariski-Samuel [1958,1960].

§ 3. Complete Varieties and Proper Morphisms

3.1. Definitions. We begin with some guiding considerations. Over the field $\mathbb{C}$, projective space $\mathbb{P}^n(\mathbb{C})$ differs from affine space $\mathbb{C}^n$ by its being compact in the classical topology. One has the intuitive feeling that – even in the abstract situation – $\mathbb{P}^n$ is 'more compact' than $\mathbb{A}^n$. Can one assign a precise meaning to this? The point is that, in the Zariski topology, every variety is quasi-compact. Hence some other approach is needed. It is based on the concept of completion. An embedding $X \subset \overline{X}$ is called a completion if X is open and dense in $\overline{X}$. Then $\mathbb{A}^n$ has some nontrivial completions, for instance, $\mathbb{A}^n \subset \mathbb{P}^n$, whereas $\mathbb{P}^n$ can no longer be completed, at least not by a separated variety. An even stronger restriction on X is to require that its image by any morphism, or in any algebraic correspondence $Z \subset X \times Y$, should be closed. This finally leads us to the definition.

Definition. A variety X is said to be *complete* if it is separated and if, for every variety Y, the projection $X \times Y \to Y$ is a closed morphism.

At this point we suspect some analogy with the closedness property of finite morphisms. As a matter of fact, there is a more general concept, which includes as special cases the notions of complete variety and of finite morphism. A morphism of varieties $f\colon X \to Y$ is said to be *proper* if it is separated (that is, the diagonal immersion $X \to X \times_Y X$ is closed) and universally closed (that is, for every base extension $Y' \to Y$ the morphism $f'\colon X \times_Y Y' \to Y'$ is closed). We remark that the proper mappings of topological spaces have very similar properties.

3.2. Properties of Complete Varieties. On the whole, we shall lay emphasis on the properties of complete varieties, since those of proper morphisms can be derived rather easily.

a) A closed subvariety of a complete variety is complete. If $X \to Y$ is a proper morphism, and Y a complete variety, then X, too, is complete.

b) A direct product of complete varieties is complete. Properness is preserved under composition and base extension.

c) The image of a complete variety by a regular mapping is complete. In particular, a regular function on a complete, connected variety, is constant. Indeed, its image is a closed, connected subset of $\mathbb{A}^1$; and it is not $\mathbb{A}^1$ (which is not complete); hence it is just one point. Over the field of complex numbers, this result can be obtained from the Maximum Principle.

d) A variety is complete if and only if all its irreducible components are.

e) A variety over $\mathbb{C}$ is complete if and only if it is compact in the classical topology.

A less simple fact is Nagata's theorem:

Theorem. *Every separated variety can be embedded as an open subvariety of a complete variety.*

3.3. Projective Varieties Are Complete. The most important examples of complete varieties are provided by projective varieties.

Theorem. *Every projective variety is complete.*

Corollary. *A regular function on a connected, projective variety is constant.*

In view of property *a*), it is enough to establish the completeness of $\mathbb{P}^n$. After blowing up a point in $\mathbb{P}^n$, we are reduced to proving the completeness of $\widetilde{\mathbb{P}^n}$. Now, this variety is a $\mathbb{P}^1$-bundle over $\mathbb{P}^{n-1}$. Hence it all comes down to checking that $\mathbb{P}^1$ is complete.

Let Z be a closed subset of $\mathbb{P}^1 \times Y$. We have to show that $p(Z)$ is closed, where p denotes projection onto Y. On replacing Y by $\overline{p(Z)}$, we may assume that Z dominates Y. Now, the intersection of Z with the section $\{\infty\} \times Y$ is of the form $\{\infty\} \times F$, where F is closed in Y. On replacing Y by $Y - F$, we may further assume that Z does not meet the section $\{\infty\} \times Y$. Then the morphism $Z \to Y$ is finite (cf. Sect. 2.4), and so $p(Z) = Y$ (by the theorem of Sect. 2.3). This completes the proof of the theorem.

We have used here the properties of finite morphisms coming from § 2. Another method of proving that $\mathbb{P}^n$ is complete, is by using the Hilbert Nullstellensatz (see Shafarevich [1972] or Mumford [1976]), or the theory of resultants. One can show in much the same way that projective morphisms are proper. In particular, the blowings-up of Sect. 1.7 are proper morphisms.

3.4. Example of a Complete Nonprojective Variety. Although the majority of complete varieties that are encountered in practice are projective, there exist also some nonprojective ones. We give the simplest example of such a variety, leaving out the details. We start from $\mathbb{P}^1 \times \mathbb{P}^1$, on which we blow up the point $(0,0)$. We thus obtain a ruled surface $\varphi\colon Y \to \mathbb{P}^1$ with one degenerate fibre above 0. This fibre is made up of two components, F and G, which are isomorphic to $\mathbb{P}^1$ (see Fig. 12).

We now take one more copy of this surface, say, $\varphi'\colon Y' \to \mathbb{P}^1$, with degenerate fibre ${\varphi'}^{-1}(0) = F' \cup G'$, and glue Y and Y' together by identifying the

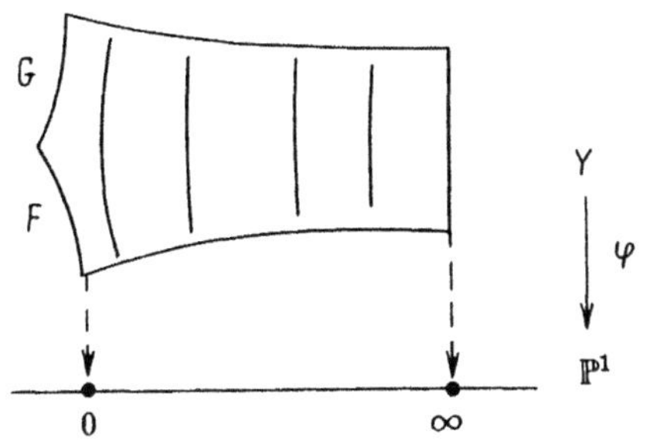

Fig. 12

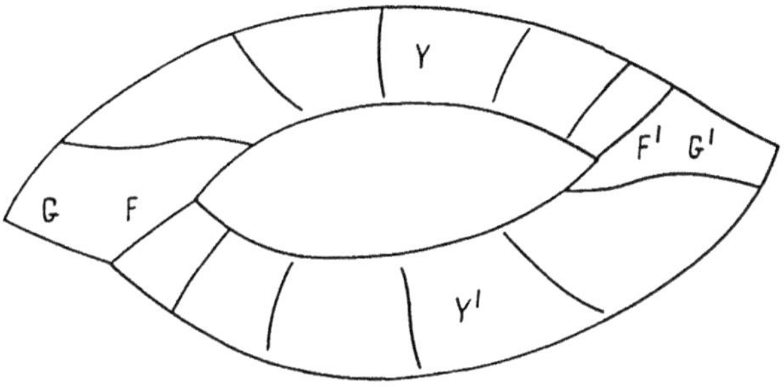

Fig. 13

curve F with the fibre ${\varphi'}^{-1}(\infty)$, and F' with $\varphi^{-1}(\infty)$. The resulting surface X is drawn in outline in Fig. 13.

It consists of two components, Y and Y', each of which is projective, so that X is complete. However, it is impossible to embed X in $\mathbb{P}^n$. Otherwise, we could take a hyperplane H in $\mathbb{P}^n$ meeting F, G, F', and G' transversally in, respectively, ν, μ, ν', and μ' points. As the fibre $\varphi^{-1}(0) = F + G$ is 'continuously deformable' into the fibre $\varphi^{-1}(\infty) = F'$, we see that $\nu + \mu = \nu'$. Similarly, $\nu' + \mu' = \nu$, whence $\mu + \mu' = 0$. As a consequence, the hyperplane H does not intersect the curves G and G', which therefore lie in $\mathbb{P}^n - H$. But $\mathbb{P}^n - H = \mathbb{A}^n$, being an affine variety, cannot contain any complete curves.

We remark that the surface X has some singularities along the curves F and F'. This is not purely casual: it can be shown that every smooth, complete surface is projective, just like any complete curve. In dimension 3 there exist some smooth, complete nonprojective varieties. It is interesting to note the following. Suppose we identify the curve F on Y with the fibre $\varphi^{-1}(\infty)$, for instance in the category of analytic spaces. What we obtain is an object which is not even an algebraic variety, but only an algebraic space in the sense of Artin and Moĭshezon (cf. Sect. 2.9 of Chap. 1).

Nevertheless, complete varieties are close to projective varieties. This is evidenced by *Chow's Lemma:*

Theorem. *A complete variety is the image of a projective variety under a proper birational morphism.*

Thus the surface X considered above is the image of the disjoint union of the two surfaces Y and Y'. This theorem of Chow allows us to reduce many questions about complete varieties to projective ones.

3.5. The Finiteness Theorem. A natural topic to consider in connection with completeness, is Serre's *Finiteness Theorem.*

Theorem. *Let F be a coherent sheaf on a complete variety X. Then the space $H^0(X, F) = F(X)$ of global sections of F is finite-dimensional.*

For the sake of brevity, we shall denote by $\mathcal{K}$ the class of coherent sheaves whose space of global sections is finite-dimensional. We also make the following trivial observation: if G is a subsheaf of F, and both G and F/G belong to $\mathcal{K}$, then $F \in \mathcal{K}$.

By induction we may suppose the theorem true for any subvariety of X which is not X itself. It follows readily that the theorem holds for any sheaf whose support is different from X. Indeed, let Y be the support of F, and $I = I(Y)$ the ideal sheaf of Y in X. Then the descending chain of sheaves $F \supset IF \supset I^2F \supset \ldots$ breaks off at some point, for a local section of F is equal to 0 outside Y and it is annihilated by I^r if r is large enough. Now, the quotient sheaves $I^kF/I^{k+1}F$ are annihilated by I. Hence they can be regarded as coherent sheaves on Y. By induction, these sheaves prove to be in $\mathcal{K}$, so that also $F \in \mathcal{K}$.

As for sheaves whose support is X, it is easy to produce one such sheaf which is in $\mathcal{K}$, namely, $F = \mathcal{O}_X$. In fact, the sections of $\mathcal{O}_X$ are the regular functions, and these are constant on each connected component of X (see property c) in Sect. 3.2). It follows that $H^0(X, \mathcal{O}_X) \simeq K^{\pi_0(X)}$, where $\pi_0(X)$ denotes – as usual – the set of connected components of X. It turns out that this is already sufficient for the proof of the theorem, because every sheaf differs from a free sheaf, say $\mathcal{O}_X^m$, by a sheaf with smaller support. Indeed, F is isomorphic to $\mathcal{O}_X^m$ over some open subset $U \subset X$. Now, even though the isomorphism cannot always be extended to an inclusion of $\mathcal{O}_X^m$ in F, it is always possible to find a coherent subsheaf $G \subset \mathcal{O}_X^m$ which coincides with $\mathcal{O}_X^m$ on U and which injects into F. Clearly, $G \in \mathcal{K}$. Now F/G is concentrated outside U, and hence $F/G \in \mathcal{K}$. Therefore $F \in \mathcal{K}$, as required.

It is worth adding two complements. Firstly, the finiteness theorem is true not only for the global sections of a coherent sheaf, but also for any cohomology group. Secondly, it can be generalized to any proper morphism $f\colon X \to Y$, in which case it asserts that f_*F, and also the higher direct image sheaves R^qf_*F, are coherent.

3.6. The Connectedness Theorem. The next theorem is the basis for many future connectedness assertions.

Theorem. *Suppose $f\colon X \to Y$ is a proper morphism and $f_*\mathcal{O}_X = \mathcal{O}_Y$. Then the fibres of f are connected.*

Indeed, suppose the fibre $f^{-1}(y)$ over some point $y \in Y$ is not connected, but splits into two disjoint closed sets, A and B. Consider on $f^{-1}(y)$ the function s_0, which is equal to 0 on A, and to 1 on B. And suppose we have managed to continue it to a regular function s on some neighbourhood of the fibre $f^{-1}(y)$. Then, by our assumption, s is of the form $f^*(s')$, where s' is a regular function in a neighbourhood of y. But such a function cannot take on different values in the fibre; a contradiction.

In order to verify that s_0 can be continued to a function s, we shall use a fundamental result of Grothendieck's. The assertion, in our special case, is as follows: a necessary and sufficient condition for s_0 to have a continuation on a Zariski neighbourhood of the fibre $f^{-1}(y)$, is that it can be continued on any infinitesimal neighbourhood of that fibre. Here is what this means. Let I be the ideal sheaf of the subscheme $f^{-1}(y)$ in X; one requires the existence (for each k) of a section s_k of the sheaf $\mathcal{O}_X/I^{k+1}$; moreover, these sections are to be compatible with the homomorphisms $\mathcal{O}_X/I^m \to \mathcal{O}_X/I^n$ $(m \geq n)$. Since, for our function s_0, the condition of infinitesimal continuation is evidently fulfilled, its continuation s exists. Unfortunately, we are deprived of the possibility of saying more about the proof of Grothendieck's theorem, which relies in an essential way on cohomology. So we refer to Grothendieck-Dieudonné [1961–1963] and Hartshorne [1977], or to the survey on cohomology.

3.7. The Stein Factorization. Let $f\colon X \to Y$ be a proper morphism. The sheaf of $\mathcal{O}_Y$-algebras $f_*\mathcal{O}_X$ is coherent, whence the Y-scheme

$$Y' = \operatorname{Specm}_Y(f_*\mathcal{O}_X) \xrightarrow{g} Y$$

is finite over Y. It is clear that f factorizes as

$$X \xrightarrow{f'} Y' \xrightarrow{g} Y$$

and that $f'_*\mathcal{O}_X = \mathcal{O}_{Y'}$. By the theorem of Sect. 3.6, the fibres of f' are connected and nonempty. This decomposition of a proper morphism is called the *Stein factorization:* a fibre $g^{-1}(y)$ is made up of the connected components of the fibre $f^{-1}(y)$.

A corollary of the foregoing results is *Zariski's Connectedness Theorem,* proved by him without any recourse to cohomology:

Theorem. *Let $f\colon X \to Y$ be a proper, dominant morphism. Assume that Y is normal and that the fibres of f over some nonempty open subset $U \subset Y$ are connected. Then all fibres of f are connected.*

Using the Stein factorization, we may assume further that f is finite and birational. Then the assertion of the theorem is deduced from the definition of normality.

As in Sect. 2.8, one can replace normality by the condition that Y be unibranch. This theorem constitutes the modern formulation of the classical Principle of Enriques, to the effect that – under continuous variation – the specialization of a connected variety should be connected (cf. Baldassarri [1956]).

§ 4. Dimension Theory

4.1. Combinatorial Definition of Dimension. The dimension of a variety is a first, rough numerical invariant of the variety. Evident though it may be either in the differentiable or in the analytic case, this notion is less trivial for algebraic varieties. It is of course natural to regard the space $\mathbb{A}^n$ as being n-dimensional, but one must also assign a dimension to every closed subset $X \subset \mathbb{A}^n$. The easiest way to achieve this is through the transcendence degree of the field $K(X)$ over K. However, this method – with its appeal to algebra – conceals the geometric significance of what goes on. That is why we shall start from the combinatorial definition by means of chains of subobjects. In this style, the dimension of a vector space is the length of a maximal flag of subspaces.

The *dimension*, $\dim_x X$, *of a variety* X *at a point* $x \in X$ is defined as the greatest length r of a chain $\{x\} = X_0 \subset X_1 \subset \ldots \subset X_r$ of distinct irreducible closed subsets of X. Clearly, $\dim_x X$ depends only on the local structure of X in a neighbourhood of x. Moreover, $\dim_x X$ is the maximum of the dimensions of the irreducible components of X passing through x. This explains why one frequently limits oneself to irreducible affine varieties when studying dimension.

From this definition it is visible that the dimension of $\mathbb{A}^n$ is the same at all points, since $\mathbb{A}^n$ is homogeneous, and that it is $\geq n$; but it is not at all clear why it is also $\leq n$.

4.2. Dimension and Finite Morphisms

Proposition. *Let $f\colon X \to Y$ be a finite morphism of irreducible varieties, $x \in X$, and $y = f(x)$. Then $\dim_x X \leq \dim_y Y$, and equality holds if f is dominant.*

We begin by showing that, for an irreducible subvariety $X' \subset X$ which is properly contained in X, one has $f(X') \neq Y$. The assertion $\dim X \leq \dim Y$ follows by induction. We shall assume Y (and hence also X and X') to be affine. Let g be a nonzero function on X that vanishes on X'. As X is finite over Y, there is an equation of integral dependence $g^n + a_1 g^{n-1} + \ldots + a_n = 0$, with $a_i \in K[Y]$. Now, if we had $f(X') = Y$, there would lie a point $x' \in X'$ above every point $y' \in Y$. This would imply that $a_n = 0$. On dividing by g, the equation would reduce to one of smaller degree, until we got $g = 0$; a contradiction.

Suppose now that f is a dominant morphism. On projecting Y onto some affine space (cf. Sect. 2.5), we may assume that $Y = \mathbb{A}^n$. Let Y' be an irreducible subset of Y that passes through y. By the theorem of Sect. 2.8, if X' is any irreducible component of $f^{-1}(Y')$ that goes through x, then X' dominates Y'. The assertion $\dim_x X = \dim_y Y$ follows by induction.

As a corollary, we obtain that the dimension of an irreducible variety X is the same at every point. (It is denoted by $\dim X$.) In fact, we may assume that X is affine, and we can project it finitely onto $\mathbb{A}^n$.

Corollary. $\dim \mathbb{A}^n = n$.

We need to check the inequality $\dim \mathbb{A}^n \leq n$. By induction we may assume this has been proved for $\mathbb{A}^k$, provided that $k < n$. Let $Y \subset \mathbb{A}^n$ be an irreducible subvariety which is properly contained in $\mathbb{A}^n$. A suitable linear projection yields a finite morphism $Y \to \mathbb{A}^{n-1}$. Hence we get $\dim Y \leq n-1$ and $\dim \mathbb{A}^n \leq n$.

Corollary. *The dimension of an irreducible variety X is equal to the transcendence degree of the field $K(X)$ over K.*

Corollary. $\dim(X \times Y) = \dim X + \dim Y$.

4.3. Dimension of a Hypersurface. Let $Y = V(f) \subset \mathbb{A}^n$ be a hypersurface. We claim that every irreducible component of Y has dimension $n-1$. By considering the factors of f, we may assume that this polynomial is irreducible. Let $\mathbb{A}^n \to \mathbb{A}^{n-1}$ be a linear projection whose restriction to Y is finite. Then $Y \to \mathbb{A}^{n-1}$ is evidently dominant, and hence $\dim Y = n-1$.

A similar assertion holds for any variety. We define a *hypersurface* in a variety X to be the set of zeros $V(f)$ of a regular function $f\colon X \to K$. When going over to a hypersurface, the dimension can only decrease, but at most by 1.

Theorem. *Let Y be a hypersurface in X. Then, for every point $y \in Y$, we have:* $\dim_y Y \geq \dim_y X - 1$.

We can assume that X and Y are irreducible and affine. Let $\pi\colon X \to \mathbb{A}^n$ be a finite surjective morphism (see Sect. 2.5) and $Y = V(f)$. Write $\varphi = (\pi, f)\colon X \to \mathbb{A}^n \times \mathbb{A}^1$. As we saw in Sect. 2.8, the image $\varphi(X)$ is given in $\mathbb{A}^n \times \mathbb{A}^1$ by the minimal polynomial of f, say $T^m + a_1 T^{m-1} + \ldots + a_m$, with $a_i \in K[T_1, \ldots, T_n]$. Then Y is the inverse image of the hypersurface $[T = 0]$. Hence $\dim Y \geq \dim \varphi(X) \cap [T = 0]$. Now, the latter set is defined in $\mathbb{A}^n$ be the equation $a_m = 0$. By what we have just seen, its dimension is $\geq n-1$.

One deduces from this by induction that, if a subvariety $Y \subset X$ is given by n equations, then we have $\dim_y Y \geq \dim_y X - n$. We will condense the results of the theory of dimension in the next assertion, which generalizes the classical *Principle of counting constants.*

4.4. Theorem on the Dimension of the Fibres. Let $f\colon X \to Y$ be a dominant morphism of irreducible varieties. Intuitively we anticipate that the fibres of f are of dimension $\dim X - \dim Y$. Perhaps not all of them (cf. the blowing-up mapping of Sect. 1.6), but at least the typical ones. And this is true!

Theorem. *Let $f\colon X \to Y$ be as above. Then*
a) $\dim_x f^{-1}(f(x)) \geq \dim X - \dim Y$ *for every point $x \in X$;*
b) *there exists a nonempty open subset $V \subset Y$ such that, for every point $y \in V$, we have* $\dim f^{-1}(y) = \dim X - \dim Y$.

For the proof of a), we may assume that Y is affine. By projecting Y finitely onto $\mathbb{A}^n$, we reduce a) to a repeat of the theorem of Sect. 4.3. For the proof of b), we can suppose that X, too, is affine. Now, using the relative version of the normalization theorem of Sect. 2.5 (and shrinking Y if necessary), we obtain a factorization $X \xrightarrow{g} \mathbb{A}^d \times Y \to Y$ of the morphism f, where g is finite and surjective. From this we see that the fibres of f have dimension $d = \dim X - \dim Y$.

4.5. The Semi-continuity Theorem of Chevalley

Theorem. *For every morphism $f\colon X \to Y$, and every integer k, the set* $X_k = \{x \in X,\ \dim_x f^{-1}(f(x)) \geq k\}$ *is closed in X.*

For the proof we may assume that X and Y are irreducible, and f dominant. Let $d = \dim X - \dim Y$. If $k \leq d$ then, by a) of the theorem of Sect. 4.4, we have $X_k = X$. Suppose now that $k > d$, and let $V \subset Y$ be as in b) of the same theorem. If we write $Y' = Y - V$ then $X_k \subset f^{-1}(Y')$. That X_k is closed thus follows by induction from the analogous assertion for the morphism $f^{-1}(Y') \to Y'$.

We mention three useful corollaries.

1. The set of points $x \in X$ that are isolated in their fibre $f^{-1}(f(x))$ (that is, those where the morphism f is quasi-finite), is open in X.

2. If f is proper then the sets $Y_k = \{y \in Y,\ \dim f^{-1}(y) \geq k\}$ are closed in Y. Indeed, $Y_k = f(X_k)$.

3. Let $f\colon X \to Y$ be a proper morphism, and assume all its fibres are irreducible and have the same dimension. If, in addition, Y is irreducible then so is X.

We point up two more consequences of the theory of dimension.

4.6. Dimension of Intersections in Affine Space. Suppose X and X' are subvarieties of $\mathbb{A}^n$, and let $x \in X \cap X'$. Then

$$\dim_x(X \cap X') \geq \dim_x X + \dim_x X' - n.$$

By virtue of reduction to the diagonal, $X \cap X'$ is isomorphic to the intersection of $X \times X'$ with the diagonal Δ in $\mathbb{A}^n \times \mathbb{A}^n$. Now, the diagonal is defined by the n equations $T_i = T_i'$, $i = 1, \ldots, n$. So the inequality follows from Sect. 4.3 and the formula on the dimension of $X \times X'$.

Corollary. *Let $X, X' \subset \mathbb{P}^n$, and assume that $\dim X + \dim X' \geq n$. Then $X \cap X' \neq \varnothing$.*

Indeed, write $X = \mathbb{P}(C)$ and $X' = \mathbb{P}(C')$, where C and C' are cones in K^{n+1} (cf. Chap. 1, Sect. 3.5). C and C' certainly meet at 0, and hence

$$\begin{aligned}\dim_0(C \cap C') &\geq \dim_0 C + \dim_0 C' - (n+1) = \\ &= \dim X + 1 + \dim X' + 1 - (n+1) \geq 1.\end{aligned}$$

Thus $C \cap C' \neq \{0\}$ and $X \cap X' \neq \varnothing$.

If, in particular, $\dim X \geq 1$ then X meets any hyperplane $H \subset \mathbb{P}^n$. We know this from another line of thought: $\mathbb{P}^n - H \simeq \mathbb{A}^n$ is affine, which implies that every complete subvariety of it is zero-dimensional. Here is another useful special case: for $k < n$ a system of k homogeneous equations $f_1 = \ldots = f_k = 0$ in n unknowns always has a nonzero solution.

4.7. The Generic Smoothness Theorem. In Chapter 1 a variety X was defined to be smooth at a point x if its tangent cone $C_x X$ coincided with the tangent space $T_x X$. This definition is equivalent to a more familiar one, which is based on the notion of dimension:

Proposition. *A point x on a variety X is smooth if and only if we have $\dim T_x X = \dim_x X$.*

This proposition derives trivially from a more general equality, namely, $\dim C_x X = \dim_x X$. To establish it, we consider the blowing-up $\sigma\colon \widetilde{X} \to X$ at the point x. Now, $\sigma^{-1}(x)$ is defined locally in $\widetilde{X}$ by one equation (cf. Sect. 1.8), so $\dim \sigma^{-1}(x) = \dim \widetilde{X} - 1$. Furthermore, $\sigma^{-1}(x) = \mathbb{P}(C_x X)$, whence $\dim C_x X = \dim \sigma^{-1}(x) + 1 = \dim \widetilde{X} = \dim X$.

Theorem. *The set of smooth points on any algebraic variety is open and dense.*

In other words, a 'general' point of a variety is smooth. For the proof we may assume that the variety X is irreducible and affine. Applying the theorem on the dimension of the fibres to $TX \to X$, we see that the set of smooth points is open. It remains to prove that it is nonempty. This is the first place where we truly use the fact that X is a variety, and not an arbitrary algebraic scheme. Suppose $X \subset \mathbb{A}^N$ and, for every point $x \in X$, the dimension of $T_x X$ is greater than $n = \dim X$. By a general linear projection $\mathbb{A}^N \to \mathbb{A}^{n+1}$, we may assume X to be a hypersurface in $\mathbb{A}^{n+1}$. But we get a contradiction with the proposition of Chap. 1, Sect. 7.1.

Corollary. *Every homogeneous variety is smooth.*

§ 5. Unramified and Étale Morphisms

5.1. The Implicit Function Theorem. In the differentiable and analytic situations, an important role is played by the implicit function theorem. Its simplest form is as follows. Suppose f is a function on $\mathbb{C}^n$ with $f(0) = 0$ and $(\partial f/\partial T_n)(0) \neq 0$. Then there exists a neighbourhood of the origin, $U \subset \mathbb{C}^{n-1}$, and a function $g\colon U \to \mathbb{C}$ of corresponding differentiable type, such that $g(0,\dots,0) = 0$ and $f(T_1,\dots,T_{n-1}, g(T_1,\dots,T_{n-1})) = 0$. In more general form, the theorem asserts that a mapping of varieties, $f\colon X \to Y$, that induces an isomorphism $T_xX \to T_{f(x)}Y$ of the tangent spaces, is locally an isomorphism.

There is nothing similar for algebraic varieties with the Zariski topology. Take the simplest function, $f = T^2$ on $\mathbb{A}^1$. At the point 1, the derivative $df/dT = 2T$ is equal to 2. However, no Zariski open subset of $\mathbb{A}^1$ is mapped injectively under f.

As a general rule, it is extremely rare that an isomorphism of the tangent spaces implies a local isomorphism of the varieties: the Zariski neighbourhoods are much too big for that. Nevertheless one feels that this is an important class of morphisms, which deserves to be given a name. They came to be known as étale, which means: spread out. Grothendieck went one step further and proposed to declare that the étale morphisms $U \to X$ are the 'open subsets' of the variety X for the étale topology, thus restoring the validity of the implicit function theorem in the algebraic situation. The étale topology (see the survey on cohomology) is an algebraic substitute for the classical topology. It allows us to talk in the abstract situation about homotopy groups, cohomologies, Betti numbers, the Lefschetz formula, etc.

5.2. Unramified Morphisms. In the above example of the function $f = T^2$, the two points $\pm\sqrt{y}$ of the fibre $f^{-1}(y)$ merge into one, as $y \to 0$. One also says that the function T^2 ramifies above 0. A function is said to be unramified if there is no ramification point. We come now to the precise definition.

Let $f\colon X \to Y$ be a morphism of algebraic varieties, $x \in X$, $y = f(x)$, and let $\mathfrak{m}_x$ and $\mathfrak{m}_y$ be the maximal ideals of the local rings, $\mathcal{O}_{X,x}$ and $\mathcal{O}_{Y,y}$ respectively. It is easy to show that the following assertions are equivalent:

a) $d_xf\colon T_xX \to T_yY$ is injective;
a′) d_xf induces an immersion of C_xX into C_yY;
b) $f^*\colon \Omega^1_Y(y) \to \Omega^1_X(x)$ is surjective;
c) $f^*(\mathfrak{m}_y)$ generates the ideal $\mathfrak{m}_x$ in $\mathcal{O}_{X,x}$;
c′) the fibre $f^{-1}(y)$ coincides, as a scheme, with the point x in a neighbourhood of x.

For instance, the proof of *b*) ⇒ *c*) makes use of Nakayama's lemma.

When these equivalent conditions hold, we say that the morphism f is *unramified at the point* x. For instance, the function $f = T^2$ is unramified at the points other than 0, and it ramifies at the origin (provided, of course, that the characteristic of K is not equal to 2; for in characteristic 2 this function is ramified everywhere). We say that the morphism f is *unramified* if it is unramified at all points of X.

An open immersion is unramified. A closed immersion is also unramified; in addition, it is injective and finite. As a matter of fact, closed immersions are characterized by these three properties.

Proposition. *If a morphism $f\colon X \to Y$ is injective, finite, and unramified, then it is a closed immersion.*

More generally, if f is finite and the fibre $f^{-1}(y)$ coincides with x as a scheme, then f is a closed immersion in some neighbourhood of y. For the proof, we can assume that Y and X are affine, with coordinate rings A and B, respectively. Let $\mathfrak{m}$ be the maximal ideal of the point y. Saying that $f^{-1}(y)$ and x coincide as schemes, means that $B/f^*(\mathfrak{m})B \cong A/\mathfrak{m}$. Since B is finite over A, it follows from Nakayama's lemma that $f^*\colon A \to B$ is surjective. (We may have to replace Y by a neighbourhood of y.) Hence $f\colon X \to Y$ is a closed immersion.

5.3. Embedding of Projective Varieties. In § 2 we constructed some finite morphisms $\pi\colon X \to \mathbb{P}^{n-1}$ by projecting $X \subset \mathbb{P}^n$ from a point $p \in \mathbb{P}^n - X$. Using the foregoing criterion, we can now check whether π is an embedding, that is, whether X is isomorphic to its image $\pi(X)$. To do that, it is convenient to regard the tangent spaces to $X \subset \mathbb{P}^n$ as being linear subspaces of $\mathbb{P}^n$. More precisely, we denote by $\overline{T}_xX$ the unique linear variety $L \subset \mathbb{P}^n$ with the property that T_xL coincides with T_xX in $T_x\mathbb{P}^n$. We call $\overline{T}_xX$ the embedded projective tangent space.

We shall deal first with local embeddings. Let $x \in X$ and suppose $n > \max(\dim_x X + 1, \dim T_xX)$. Then we can find a centre of projection, say $p \in \mathbb{P}^n$, whose associated map $\pi\colon X \to \mathbb{P}^{n-1}$ induces a closed embedding of some neighbourhood of the point x. Indeed, if x' runs through $X - \{x\}$, the set of chords $\overline{xx'}$ generates a variety S_x of dimension $\leq \dim X + 1$. Hence $S_x \cup \overline{T}_xX$ is of smaller dimension than $\mathbb{P}^n$. Thus, if the centre of projection p is chosen outside $S_x \cup \overline{T}_xX$ then the fibre $\pi^{-1}(\pi(x)) \cong \overline{px} \cap X$ of the projection map consists of one point, namely x (because $p \notin S_x$), and coincides with it as a scheme (since $p \notin \overline{T}_xX$). It remains only to apply the above criterion.

In particular, given a point x on an arbitrary variety X, one can always embed some neighbourhood of x as a subvariety of $\mathbb{A}^r$, where $r = \max(\dim_x X + 1, \dim T_xX)$. In case x is a smooth point of X then X embeds locally in $\mathbb{A}^{n+1}$, where $n = \dim X$. Thus every irreducible variety of dimension n is birationally equivalent to a hypersurface in $\mathbb{A}^{n+1}$.

We now turn to global embeddings.

Proposition. *A smooth projective variety of dimension n can be embedded in* $\mathbb{P}^{2n+1}$.

Suppose $X \subset \mathbb{P}^N$. We consider two subsets of $\mathbb{P}^N$ which are associated with X. The first one is the *variety of secants,* $\operatorname{Sec} X$; it is the closure of the set of points lying on the secants (or chords) $\overline{xx'}$, where x and x' are distinct points of X. The other subset, $\operatorname{Tan} X$, is the union of the projective tangent spaces $\overline{T}_x X$, for all $x \in X$. Of course, $\dim \operatorname{Sec} X \leq 2n+1$ and $\dim \operatorname{Tan} X \leq 2n$. Thus, for $N > 2n+1$, one can find a point that does not lie in $\operatorname{Sec} X \cup \operatorname{Tan} X$. Again, the projection map $\pi\colon X \to \mathbb{P}^{N-1}$ centred at this point, is an injective, finite, unramified morphism, and hence a closed embedding.

For instance, every smooth curve can be embedded in $\mathbb{P}^3$; every smooth surface, in $\mathbb{P}^5$; etc. If $\dim \operatorname{Sec} X \leq 2n$, one can embed X in $\mathbb{P}^{2n}$, but this happens rather infrequently. One can also project X further, while trying to obtain singularities as simple as possible (cf. Griffiths-Harris [1978], Chap. 4, § 6).

5.4. Étale Morphisms

Definition. A morphism $f\colon X \to Y$ is said to be *étale at a point* $x \in X$ if $d_x f$ induces an isomorphism $C_x X \to C_{f(x)} Y$ of the tangent cones, viewed as schemes.

We may also say that a morphism is étale if it is smooth and unramified.

For instance, an open immersion is étale. On the other hand, if a closed immersion is étale at a point x then it is locally an isomorphism. Dimension, as well as smoothness, are preserved under étale morphisms.

A typical example. Suppose Y is an affine variety, and $X \subset \mathbb{A}^1 \times Y$ is given by the zeros of a polynomial $P \in K[Y][T]$. If a point $x \in X$ is such that the derivative $(dP/dT)(x) \neq 0$, then X is étale over Y at the point x.

As a matter of fact, every étale morphism looks locally exactly as in this example. We shall establish this under the assumption that X is finite over Y. (In view of Sect. 7.2, there is no loss of generality in assuming that.) Suppose X is embedded in $\mathbb{A}^n \times Y$. By choosing an appropriate projection $\mathbb{A}^n \to \mathbb{A}^1$, and using the fact that f is unramified at x, we can consider that X lies in $\mathbb{A}^1 \times Y$.

We shall now look at the fibre over the point $y = f(x)$. As a subscheme of $\mathbb{A}^1 \cong \mathbb{A}^1 \times \{y\}$, it is defined by the zeros of a polynomial, say $T^m + \bar{a}_1 T^{m-1} + \ldots + \bar{a}_m$, with coefficients in $K[Y]/\mathfrak{m}_y$. Hence $1, T, \ldots, T^{m-1}$ generate $K[X]$ modulo $\mathfrak{m}_y$. So, by Nakayama's lemma, they generate $K[X]$ over the ring $K[Y]$. As a result, there holds in $K[X]$ a relation of the form $P(T) = T^m + a_1 T^{m-1} + \ldots + a_m = 0$, where the $a_i \in K[Y]$. This means that X is contained in the variety of zeros of the polynomial P, say $X' \subset \mathbb{A}^1 \times Y$. Now, x is a simple root of $\overline{P}$, since f is unramified at x. It follows that $(dP/dT)(x) \neq 0$, so that X' is étale over Y at x. But then so is the closed embedding $X \subset X'$, which is therefore an isomorphism in a neighbourhood of x.

As a corollary we obtain that the set of points where a morphism $f: X \to Y$ is étale, is open in X. Moreover, one sees that the property of being étale is preserved under base extension.

5.5. Étale Coverings. A finite étale morphism is called an *étale covering.* Over the field of complex numbers, with the classical topology, such morphisms are unramified coverings, that is, locally trivial bundles with finite fibres. In particular, the number of points in a fibre $f^{-1}(y)$, is independent of $y \in Y$. We shall prove that this assertion is also true in the abstract situation.

Theorem of conservation of number. *Suppose $f: X \to Y$ is an étale covering, and Y is connected. Then the number of points in a fibre $f^{-1}(y)$ is independent of $y \in Y$.*

Indeed, arguing as in Sect. 5.4, we see that, locally on Y, the variety X can be defined in $\mathbb{A}^1 \times Y$ by one equation, say $T^m + a_1 T^{m-1} + \ldots + a_m = 0$, where the $a_i \in K[Y]$. Now, since f is étale, all the roots of the equation $T^m + a_1(y)T^{m-1} + \ldots + a_m(y) = 0$ are simple, and there are precisely m of them.

The analogy between étale coverings and the unramified coverings from topology enables one to build a purely algebraic theory of the fundamental group, which is closely related to Galois theory. Referring for details to Grothendieck [1971], we shall content ourselves here with one definition. A connected variety X is said to be *simply connected* if every étale covering $X' \to X$, where X' is connected, is an isomorphism.

For instance, as we shall see later on, $\mathbb{P}^n$ is simply connected.

5.6. The Degree of a Finite Morphism. From the proof of the theorem of conservation of number, one sees that the number of points in each fibre of an étale covering $f: X \to Y$ – which it is natural to call the degree of f – is equal to the dimension of the ring $K(X)$ over the field $K(Y)$. This suggests a more general definition of the degree.

Definition. Let $f: X \to Y$ be a finite dominant morphism. The dimension $[K(X) : K(Y)]$ of the ring of rational functions $K(X)$ over the field $K(Y)$ is called the *degree* of f and is denoted by $\deg f$.

This raises a natural question: what is the relationship between the number of points in the fibres of a finite morphism f and $\deg f$?

Theorem. *Suppose $f: X \to Y$ is a finite dominant morphism, and Y is normal. Then $\left|f^{-1}(y)\right| \leq \deg(f)$ for every point $y \in Y$.*

This is obvious if X is defined in $\mathbb{A}^1 \times Y$ as the set of zeros of a polynomial $T^m + a_1 T^{m-1} + \ldots + a_m \in K[Y][T]$. The general case reduces to that by using some linear projections and the lemma of Sect. 2.8. The normality assumption on Y is crucial, as is shown by the example of the normalization of the plane curve $C = [T_2^2 = T_1^3 + T_1^2]$ (cf. Fig. 16 or Fig. 4).

5.7. The Principle of Conservation of Number. To account for the diminished number of points in some fibre $f^{-1}(y)$, as compared to $\deg f$, one can argue that several points have merged, or agglutinated, so that one needs to take each point in the fibre with some multiplicity – more or less as is done with the roots of a polynomial in one variable. In fact, the number of roots of a polynomial, counting multiplicities, is equal to its degree. Analogously, the *Principle of conservation of number* (or *Principle of continuity*) states that, provided one gives a correct definition of the multiplicity, or of the local degree, $\deg_x(f)$ of a finite morphism f, then the following formula holds:

$$\deg(f) = \sum_{x \in f^{-1}(y)} \deg_x(f).$$

This generalizes the theorem of Sect. 5.5. We put off the discussion of the general case until the next chapter and consider here one important class of finite morphisms.

Definition. A finite morphism $f\colon X \to Y$ is said to be *locally free* if the sheaf $f_*\mathcal{O}_X$ is locally free over $\mathcal{O}_Y$. The *local degree* of such a morphism f at a point x is the number

$$\deg_x(f) = \dim_K(\mathcal{O}_{X,x}/f^*(\mathfrak{m}_y)\mathcal{O}_{X,x}).$$

The Principle of conservation of number holds in this situation. Indeed, upon replacing Y by some neighbourhood of y, we may assume that Y is affine, and that $K[X]$ is a free $K[Y]$-module of rank $d = \deg f$. Hence the $K[Y]/\mathfrak{m}_y$-module $K[X]/\mathfrak{m}_y K[X]$ also has rank (or dimension) d. But

$$K[X]/\mathfrak{m}_y K[X] = \bigoplus_{x \in f^{-1}(y)} \mathcal{O}_{X,x}/\mathfrak{m}_y \mathcal{O}_{X,x},$$

so that $d = \sum_{x \in f^{-1}(y)} \deg_x(f)$.

Example 1. An étale covering is a locally free morphism with the property that all $\deg_x f$ are equal to 1. We thus regain the theorem of Sect. 5.5. Conversely, if f is locally free and $\deg_x(f) = 1$ then f is étale at the point x.

Example 2. Let $f\colon X \to Y$ be a finite dominant morphism, where Y is a smooth curve. Then f is a locally free morphism.

Indeed, let $u \in K[Y]$ be a generator of the maximal ideal of a point $y \in Y$. Let $s_1, \ldots, s_m$ be elements of $K[X]$ which yield a basis of $K[X]$ modulo u. By Nakayama's lemma, we can assume that $s_1, \ldots, s_m$ generate the $K[Y]$-module $K[X]$. We are going to prove that they are independent over $K[Y]$. Suppose $\Sigma a_j s_j = 0$. As the s_j are independent modulo u, all of the a_j

are divisible by it and we have, say, $a_j = ua'_j$. Therefore $u \Sigma a'_j s_j = 0$, whence $\Sigma a'_j s_j = 0$ and the a'_j are divisible by u, etc. In the end we see that the a_j are divisible by an arbitrary power of u, which means that they are equal to 0.

Example 3. We give an example in which our naïve definition of multiplicity does not work. Let V be the plane in $\mathbb{A}^4$ with equations $T_1 = T_3 = 0$, and V' the plane with equations $T_2 = T_4 = 0$. Let $X = V \cup V'$, and $f\colon X \to \mathbb{A}^2$ be the map defined by the two functions $T_1 + T_2$ and $T_3 + T_4$. The restriction of f to either V or V' is an isomorphism, so the degree of f is equal to 2, which is also the cardinality of almost every fibre. Nevertheless, $\deg_0(f) = 3$.

Indeed, X is defined in $\mathbb{A}^4$ by the ideal $(T_1T_2,\ T_1T_4,\ T_3T_2,\ T_3T_4)$, and the subscheme $f^{-1}(0)$ by the ideal $(T_1T_2,\ T_1T_4,\ T_3T_2,\ T_3T_4,\ T_1 + T_2,\ T_3 + T_4)$. Hence the corresponding quotient ring is 3-dimensional, with basis $1, T_1, T_3$.

§ 6. Local Properties of Smooth Varieties

6.1. Smooth Points. In Chapter 1 we defined the notion of a smooth point on a variety X by the condition that $C_xX = T_xX$, where C_xX is the tangent cone, and T_xX the tangent space, to X at the point x. In view of Sect. 4.7, this is equivalent to requiring that the dimension of T_xX (or of its dual vector space $\Omega^1_X(x)$) be equal to $\dim_x X$. Finally, being smooth at x is equivalent to the existence of a neighbourhood of x, say $U \subset X$, on which there is an étale morphism $U \to \mathbb{A}^n$. Indeed, if $u_1, \ldots, u_n$ are functions in a neighbourhood of x with the property that $du_1, \ldots, du_n$ form a basis of $\Omega^1_X(x)$ then $du\colon T_xX \to T_{u(x)}\mathbb{A}^n$ is an isomorphism; so $u\colon U \to \mathbb{A}^n$ is étale in a neighbourhood of x.

It is clear that the set of smooth points on a variety X is open; according to Sect. 4.7, it is everywhere dense. Thus, a typical – or 'general' – point of X is smooth. So we naturally ask ourselves what a variety looks like, in a neighbourhood of a smooth point. Our intuition leads us to expect that it has some resemblance to the affine space T_xX. Of course, as we explained in Sect. 5.1, 'similar' cannot be taken to mean locally isomorphic. It can only mean that a smooth variety has, locally, some of the important qualitative properties of affine space.

6.2. Local Irreducibility. Among all properties of $\mathbb{A}^n$, the simplest is irreducibility. We wish to prove that a smooth point $x \in X$ is contained in some irreducible neighbourhood $U \subset X$. To this end we take two functions, a and b, on X such that $ab = 0$. Suppose $a \in \mathfrak{m}^i$ and $b \in \mathfrak{m}^j$, where $\mathfrak{m}$ is the maximal ideal of x in the ring $K[X]$, and let $\bar{a}, \bar{b}$ be their images in $\mathfrak{m}^i/\mathfrak{m}^{i+1}$, respectively, $\mathfrak{m}^j/\mathfrak{m}^{j+1}$. Seeing that $ab = 0$, we also have $\bar{a}\,\bar{b} = 0$ in the ring $\mathrm{gr}(K[X], \mathfrak{m})$. But, for a smooth point, $\mathrm{gr}(K[X], \mathfrak{m})$ is a polynomial ring and has no zero-divisors. Hence $\bar{a} = 0$ (say) and $a \in \mathfrak{m}^{i+1}$. Since we can repeat this argument, in the end we find that $a \in \mathfrak{m}^\infty = \bigcap_{i \geq 0} \mathfrak{m}^i$. It remains to show that $\mathfrak{m}^\infty = (0)$. Now, the ideal $\mathfrak{m}^\infty$ is of finite type, and $\mathfrak{m} \cdot \mathfrak{m}^\infty = \mathfrak{m}^\infty$. By Nakayama's lemma, it follows that $\mathfrak{m}^\infty = (0)$, as required.

6.3. Factorial Varieties. Affine space has yet a more subtle property. Namely, suppose Y is a subvariety of $\mathbb{A}^n$ of dimension $n-1$. (One also says that Y has *codimension* 1; more generally, the *codimension* of a subvariety Y of X is defined as the number $\dim X - \dim Y$.) Then Y is a hypersurface in $\mathbb{A}^n$. Moreover, the ideal $I(Y)$ in $K[\mathbb{A}^n]$ is principal.

Indeed, we may assume Y to be irreducible. Then there exists an irreducible polynomial f that vanishes on Y. We shall prove that the ideal $I(Y)$ is generated by f. Let g be another polynomial that vanishes on Y. Then some power of g is divisible by f. Since unique factorization holds in polynomial rings, g itself is divisible by f, as we wanted to show.

The analogous property holds for any affine variety X whose coordinate ring $K[X]$ is a unique factorization domain (for further details on these rings, see Bourbaki [1961–1965]). But not by any means for every variety! However, if X is smooth, the property is fulfilled locally at any point x. This means that, for every subvariety $Y \subset X$ of codimension 1, one can find an affine neighbourhood U of x such that the ideal $I(Y \cap U)$ is principal in the ring $K[U]$. When this happens, we shall say that the variety X is *factorial* in a neighbourhood of x.

Theorem. *Every variety is factorial in a neighbourhood of a smooth point.*

For the proof we can assume (cf. Sect. 5.3) that the variety X is embedded in $\mathbb{P}^{n+1}$, where $n = \dim X$. Furthermore, we may suppose that Y is irreducible and passes through x, which is a smooth point of X. Now we pick a point $p \in \mathbb{P}^{n+1}$ in such a way that:

a) $p \notin X$;
b) $p \notin \overline{T}_x X$;
c) the line $\overline{px}$ meets Y only at x.

Clearly, such a point exists. We shall project everything from it onto $\overline{T}_x X$. Going over to an affine chart $\mathbb{A}^{n+1} \subset \mathbb{P}^{n+1}$, we obtain a finite, surjective morphism $\pi\colon X \to \mathbb{A}^n$, which is étale at the point x. Further, Y projects onto a closed subset $\pi(Y) \subset \mathbb{A}^n$ of codimension 1, and this projection is an isomorphism in some neighbourhood of x (cf. Sect. 5.3). Since $\mathbb{A}^n$ is factorial, the variety $\pi(Y)$ is defined by just one irreducible polynomial $f \in K[T_1, \dots, T_n]$. It remains to show that $\pi^*(f) \in K[X]$ defines Y in a neighbourhood of x. Let $Y_1, \dots, Y_k$ be the components of $\pi^{-1}(\pi(Y))$, other than Y. If we can show that the Y_i do not pass through x, then $X - \bigcup_{i=1}^{k} Y_i$ will be a neighbourhood of x in which Y is defined by the function $\pi^*(f)$.

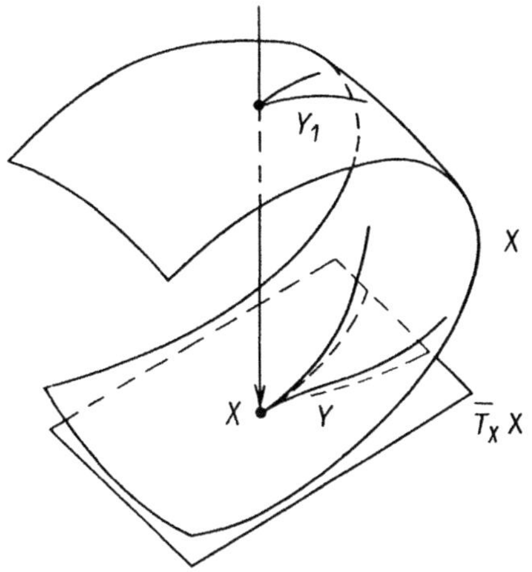

Fig. 14

This is where we use the fact that X is smooth. If $T_1, \ldots, T_n, T$ are coordinates on $\mathbb{A}^{n+1}$, the variety X is given by the zeros of an irreducible polynomial $F = T^m + a_1T^{m-1} + \ldots + a_m$, where the $a_i \in K[T_1, \ldots, T_n]$ and $\partial F/\partial T(x) \neq 0$. We shall suppose that x is the origin; thus $a_m(0) = 0$ and $a_{m-1}(0) \neq 0$. Let now $A = K[T_1, \ldots, T_n]/(f)$ be the ring of functions on $\pi(Y)$. Then $\pi^{-1}(\pi(Y))$ is defined by the zeros of the polynomial $\overline{F} = T^m + \bar{a}_1T^{m-1} + \ldots + \bar{a}_m$, where $\bar{a}_i$ denotes the class of a_i in A. As $Y \to \pi(Y)$ is a local isomorphism, there exist a neighbourhood U of 0 in $\pi(Y)$ and a regular function g on U, such that $(t, g(t)) \in Y$ for every $t \in U$. Then $\overline{F}(g) = 0$, and hence $\overline{F}(T) = \overline{G}(T)(T - g)$, with $\overline{G} \in K[U]$. Now, since $a_{m-1}(0) \neq 0$, we have $\overline{G}(0) \neq 0$; but $\overline{G}$ vanishes on $\bigcup Y_i$. Therefore none of the Y_i passes through $x = 0$.

In particular, we see that a smooth variety is normal (cf. Sect. 2.7).

6.4. Subvarieties of Higher Codimension. Let now $Y \subset X$ be a subvariety of codimension $p > 1$. Is it possible to define Y locally by means of p equations? (In the affirmative, Y is called a *complete intersection.*) Here one can think of Y as being given either set-theoretically, as $Y = V(f_1, \ldots, f_p)$, or as a scheme, that is, $I(Y) = (f_1, \ldots, f_p)$. The latter variant is more precise, and the former can be deduced from it. There are, however, some varieties which are not complete intersections, even set-theoretically. For instance, let $Y \subset \mathbb{A}^4$ be the plane with equations $T_1 = T_3 = 0$, and $Y' \subset \mathbb{A}^4$ the plane with equations $T_2 = T_4 = 0$. Then $Y \cup Y'$ cannot be given in $\mathbb{A}^4$ by two equations (cf. Chap. 3, § 1, and also Hartshorne [1977], Chapter III, Exercise 4.9).

Nevertheless, if the subvariety $Y \subset X$ is smooth then it is locally a complete intersection in the scheme-theoretic sense. This can be derived from the *Jacobian criterion for simplicity:*

Proposition. *A subvariety Y, of codimension p in a smooth variety X, is smooth at a point $x \in Y$ if and only if it can be described, in a neighbourhood of x, as the set of zeros of p functions $f_1, \dots, f_p$, whose differentials $d_x f_i$ are linearly independent.*

To begin with, we shall treat the case of one function f on X with $f(x) = 0$ and $d_x f \neq 0$, and check that the subscheme $Y = f^{-1}(0)$ is smooth at the point x. Indeed, $T_x Y$ is defined as the kernel of the nonzero linear map $d_x f\colon T_x X \to K$, so that its dimension is equal to $\dim T_x X - 1$. On the other hand, $\dim_x Y = \dim_x X - 1$. Hence $\dim T_x Y = \dim_x Y$ and Y is smooth. That Y is a variety in a neighbourhood of x, can be established as in Sect. 6.2. To show in the general case that Y is smooth, one proceeds by induction.

Conversely, suppose Y is smooth at x. The space $\Omega^1_Y(x)$ is obtained from $\Omega^1_X(x)$ via factorization by the subspace generated by the differentials of functions in $I(Y)$ (cf. Chap. 1, Sect. 7.6). Hence we can find p functions, say $f_1, \dots, f_p$, in $I(Y)$ with linearly independent differentials at x. By the above, the variety $Y' = V(f_1, \dots, f_p)$ is smooth at x, irreducible, and of codimension p. Since $Y \subset Y'$ also has codimension p, we conclude that $Y = Y'$.

6.5. Intersections on a Smooth Variety

Proposition. *Let X be a smooth n-dimensional variety, Y and Y' two subvarieties. Then for every point $x \in Y \cap Y'$ we have*

$$\dim_x Y \cap Y' \geq \dim_x Y + \dim_x Y' - n. \qquad (*)$$

As in Sect. 4.6, we invoke reduction to the diagonal. Now, the diagonal Δ in $X \times X$, being a smooth variety, is defined locally by n equations.

If equality holds in the relation $(*)$, one says that the subvarieties Y and Y' *intersect properly* (or: *correctly*) at x. An improper intersection indicates that Y and Y' are in special position with respect to each other (for instance, $Y = Y'$). An even more stringent condition on the position of the subvarieties is transversality. We say that Y and Y' *meet transversally* at a point $x \in X$ if they are smooth at x and the vector subspaces $T_x Y$ and $T_x Y'$ are in general position in $T_x X$. When such is the case, the variety $Y \cap Y'$ is also smooth at x, and its codimension is equal to the sum of the codimensions of Y and Y'.

Note that the assertion $(*)$ does not hold for arbitrary varieties. For let Y and Y' be the two planes in $\mathbb{A}^4$ introduced in Sect. 6.4. Both of them lie in the three-dimensional variety $X \subset \mathbb{A}^4$ defined by the equation $T_1 T_2 = T_3 T_4$. Nonetheless, their intersection is reduced to the origin.

6.6. The Cohen-Macaulay Property. The local ring of a smooth point possesses one further important property, which was discovered by Macaulay in the case of polynomial rings, and by Cohen for arbitrary regular rings (see Serre [1965] or Zariski-Samuel [1958,1960] for further details).

Let $A = \mathcal{O}_{X,x}$ be the local ring of a smooth point $x \in X$. A sequence of elements $a_1, \ldots, a_n$ in the maximal ideal of A is said to be *regular* if, for every $i = 1, \ldots, n$, the element a_i is not a zero-divisor in $A/(a_1, \ldots, a_{i-1})$. (For $i = 1$, this means that a_1 is not a zero-divisor in A.) For such a sequence, we have $\dim_x V(a_1, \ldots, a_i) = \dim_x X - i$. Indeed, a_i does not vanish on any component of $V(a_1, \ldots, a_{i-1})$, and hence the dimension of $V(a_1, \ldots, a_i)$ decreases exactly by one (see Sect. 4.3). One of the remarkable properties of smooth points is that the converse is true as well.

Theorem. *With the preceding assumptions, the following are equivalent:*
a) $\dim_x V(a_1, \ldots, a_n) = \dim_x X - n$;
b) $\dim_x V(a_1, \ldots, a_i) = \dim_x X - i$ *for* $i = 1, \ldots, n$;
c) *the sequence* $(a_1, \ldots, a_n)$ *is regular in* A.

We have already proved that *c)* ⇒ *b)*. The equivalence of *a)* and *b)* follows from the theory of dimension. The main point is to show that *a)* ⇒ *c)*, and for this we may assume that $n = \dim X$. The dimension of the variety $V(a_1, \ldots, a_{n-1})$ is equal to 1. Hence there is a function f with $d_x f \neq 0$ such that $V(a_1, \ldots, a_{n-1}) \cap V(f) = \{x\}$. According to Sect. 6.4, the variety $V(f)$ is smooth and of smaller dimension. By induction, the sequence $a_1, \ldots, a_{n-1}$ is regular in $A/(f)$, that is, $(f, a_1, \ldots, a_{n-1})$ is regular in A. It is fairly easy to verify that – in any noetherian local ring – the regularity of a sequence (a, b) is equivalent to that of (b, a). It follows that also $(a_1, f, \ldots, a_{n-1})$ is regular, and further on until we get that $(a_1, \ldots, a_{n-1}, f)$ is regular.

Let now $\bar{A} = A/(a_1, \ldots, a_{n-1})$; we have to show that multiplication by a_n in $\bar{A}$ is injective. By assumption, $V(a_1, \ldots, a_n) = \{x\}$. Since $f(x) = 0$, it follows that some power of f lies in the ideal $(a_1, \ldots, a_n)$, so that, say, $f^r = g a_n$ modulo $(a_1, \ldots, a_{n-1})$. Now, multiplication by f (and hence also by f^r) is injective in $\bar{A}$. Hence so is multiplication by a_n. The theorem is proved.

The *Cohen-Macaulay property* of the above theorem holds not only for smooth points. For instance, every hypersurface (or scheme-theoretic complete intersection) in a smooth variety has that property. By contrast, the union of two planes $Y \cup Y'$ of the example given in Sect. 6.4 is not Cohen-Macaulay. We may also quote the following fact (cf. Example 2 in Sect. 5.7):

Proposition. *Suppose* $f\colon X \to Y$ *is a finite dominant morphism, and* Y *is a smooth variety. The following are equivalent:*
a) *the morphism* f *is locally free* (cf. Sect. 5.7);
b) *the variety* X *has the Cohen-Macaulay property.*

§ 7. Application to Birational Geometry

7.1. Fundamental Points. We shall examine what can be said about the structure of the set of indeterminacy points of a rational map $f\colon X \dashrightarrow Y$.

To that effect, we shall identify the map f with its graph (cf. Sect. 1.5), which is a closed subvariety $\Gamma \subset X \times Y$, whose projection $p\colon \Gamma \to X$ is a birational morphism. The map f is defined at a point $x \in X$ if p is an isomorphism above some neighbourhood of x. Otherwise, we say that x is a *point of indeterminacy,* or a *fundamental point* of f.

We shall restrict attention to the case where X is normal. There is a marked difference in the structure of the set of fundamental points of f, according as Y is affine or complete. We begin with the former case, in which the statement is reminiscent of Hartogs's Continuation principle of Chap. 1, § 2.

Proposition. *Let X be a normal variety, $F \subset X$ a closed subset of codimension ≥ 2, and Y an affine variety. Then every morphism $f\colon (X - F) \to Y$ extends to a morphism $\bar{f}\colon X \to Y$.*

For the proof we may assume that $Y = \mathbb{A}^1$. We shall regard f as a rational map from X to $\mathbb{P}^1$ and write $\Gamma \subset X \times \mathbb{P}^1$ for its graph. The set $\Gamma \cap (X \times \{\infty\})$ is contained in $F \times \{\infty\}$. Hence its dimension is less than $\dim X - 1$. On the other hand, $X \times \{\infty\}$ is defined locally in $X \times \mathbb{P}^1$ by one equation, so that $\dim(\Gamma \cap (X \times \{\infty\})) \geq \dim \Gamma - 1 = \dim X - 1$. This means that Γ does not meet $X \times \{\infty\}$. Therefore the morphism $\Gamma \to X$ is finite. As it is birational and X is normal, it is an isomorphism.

7.2. Zariski's Main Theorem. The set of indeterminacy points of a map into an affine variety has codimension 1. On the contrary, a map into a complete variety is undefined only in codimension ≥ 2.

Theorem. *Let $f\colon X \dashrightarrow Y$ be a rational map of a normal variety X into a complete variety Y. If x is a fundamental point of f, then its image $f(x)$ has positive dimension.*

Thus a rational map can be pictured as blowing up each point of indeterminacy into a variety of dimension ≥ 1. We shall verify the theorem when Y is projective, and even only for $Y = \mathbb{P}^n$. Suppose the contrary, that $f(x)$ is finite, and let $H \subset \mathbb{P}^n$ be a hyperplane that avoids $f(x)$. The set $p(\Gamma \cap (X \times H))$ is closed and does not contain x. Its complement is a neighbourhood of x over which the morphism $\Gamma \to X$ is finite, and hence an isomorphism. Then f is defined at the point x.

Besides that, as we remember, $f(x)$ is connected (cf. § 3). In reality, Zariski proved a more subtle fact, which Grothendieck rephrased as follows: for every separated, quasi-finite morphism $X \to Y$, there exists a decomposition $X \to X' \to Y$, where $X \to X'$ is an open immersion, and $X' \to Y$ a finite morphism.

Corollary. *With the assumptions of the theorem, the set F of fundamental points of f has codimension ≥ 2 in X.*

Indeed, on the one hand, the theorem implies that $\dim p^{-1}(F) > \dim F$. On the other hand, $p^{-1}(F)$ is a proper subvariety of Γ.

7.3. Behaviour of Differential Forms under Rational Maps. Let $\omega \in H^0(Y, \Omega_Y^p)$ be a regular differential form on Y, and $f\colon X \dashrightarrow Y$ a rational map satisfying the conditions of the theorem of Sect. 7.2. It is a striking fact that the form $f^*(\omega)$ is also regular!

Indeed, let Γ be the graph of f, and p, q the projections of Γ into X and Y, respectively. Then the form $\omega' = q^*(\omega)$ is regular on Γ. By the corollary of Sect. 7.2, p is an isomorphism outside a subset $F \subset X$ of codimension ≥ 2, so that $f^*(\omega) = (p^{-1})^*\omega'$ is regular on $X - F$. And then, by the Continuation principle of Sect. 7.1, it is regular everywhere on X.

As a corollary we get that, for a smooth complete variety X, the dimension of the space $H^0(X, \Omega_X^p)$ of regular differential p-forms is a birational invariant. This applies equally to all sheaves that are built from Ω_X^p by means of covariant tensor operations. There is one especially important case.

Suppose X is a smooth n-dimensional variety. Then the sheaf Ω_X^1 is locally free of rank n. Its n-th exterior power $\Omega_X^n = \Lambda^n \Omega_X^1$, which is an invertible sheaf, is called the *canonical sheaf* on X and is denoted by ω_X. The dimension of $H^0(X, \omega_X)$ is called the *geometric genus* of X and is denoted by $p_g(X)$. As we saw in Chap. 1, § 7, $p_g(\mathbb{P}^n) = 0$. It was also shown there that, for a smooth plane cubic curve, the genus is ≥ 1. It follows from this that it is not rational.

7.4. The Exceptional Variety of a Birational Morphism. Let $p\colon \Gamma \to X$ be a proper birational morphism, where X is normal, and let F be the set of fundamental points of p^{-1}. Then $p^{-1}(F)$ is called the *exceptional subvariety* of p. As was shown in Sect. 7.2, its dimension is greater than that of its image F. In other words, when we have a birational morphism, there is a subvariety which 'contracts' (or 'collapses') into a variety of smaller dimension. If X is smooth, we can make this statement more precise:

Theorem (van der Waerden). *Suppose the variety X is locally factorial (for instance, smooth), and $p\colon \Gamma \to X$ is a birational morphism. Then every component of the exceptional variety of p has codimension 1 in Γ.*

Let $x \in X$ be a fundamental point of p and let $z \in p^{-1}(x)$. The regular functions on Γ can be viewed as rational functions on X. Now, since x is a fundamental point, one can find a regular function u on Γ, with $u(z) = 0$, which is non-regular at x. Let $u = a/b$ be an irreducible representation, where a and b are regular at x. Finally, define $Z \subset \Gamma$ as the set of zeros of the function $p^*(b)$; its codimension is equal to 1. As $p^*(a) = u\,p^*(b)$, the function $p^*(a)$ also vanishes on Z. Therefore $p(Z)$ is included in the zeros of both a and b; so it is of codimension > 1. It follows that the variety Z is exceptional.

Example. If X is not smooth, the exceptional variety can be of codimension > 1. Take X to be the cone in $\mathbb{A}^4$ which is defined by the equation $T_1T_2 = T_3T_4$. This is a normal variety, as can be seen from the fact that the ring $K[X]$ is the intersection of two normal rings, $K[T_1, T_4/T_1, T_3/T_1]$ and $K[T_2, T_4/T_2, T_3/T_2]$, in the field $K(X)$. On the other hand, we may consider the rational function $f = T_1/T_3 = T_4/T_2$ on X, and its graph $p: \Gamma \to X$. If U and V are homogeneous coordinates on $\mathbb{P}^1$, this graph is defined by the equations $UT_3 = VT_1$ and $UT_2 = VT_4$. The projection $p: \Gamma \to X$ is an isomorphism over $X - \{0\}$, whereas the exceptional variety $p^{-1}(0) = \{0\} \times \mathbb{P}^1$ has codimension 2 in Γ.

In particular, the variety X is not factorial.

7.5. Resolution of Singularities. A *resolution of singularities* of a variety X is a proper birational morphism $X' \to X$ with X' smooth. For instance, the morphism $\Gamma \to X$ of the preceding example resolves the singularity of the cone X. This notion brings up two questions, about the existence and the uniqueness of a solution. The former is more fundamental, and it appears to have a positive answer. At any rate, the answer is positive if K is of characteristic 0 (Hironaka's theorem) or if $\dim X \leq 2$ (Abhyankar [1968]).

As to uniqueness, the answer depends on the dimension of X. The desingularization of a curve is uniquely determined up to isomorphism (cf. Sect. 7.2) and is identical with its normalization. If the dimension is ≥ 2, there is not a unique smooth model. In fact, if we blow up a point on a smooth variety, what we get is again a smooth variety.

Proposition. *Let $\sigma: \widetilde{X} \to X$ be the blowing-up of a smooth variety along a smooth subvariety Y. Then $\widetilde{X}$ is smooth.*

Indeed, in this case $C_{Y|X} = N_{Y|X}$ is a locally trivial vector bundle on Y. Hence also the exceptional variety $E = \sigma^{-1}(Y) = \mathbb{P}_Y(N_{Y|X})$ is smooth. On the other hand, E is defined locally by a single equation. It is easy to deduce from this that $\widetilde{X}$ is smooth at the points of E. Now, $\widetilde{X}$ is also smooth at the remaining points, where it is isomorphic to X.

Still, for surfaces, there is a uniquely defined minimal resolution of singularities (cf. Shafarevich [1972], Griffiths-Harris [1978], Hartshorne [1977] or Mumford [1976]). In dimension 3 and higher, this is no longer so. Let again X be the cone in $\mathbb{A}^4$ which is defined by the equation $T_1T_2 = T_3T_4$. The desingularization $p: \Gamma \to X$ (the graph of T_1/T_3) is minimal. But the same can be said of $p': \Gamma' \to X$ (the graph of T_1/T_4), given in $X \times \mathbb{P}^1$ by the equations $U'T_4 = V'T_1$, $U'T_2 = V'T_3$. The two models, Γ and Γ', are not isomorphic as X-varieties. Indeed, the plane $\mathbb{A}^2 \subset \Gamma$ with equations $U = 0$, $V = 1$, and $T_1 = T_4 = 0$ is carried into the surface $\widetilde{\mathbb{A}}^2 \subset \Gamma'$ with equations $U'T_2 = V'T_3$, $T_1 = T_4 = 0$. The picture looks as in Fig. 15.

7.6. A Criterion for Normality. We begin with the following property of a normal variety:

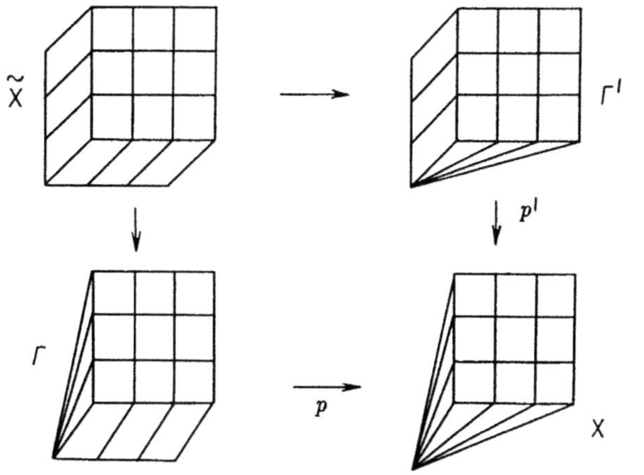

Fig. 15. The top left variety $\widetilde{X}$ is yet another desingularization of X, isomorphic to $\Gamma \times_X \Gamma'$. It is the blow-up of X at the origin.

Proposition. *The set* Sing X *of singular points of a normal variety* X *has codimension* ≥ 2.

By restricting X, we may assume the subvariety Sing X to be smooth. Now, X being singular at all points of Sing X, the latter variety is not defined by one equation (even locally) in X. Hence the blowing-up $\sigma\colon \widetilde{X} \to X$ along Sing X is not an isomorphism. Thus Sing X is fundamental for σ^{-1} and, according to Sect. 7.2, it is of codimension ≥ 2.

Corollary. *A normal curve is smooth.*

Thus we see that a normal variety has two important properties: the Hartogs property (cf. Sect. 7.1) and smoothness in codimension 1. We claim that the converse is also true. Indeed, let f be a rational function on X which is integral over X. Since $X -$ Sing X is normal, f is regular outside Sing X. But the codimension of Sing X is greater than 1. So, by the Hartogs property, f is regular everywhere.

Further, we remark that the Cohen-Macaulay property implies the Hartogs continuation property. Indeed, suppose the codimension of F in X is greater than 1, and a/b is a rational function which is regular outside F. Let g be a function on X such that $V(g) \supset F$ and $V(g) \cap V(b)$ has codimension 2 in X. Since the function a/b is regular on $X - V(g)$, it follows that b divides $g^r a$ for some integer r. In view of the Cohen-Macaulay property, g is not a zero-divisor modulo b. Hence b divides a and the function a/b is regular everywhere.

In particular, if X is a hypersurface in a smooth variety, and Sing X has codimension ≥ 2, then X is normal. This yields another proof that the quadratic cone X of the Example in Sect. 7.5 is normal. One can check by a similar argument that the pair of planes $V \cup V'$ of Example 3 of § 5 is not a complete intersection, even in the set-theoretical sense.

Chapter 3

Geometry on an Algebraic Variety

Geometry is concerned with the properties of geometric figures and their relative positions. The figures of algebraic geometry are the subvarieties (or the *algebraic cycles*) of some fixed algebraic variety, mostly of projective space. Usually these figures admit some continuous variations, and the central theme of this chapter will be the notion of an algebraic family of figures. The simplest of them are the *divisors* (subvarieties of codimension 1), which are defined locally by a single equation. *Linear systems* of divisors provide the most important examples of algebraic families. One can build more complicated figures from simpler ones by means of operations like intersection or union: this is the domain of *intersection theory.*

Continuous variations preserve certain invariants, for example the *degree* of a projective figure. In this case, all figures of the same dimension and of given degree are parametrized by the points of an algebraic variety, known as the *Chow variety.*

§ 1. Linear Sections of a Projective Variety

1.1. External Geometry of a Variety. The simplest projective variety is projective space $\mathbb{P}^n$. The simplest subvarieties of $\mathbb{P}^n$ are the linear subvarieties, that is, those of the form $\mathbb{P}(\lambda)$, where λ is a vector subspace of K^{n+1}. The problems connected with such varieties are elementary and have more to do with linear algebra than with projective geometry. We recall that they are parametrized by the Grassmann varieties.

Let now $X \subset \mathbb{P}^n$ be any projective variety. So far, we have been concerned with the intrinsic properties of X, such as dimension, singular points, etc. From now on we shall be more interested in its 'external geometry', that is, in the properties of X that relate to the embedding $X \subset \mathbb{P}^n$. In other words, we shall consider properties of the pair $(X, \mathbb{P}^n)$, or of the cone $C \subset K^{n+1}$ such that $X = \mathbb{P}(C)$. It is natural to begin our investigation by studying how X interacts with the linear subvarieties of $\mathbb{P}^n$. This can be useful even if we want to examine the intrinsic properties of X. For instance, in the proof of the nonrationality of a smooth cubic hypersurface in $\mathbb{P}^4$, the 2-dimensional family of lines lying on the cubic plays an essential role (cf. Tyurin [1979]).

See also the more elementary example in § 3, where it is shown that a smooth cubic curve in $\mathbb{P}^2$ is not rational.

One uses the intersection of X with 'general', or typical, linear subvarieties to construct some projective invariants, the most important one being the degree. Here, and further on in Chapter 3, a 'general' linear subspace means an element of an open dense subset of the corresponding Grassmannian. The linear subvarieties that are situated in a special way with respect to X (tangent, or secant, or lying on X) also carry some important information and allow us to attach some new varieties to X. We shall discuss here a few general facts about intersections with linear varieties, the general philosophy being that a linear section $X \cap L$ inherits many of the properties of X. It is customary to subdivide statements of this nature into Bertini type theorems about general sections and Lefschetz type theorems on arbitrary sections. For instance, as we saw in Chapter 2, if $\dim L = n - \dim X$ then $X \cap L$ is nonempty; for general L, it is even finite.

1.2. The Universal Linear Section. It is often useful to consider all linear sections at once. To that effect we fix an integer $m \geq 0$ and denote by G the Grassmannian $G(n+1-m, n+1)$ of vector subspaces of codimension m in K^{n+1} (equivalently, of linear subvarieties of codimension m in $\mathbb{P}^n$). There is on G the universal vector (sub-)bundle $S \subset K^{n+1} \times G$ (Chapter 1), and the associated projective subbundle $\mathbb{P}(S) \subset \mathbb{P}^n \times G$. The fibre of $\mathbb{P}(S)$ over a point $\lambda \in G$ (that is, over a vector subspace $\lambda \subset K^{n+1}$) is the linear subspace $\mathbb{P}(\lambda) \subset \mathbb{P}^n$.

Let now $X \subset \mathbb{P}^n$ be a projective variety (usually irreducible). We form the incidence variety IX as the intersection of $X \times G$ with $\mathbb{P}(S)$ in $\mathbb{P}^n \times G$. It consists of the pairs $(x, L) \in X \times G$ such that $x \in L$. Let p and q denote the projections of IX onto X, respectively, G. The bundle $q\colon IX \to G$ will be called the *universal linear section* of X of codimension m. Its fibre over a point $L \in G$ is isomorphic, as a scheme, to $X \cap L$. The other projection, $p\colon IX \to X$, plays an auxiliary role. For $x \in X$, the fibre $p^{-1}(x)$ consists of the linear varieties $L \in G$ that pass through x, and it is isomorphic to $G(n-m, n)$. In particular, the fibres of p are irreducible and of dimension $m(n-m)$.

We shall discuss in more detail the case where $m = r = \dim X$, in which we intersect X with linear spaces of complementary dimension. We denote by U (respectively, U_0) the subset of G consisting of those L which meet X in a finite number of points (respectively, transversally); clearly, $U_0 \subset U$.

Theorem. *The sets U and U_0 are open and dense in G. For every $L \in U$, the number of points $|X \cap L|$ of the set $X \cap L$ does not exceed* $\deg(q)$; *and it is equal to* $\deg(q)$ *for $L \in U_0$.*

Indeed, in this case we have $\dim IX = \dim X + \dim G(n-m, n) = r + r(n-r) = \dim G$. Moreover, as already mentioned, $X \cap L \neq \varnothing$ for any $L \in G$, so that $q\colon IX \to G$ is surjective. That U is open and dense now follows from the theorem of Chap. 2, Sect. 4.5, and the inequality $|X \cap L| \leq \deg q$ from the theorem of Chap. 2, Sect. 5.6. We now move on to transversality. Let $F \subset IX$ consist of those pairs (x, L) for which L is not transversal to X at the point x. This subset is closed by virtue of the theorem on the dimension of the fibres, and nowhere dense since X contains some smooth points. Therefore $q(F)$ is closed in G, and $\dim q(F) \leq \dim F < \dim IX = \dim G$. This shows that $U_0 = G - q(F)$ is open and dense. Since q is étale over U_0, it follows from the theorem of conservation of number that $|X \cap L| = \deg q$ for $L \in U_0$.

One can prove in the same way that a general $L \subset \mathbb{P}^n$ of codimension $m > r = \dim X$ does not meet X. More precisely, in this case the projection map $q\colon IX \to G$ is (birationally) an immersion and $q(IX)$ has codimension $m - r$ in G.

Similarly, if L is a general linear subspace of codimension $m < \dim X$ then L meets X transversally at all smooth points of X. We shall analyse a special case of this situation in further detail.

1.3. Hyperplane Sections. For $m = 1$, $G = \mathbb{P}^*$ is the dual projective space to $\mathbb{P} = \mathbb{P}^n$. Suppose, further, that the r-dimensional variety X is smooth. Then IX is also smooth, with dimension $r + n - 1$, and $q\colon IX \to \mathbb{P}^*$ is a morphism of smooth varieties. As we have already said, the generic fibre of q, which is just the intersection of X with a general hyperplane H, is smooth. Let us examine the degeneracy locus of q.

The points of IX where the morphism q is not smooth are the pairs (x, H) such that the hyperplane H is tangent to X at x, that is, $H \supset \overline{T}_x X$. These points form a set CX, which is called the *conormal set* of X, and its image $q(CX) = X^* \subset \mathbb{P}^*$ is the *dual variety* to X.

This terminology seems to imply that the dual variety of X^* is again X. This is nearly always true, but one must make two qualifications. First of all, X^* usually has some singularities, so that the definitions need to be slightly generalized. And then, even if X^* is smooth, there are cases in which reflexivity fails to hold.

Example 1. Suppose K is of characteristic 2 and let $C = [T_0 T_2 = T_1^2]$, which is a smooth conic in $\mathbb{P}^2$. Then the tangents to C all pass through one point, namely, $P = (0, 1, 0)$ (cf. Example 2 in Chap. 1, Sect. 2.6). Hence C^* is a straight line in $\mathbb{P}^{2^*}$, and $C^{**} = P$.

This example is interesting for one more reason: the map $q\colon IC \to \mathbb{P}^{2^*}$ is a double covering, ramified over the line C^*. Its restriction to $\mathbb{P}^{2^*} - C^*$ yields an unramified covering of $\mathbb{A}^2$; so $\mathbb{A}^2$ is not simply connected!

On the other hand, in characteristic zero the equality $X^{**} = X$ is always satisfied (see Kleiman [1977]).

By looking at the first projection $p\colon CX \to X$, one readily sees that CX is a smooth variety of dimension $n - 1$.

Example 2. If X is a hypersurface in $\mathbb{P}$ then CX and X are isomorphic. The map $q \circ p^{-1}: X \to \mathbb{P}^*$ is called the *Gauss mapping*. If X is given by a form $F(T_0, \ldots, T_n)$, then to a point x the Gauss mapping associates the point $(\partial F/\partial T_0(x), \ldots, \partial F/\partial T_n(x))$.

Considering that CX has smaller dimension than $\mathbb{P}^*$, we expect in general the map $CX \to X^*$ to be a birational equivalence, whence $\dim X^* = n - 1$. We quote several results, referring to Grothendieck, Deligne & Katz [1973] and to Kleiman [1977] for details:

Proposition. a) *The morphism $CX \to \mathbb{P}^*$ is unramified at a point (x, H) if and only if x is a nondegenerate quadratic singularity of $X \cap H$.*

b) *If the morphism $CX \to \mathbb{P}^*$ is not everywhere ramified (so, for instance, if we are in characteristic zero and* $\dim X^* = n - 1$*), then the map $CX \to X^*$ is an isomorphism over X^* –* Sing X^*. *For every point H which is smooth on X^*, the section $X \cap H$ has exactly one (quadratic, nondegenerate) singular point.*

When X is a curve, the morphism $q: IX \to \mathbb{P}^*$ is finite, so that X^* has codimension 1 in $\mathbb{P}^*$. This is a special case of the *Zariski-Nagata theorem on the purity of the branch locus* (cf. Grothendieck [1968]):

Theorem. *Suppose $f: Y \to Z$ is a finite dominant morphism, where Z is smooth and Y normal. Then the branch locus of f has pure codimension* 1 *in Z.*

It is worth comparing this statement with Example 3 of Chap. 2, Sect. 5.7, where ramification occurs in codimension 2.

Example 3. It may happen that X^* has codimension greater than 1. In this case, $H \in X^*$ is tangent to X along a subvariety (in general linear) of positive dimension. The simplest nontrivial example is the image of $\mathbb{P}^1 \times \mathbb{P}^2$ in $\mathbb{P}^5$ under the Segre embedding; it is self-dual.

1.4. The Connectedness Theorem. If the codimension of L is less than $\dim X$ then $X \cap L$ is connected. It is convenient to prove a slightly more general assertion.

Theorem. *Let $f: X \to \mathbb{P}^n$ be a proper morphism, where X is an irreducible variety, and let $L \subset \mathbb{P}^n$ be a linear subvariety of codimension* $< \dim X$. *Then $f^{-1}(L)$ is connected.*

We give a sketch of the argument. By appealing to the Stein factorization, we can suppose that f is finite. Using the Connectedness principle of Enriques and Zariski (cf. Chap. 2, Sect. 3.7), we may also assume L to be general. If now we project $f(X)$ linearly, we may further assume that f is surjective. Finally, we can suppose that $\dim L = 1$. We shall pick a point $p \in \mathbb{P}^n$ and show that, given any line L passing through p, the curve $f^{-1}(L)$ is connected. Let $\pi: \mathbb{P}^n \dashrightarrow \mathbb{P}^{n-1}$ denote projection from p. We have the commutative diagram

$$\begin{array}{ccccc} X & \xleftarrow{\sigma_X} & \widetilde{X} & & \\ \downarrow{\scriptstyle f} & & \downarrow{\scriptstyle \tilde{f}} & \searrow^{\pi_X} & \\ \mathbb{P}^n & \xleftarrow{\sigma} & \widetilde{\mathbb{P}^n} & \xrightarrow{\pi} & \mathbb{P}^{n-1} \end{array}$$

where σ (respectively, σ_X) denotes the blowing-up of $\mathbb{P}^n$ at p (respectively, the blowing-up of X at the points of $f^{-1}(p)$). We must prove that the curves $\pi_X^{-1}(y)$ are connected for $y \in \mathbb{P}^{n-1}$. The trick is that the morphism π_X has some sections. In fact, for a point $x \in f^{-1}(p)$, a section is given by $\sigma_X^{-1}(x) = \{x\} \times \sigma^{-1}(p)$. Thus, if we write the Stein factorization of π_X as $\pi_X : \widetilde{X} \to Z \to \mathbb{P}^{n-1}$, then Z will also have a section over $\mathbb{P}^{n-1}$. On the other hand Z, being the image of the irreducible variety $\widetilde{X}$, is irreducible. Hence $Z \to \mathbb{P}^{n-1}$ is an isomorphism, and the fibres $\pi_X^{-1}(y)$ are connected.

Fulton and Hansen have enunciated the connectedness theorem in a more powerful form:

Theorem. *Let X be a complete irreducible variety, and $f: X \to \mathbb{P}^n \times \mathbb{P}^n$ a morphism with the property that* $\dim f(X) > n$. *Then $f^{-1}(\Delta)$ is connected, where Δ denotes the diagonal in $\mathbb{P}^n \times \mathbb{P}^n$.*

This statement reduces to the preceding one by means of the forthcoming simple but useful construction.

1.5. The Ruled Join. Let V and V' be two vector spaces over K, embedded in $V \times V'$ as $V \times \{0\}$ and $\{0\} \times V'$. Then $\mathbb{P}(V)$ and $\mathbb{P}(V')$ are nonintersecting linear subvarieties of $\mathbb{P}(V \times V')$, and every point that belongs to neither of them lies on a unique line which meets both $\mathbb{P}(V)$ and $\mathbb{P}(V')$.

More generally, if X and X' are varieties in $\mathbb{P}(V)$, respectively $\mathbb{P}(V')$, then we can join the points of X to those of X' by means of straight lines, which span some subvariety of $\mathbb{P}(V \times V')$. This variety, which can be named the *ruled join* of X and X', will be denoted by $X * X'$. For instance, the join of X and a point is the cone with vertex at that point and base X. If now $X = \mathbb{P}(C)$ and $X' = \mathbb{P}(C')$, where C and C' are cones in V, respectively V', then $X * X' = \mathbb{P}(C \times C')$.

As we have already said, $\mathbb{P}(V \times V') - \big(\mathbb{P}(V) \cup \mathbb{P}(V')\big)$ has a natural projection onto $\mathbb{P}(V) \times \mathbb{P}(V')$ with fibre $\mathbb{A}^1 - \{0\}$. Blowing up $\mathbb{P}(V)$ and $\mathbb{P}(V')$ actually yields a morphism

$$\rho: \mathbb{P}(V \times V')^{\sim} \to \mathbb{P}(V) \times \mathbb{P}(V'),$$

with fibre $\mathbb{P}^1$, having two canonical sections.

Let us revert to the Fulton-Hansen theorem. Suppose $V = V' = K^{n+1}$, and let δ be the diagonal in $K^{n+1} \times K^{n+1}$. Then $\mathbb{P}(\delta) \subset \mathbb{P}^{2n+1} = \mathbb{P}^n * \mathbb{P}^n$ projects isomorphically under ρ onto the diagonal $\Delta \subset \mathbb{P}^n \times \mathbb{P}^n$. We form the Cartesian square

$$\begin{array}{ccc} \widetilde{X} & \longrightarrow & X \\ \big\downarrow \tilde{f} & & \big\downarrow f \\ \widetilde{\mathbb{P}}^{2n+1} & \xrightarrow{\ \rho\ } & \mathbb{P}^n \times \mathbb{P}^n \end{array}$$

Since $f^{-1}(\Delta) = \tilde{f}^{-1}(\mathbb{P}(\delta))$, everything follows from the former version of the connectedness theorem.

1.6. Applications of the Connectedness Theorem

Theorem (Bertini). *Let X be an irreducible subvariety of $\mathbb{P}^n$. Then, for a general linear subspace $L \subset \mathbb{P}^n$ of codimension $< \dim X$, the intersection $X \cap L$ is irreducible. Thus, if $\dim X \geq 2$, a general hyperplane section is irreducible.*

We shall verify this in the limiting case in which the codimension of L is equal to $\dim X - 1$. On applying the connectedness theorem to the normalization $f\colon X^\nu \to \mathbb{P}^n$, we obtain that the curve $f^{-1}(L)$ is connected for general L. On the other hand, X^ν is smooth in codimension 1 (§ 7 of Chap. 2). It follows that, for general L, the curve $f^{-1}(L)$ is smooth, and hence irreducible. This means that $X \cap L$, as the image of $f^{-1}(L)$, is also irreducible.

Theorem. *If X, $Y \subset \mathbb{P}^n$ are irreducible and $\dim X + \dim Y > n$, then $X \cap Y$ is connected.*

In fact, $X \cap Y$ is the intersection of $X \times Y \subset \mathbb{P}^n \times \mathbb{P}^n$ with the diagonal.

Theorem. *Let X be an irreducible variety, and $f\colon X \to \mathbb{P}^n$ a finite, unramified morphism. If $2\dim X > n$ then f is a closed immersion.*

Indeed, saying that f is unramified means that the diagonal $\Delta_X \subset X \times X$ is open and closed in $X \times_{\mathbb{P}^n} X = (f \times f)^{-1}(\Delta_{\mathbb{P}^n})$, which is connected, by the Fulton-Hansen theorem. Hence $\Delta_X = X \times_{\mathbb{P}^n} X$, f is injective, and it remains only to apply the criterion of Chap. 2, § 5.

Corollary. *Every subvariety of $\mathbb{P}^n$ with dimension $> n/2$ is simply connected, in particular $\mathbb{P}^n$ itself.*

It is worth noting that $\mathbb{A}^n$ is simply connected only in characteristic zero!

Theorem. *Let X be a normal projective variety of dimension ≥ 2. If a hyperplane section of X is simply connected then so is X.*

Indeed, let $f\colon X' \to X$ be an étale covering, where X' is connected. X being normal, X' is irreducible. In view of the connectedness theorem, $f^{-1}(H)$ is also connected. Hence the degree of X' over X is equal to 1, since by assumption $X \cap H$ is simply connected.

Theorem (F. L. Zak [1981]). *Let $X \subset \mathbb{P}^n$ be an irreducible variety which is not contained in any hyperplane. Let $L \subset \mathbb{P}^n$ be a nontrivial linear subspace which is tangent to X along a subvariety Y (so that $\overline{T}_y X \subset L$ for all $y \in Y$). Then* $\dim Y \leq \dim L - \dim X$.

To prove this, we choose a linear subvariety $M \subset \mathbb{P}^n$ of dimension $n-1-\dim L$ which is skew to L. Let $\pi_M : X \to L$ denote linear projection with centre M. Since X and L have a contact along Y, this projection is unramified at the points of Y. It follows that Y is a connected component of the set $\pi_M^{-1}(Y) = X \cap (Y * M)$. Suppose now that $\dim Y > \dim L - \dim X$. Then $\dim(Y * M) + \dim X > n$, and hence $X \cap (Y * M)$ would be connected, by virtue of the connectedness theorem. Therefore $Y = X \cap (Y * M)$. Now, since M was arbitrary, it would follow that $X = Y \subset L$. A contradiction!

Corollary. *If X is smooth then* $\dim X^* \geq \dim X$.

Corollary. *Let X be a smooth subvariety of $\mathbb{P}^n$. Then every hyperplane section of X is reduced if* $2\dim X > n$, *and normal if* $2\dim X > n+1$.

(Apply the normality criterion of Chap. 2, § 7).

For further applications of the connectedness theorem, see Fulton-Lazarsfeld [1981], a survey which also contains references to the literature concerning properties apt to be inherited by the hyperplane sections.

§ 2. The Degree of a Projective Variety

2.1. Definition of the Degree. The degree of a projective variety $X \subset \mathbb{P}^n$ is, in order of importance, the second numerical character of X (after dimension). It reflects its position in $\mathbb{P}^n$. If $r = \dim X$ (as a rule it is understood that all components of X have dimension r), then the *degree* $\deg X$ is the number of intersection points of X with a general linear variety L of dimension $n-r$ (cf. the theorem of Sect. 1.2). If X meets an $(n-r)$-plane L in more than $\deg X$ points then $X \cap L$ is infinite.

Example 1. Suppose X is a linear variety in $\mathbb{P}^n$. Then $\deg X = 1$. The converse is also true.

Example 2. Let X be a hyperplane, defined by an irreducible homogeneous polynomial $F(T_0, \ldots, T_n)$. Then $\deg X = \deg F$.

Example 3. Let $C \subset \mathbb{P}^n$ be an irreducible curve of degree 2. Then C lies in some plane. Indeed, pick three general points x, y, and z on C and consider the plane $L = \overline{xyz}$ through them. As $L \cap C$ is infinite, we have $C \subset L$. Note that the irreducibility of C is essential, as can be seen from the example of two skew lines in $\mathbb{P}^3$.

Example 4. The degree of the intersection of X with a hyperplane H is $\leq \deg X$, and it is equal to $\deg X$ if H is general.

Example 5. Let $X * Y$ be the ruled join of X and Y. Then

$$\deg X * Y = \deg X \cdot \deg Y.$$

Indeed, let $L = \mathbb{P}(\lambda)$ be a general linear variety which meets X in $\deg X$ points; and similarly, $L' = \mathbb{P}(\lambda')$ for Y. Then the linear variety

$$L * L' = \mathbb{P}(\lambda \times \lambda')$$

meets $X * X'$ in the one-dimensional variety $(X \cap L) * (X' \cap L')$, which is composed of $\deg X \cdot \deg X'$ straight lines. Now everything should be clear.

Over $K = \mathbb{C}$ the degree can be interpreted as a volume; cf. Griffiths-Harris [1978] and Mumford [1976].

2.2. Theorem of Bézout. If all the irreducible components, say $X_1, \ldots, X_s$, of a variety X have the same dimension then $\deg X = \deg X_1 + \ldots + \deg X_s$. This can be thought of as the additiveness of degree under unions. The most famous theorem on the degree, the *Bézout theorem*, states that it is multiplicative under intersections. Of course, we must either suppose that the varieties to be intersected are in transversal position or assign some multiplicities to the intersections. The question of multiplicities will be discussed in the section devoted to general intersection theory. We state now the Bézout theorem in a form that goes back to Fulton and MacPherson (Fulton [1984]).

Theorem. *Let $X_1, \ldots, X_s$ be pure-dimensional varieties in $\mathbb{P}^n$, and let $Z_1, \ldots, Z_t$ be the irreducible components of $X_1 \cap \ldots \cap X_s$. Then*

$$\sum_{j=1}^{t} \deg Z_j \le \prod_{i=1}^{s} \deg X_i.$$

For instance, if the intersection of n hypersurfaces $X_1, \ldots, X_n$ in $\mathbb{P}^n$ is finite, then it consists of at most $d_1 \cdot \ldots \cdot d_n$ points, where $d_i = \deg X_i$. This is precisely what Bézout had established.

For the proof we can assume that the X_i are irreducible. Further, for simplicity, we shall limit ourselves to the intersection of two varieties. We shall make use of the projective variant of reduction to the diagonal. Let δ be the diagonal in $K^{n+1} \times K^{n+1}$. Then the linear variety $\mathbb{P}(\delta)$ in $\mathbb{P}^n * \mathbb{P}^n = \mathbb{P}^{2n+1}$ intersects the ruled join $X_1 * X_2$ precisely in $X_1 \cap X_2$. Now, in view of Example 5 above, we can assume X_2 to be linear. Further, since a linear variety can be expressed as an intersection of hyperplanes, we may assume that X_2 is a hyperplane. Now everything is clear (cf. Example 5).

When the intersections are transversal, one can state more. We shall say that two varieties $X, Y \subset \mathbb{P}^n$ *have simple intersection* (or: intersect with multiplicity 1) if on each component of $X \cap Y$ there is a point where X and Y meet transversally. In this situation, X and Y intersect properly in the sense of dimension theory (cf. Chap. 2, § 6). Now, the foregoing considerations in the case of a simple intersection yield the equality

$$\deg(X \cap Y) = \deg X \cdot \deg Y.$$

Example. Let $C \subset \mathbb{P}^3$ be the normal rational curve of degree 3. It is not a complete intersection, that is, a simple intersection of two surfaces. Indeed, by the Bézout theorem, one of them would have to be a plane and C does not lie in any plane.

Another simple consequence of the Bézout theorem can surprise one only because it did not emerge before:

Corollary. *The automorphisms of $\mathbb{P}^n$ carry hyperplanes into hyperplanes and are induced by the linear automorphisms of K^{n+1}.*

Indeed, let H be a hyperplane in $\mathbb{P}^n$, and L a transversal line. If φ is any automorphism of $\mathbb{P}^n$ then $\varphi(H)$ and $\varphi(L)$ again meet transversally in a unique point, whence $\deg \varphi(H) = 1$ and $\varphi(H)$ is a hyperplane.

2.3. Degree and Codimension. The degree of a variety places some restrictions on the dimension of its linear hull. Suppose, for example, that C is an irreducible curve. We may pick $\deg C + 1$ points on C and draw through them a linear space L of dimension $\leq \deg C$. As C and L intersect in all of these points, C must lie in L. Thus, every irreducible curve lies in a space of dimension $\leq \deg C$. On taking hyperplane sections, it is easy to deduce:

Proposition. *Every irreducible projective variety X lies in a space of dimension $< \dim X + \deg X$.*

For instance, a variety of degree 2 lies in a space of dimension $\dim X + 1$. We have already seen that for conics. We can also say that the codimension of an irreducible variety in its linear hull is less than its degree.

Example 1. Let $C = v_n(\mathbb{P}^1) \subset \mathbb{P}^n$ be the rational normal curve of degree n. It is clear that $\langle C \rangle = \mathbb{P}^n$; so the estimate of the proposition is sharp. On the other hand, consider a nondegenerate curve C of degree n in $\mathbb{P}^n$. (By nondegenerate, we mean that it does not lie in any hyperplane.) Pick $n-1$ points on it, say $x_1, \ldots, x_{n-1}$. Then every hyperplane H passing through $x_1, \ldots, x_{n-1}$ meets C in one more point $x_H \in C$. This sets up an isomorphism of C with $\mathbb{P}^1$. It can be shown that C is of the form $v_n(\mathbb{P}^1)$.

Example 2. Let $S \subset \mathbb{P}^5$ be the Veronese surface, that is, the image of the plane $\mathbb{P}^2$ under the Veronese embedding $v \colon \mathbb{P}^2 \to \mathbb{P}^5$. We may express a general 3-plane in $\mathbb{P}^5$ as the intersection of two hyperplanes, say H and H'. The inverse images of H and H' under v are two conics in $\mathbb{P}^2$, which meet in 4 points. Therefore the degree of S is equal to 4. This is also the least possible degree for a surface in $\mathbb{P}^5$.

The Veronese surface S has one further interesting property: its variety of secants $\operatorname{Sec} S$ has dimension 4, instead of 5, as one would expect. The reason is that v carries every line $l \subset \mathbb{P}^2$ into a curve of degree 2. Hence the linear hull $\langle v(l) \rangle$ is a 2-plane. Now, $\operatorname{Sec} S$ is spanned by the planes $\langle v(l) \rangle$ as l runs through the 2-dimensional family of lines in $\mathbb{P}^2$, and so $\dim \operatorname{Sec} S = 4$. One can associate with S yet another 2-dimensional family of planes, namely, the tangent planes to S. They span a 4-dimensional variety $\operatorname{Tan} S$. Now, since $\operatorname{Tan} S$ is contained in $\operatorname{Sec} S$, these two varieties coincide.

The varieties of least possible degree have rather particular shapes and can be described completely (Semple-Roth [1949], Iskovskikh [1979]). Varieties of small degree and codimension are examined in Hartshorne [1974].

2.4. Degree of a Linear Projection. Varieties of large codimension can be simplified by using some linear projections into a space of smaller dimension. How does the degree behave then? For instance, if $X \subset \mathbb{P}^n$ is not a hypersurface then a general projection preserves the degree. More generally, if X is irreducible and the centre of projection does not lie on X, then the projection map $\pi \colon X \to \pi(X)$ is finite and $\deg X = \deg(\pi) \cdot \deg \pi(X)$.

More interesting things happen when the centre of projection lies on X. Then the image, $\pi(X)$, of the projection must be understood as being $\tilde{\pi}(\widetilde{X})$, where $\widetilde{X}$ is the blow-up of X at the point p. The degree of the image diminishes in this process!

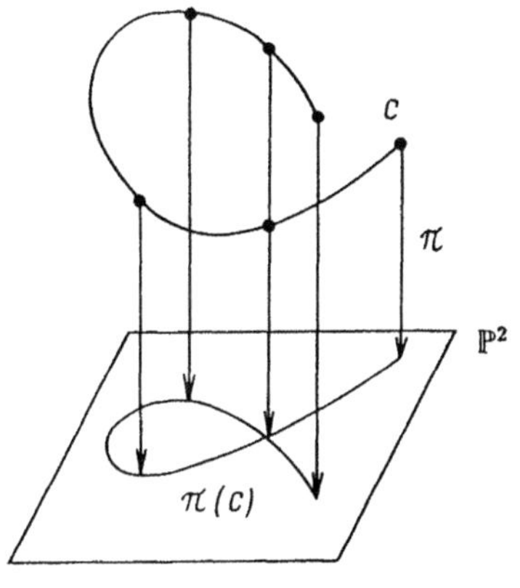

Fig. 16

Example. Let $C \subset \mathbb{P}^3$ be the rational normal cubic curve, given parametrically as $\{(1, t, t^2, t^3)\}$, $t \in \mathbb{P}^1$. If we project it from a point $p \notin C$, its image will be a plane cubic, necessarily singular (cf. Fig. 16), since a smooth cubic is nonrational. However, projecting from the point $p = (1,0,0,0) \in C$ yields the plane conic $\{(1, t, t^2)\}$, $t \in \mathbb{P}^1$.

It is intuitively clear that, if p is a smooth point of X, the degree of the image must go down by one. In the general case, the degree decreases by an integer $\operatorname{mult}_p X > 0$, which is called the *multiplicity* of p on X. Intuitively,

$\operatorname{mult}_p X$ is the intersection multiplicity at p of the variety X with a general linear variety of complementary dimension, passing through p. This can be given a precise meaning by using the universal linear section and the notion of local degree of a finite morphism introduced in Chap. 2, Sect. 5.7. The following holds (for details, see Mumford [1976]):

Proposition. *Let X be an irreducible variety which is not a cone with vertex p, and let π be projection from the point p. Then*

$$\deg X = \operatorname{mult}_p X + \deg(\pi) \cdot \deg \pi(X).$$

2.5. The Hilbert Polynomial. In principle all projective invariants of an embedding $X \subset \mathbb{P}^n$ are determined by the cone $C \subset K^{n+1}$ such that $X = \mathbb{P}(C)$, or by its coordinate ring $R = K[C]$. What then corresponds to the degree of X?

The ring R, being the quotient of the polynomial ring $K[T_0, \dots, T_n]$ by the homogeneous ideal $I(C)$, has a natural structure as a graded ring $R = \bigoplus_{k \geq 0} R_k$. The simplest invariants associated with this structure are the dimensions of the homogeneous components R_k as vector spaces over K. It turns out that, for large k, these numbers behave in a fairly regular way.

Theorem (Hilbert). *There exists a polynomial $P_R \in \mathbb{Q}[T]$, of degree $\leq n$, such that*

$$\dim R_k = P_R(k)$$

for all sufficiently large integers k.

A similar statement holds for any graded $K[T_0, \dots, T_n]$-module of finite type. It is proved fairly simply by induction on n (see Manin [1970], Hartshorne [1977], Mumford [1976], Serre [1965] or Zariski-Samuel [1958, 1960]). This polynomial P_R (or P_X) is called the *Hilbert polynomial* of the graded ring R (or: of the projective variety X).

Example. Let $R = K[T_0, \dots, T_n]$. Then, for $k \geq 0$,

$$\dim R_k = \text{number of monomials of degree } k \text{ in } T_0, \dots, T_n = \binom{k+n}{n}.$$

Hence $P_{\mathbb{P}^n} = \dfrac{1}{n!}(T+n) \cdot \ldots \cdot (T+1) = \dfrac{T^n}{n!} + \dots$.

If $X \subset \mathbb{P}^n$ is a hypersurface given by a homogeneous polynomial F of degree d then, for $k \geq d$,

$$\dim\bigl(K[T_0, \dots, T_n]/(F)\bigr)_k = \binom{k+n}{n} - \binom{k+n-d}{n},$$

whence $P_X = \dfrac{d}{(n-1)!} T^{n-1} + \dots$.

In both cases, the degree of the polynomial P_X is equal to $\dim X$, and $\deg X$ appears in the leading coefficient of P_X. This is not a mere coincidence.

Theorem. *Let $X \subset \mathbb{P}^n$ be a variety of dimension r and degree d. Then* $P_X = \frac{d}{r!}T^r + o(T^r)$.

This is established by passing to a hyperplane section (see, for example, Mumford [1976], § 6B). Then, using the fact that $K[C \times C'] = K[C] \otimes K[C']$, one can once more derive the formula for the degree of the ruled join, and thereby also the Bézout theorem.

2.6. The Arithmetic Genus. Of course, all the coefficients of P_X, not only its leading coefficient, are projective invariants of $X \subset \mathbb{P}^n$ and allow of some geometric interpretation (cf. Baldassarri [1956]). The most important of them is the constant term of P_X, that is, $P_X(0)$. For historical reasons one mostly uses the number

$$p_a(X) = (-1)^{\dim X}(P_X(0) - 1),$$

which is called the *arithmetic genus* of X. It is always an integer and it depends only on X, not on the embedding of X in $\mathbb{P}^n$. This follows from the cohomological interpretation of $P_X(0)$. As a matter of fact, for smooth varieties over a field of characteristic 0, the arithmetic genus is even a birational invariant. (See Baldassarri [1956], Hartshorne [1977] or Mumford [1976].)

For instance, the arithmetic genus of $\mathbb{P}^n$ is equal to 0, like the genus of all hypersurfaces of degree d in $\mathbb{P}^n$ with $d \leq n$. The genus of a plane cubic curve is equal to 1, and hence a smooth cubic curve is not rational, as we already knew.

§ 3. Divisors

3.1. Cartier Divisors. In the preceding two sections we were concerned with geometry in $\mathbb{P}^n$. We shall now deal with varieties in general, starting from their simplest figures, which are given (locally) by one equation.

Suppose X is a smooth variety, and Y a subvariety of codimension 1 in X. As we know from Chap. 2, § 6, Y is defined locally by one equation. That is, there exist an open covering (U_i) of X and, on each U_i, a regular function g_i, such that $Y \cap U_i$ (as a subscheme of U_i) is defined by the equation $g_i = 0$. On the intersections $U_i \cap U_j$ the functions g_i and g_j define the same subscheme. Hence g_i/g_j and g_j/g_i are regular on $U_i \cap U_j$. This brings us to a general definition:

Definition. A *Cartier divisor* on a variety X is a family (U_i, g_i), $i \in I$, where the U_i are open subsets of X which cover X, and the g_i are rational functions on the U_i such that, on each intersection $U_i \cap U_j$, g_i/g_j is regular. The functions g_i are called the *local equations* of the divisor.

More precisely, a Cartier divisor is an equivalence class of such data. Two collections, (U_i, g_i) and (U'_j, g'_j), are equivalent if their union is still a divisor. Cartier divisors can be added, by multiplying their local equations. Thus they form a group, which is denoted by $\mathrm{Div}(X)$.

Example 1. If each local equation g_i is regular on U_i then we say that the divisor $\mathcal{D}$ is *effective,* and we write $\mathcal{D} \geq 0$. The subschemes $[g_i = 0]$ of the U_i can then be glued together into a subscheme of X, which is also denoted by $\mathcal{D}$. Effective Cartier divisors thereby identify with the subschemes of X that are locally given by one equation.

Example 2. A nonzero rational function $f \in K(X)^*$ determines a Cartier divisor (X, f), which is said to be *principal* and is denoted by $\mathrm{div}(f)$. Principal divisors form a subgroup of $\mathrm{Div}(X)$.

3.2. Weil Divisors. The *support* of a Cartier divisor (U_i, g_i), $i \in I$, is the set of points x where the local equations g_i either vanish or become infinite. This closed subset of codimension 1 allows us to get a more geometric picture of the divisor. But one can go further and assign some multiplicities to the various components of the support, so as to reflect the order of the zero or pole of the local equation.

Definition. A *Weil divisor* on X is a finite formal sum $\Sigma\, n_i F_i$, with integer coefficients, where the F_i are irreducible subvarieties of X of codimension 1. The group formed by these divisors is denoted by $\mathcal{Z}(X)$.

Thus, to every Cartier divisor $\mathcal{D}$ we wish to associate a Weil divisor $[\mathcal{D}] = \Sigma\, \mathrm{ord}_F(\mathcal{D}) \cdot F$. To this end we need to define the order, $\mathrm{ord}_F(\mathcal{D})$, of the divisor $\mathcal{D}$ along any irreducible subvariety $F \subset X$ of codimension 1. For X normal, this is done as follows. Let g be a local equation for $\mathcal{D}$ at a general point of F. By shrinking X, we may assume that the ideal $I(F)$ is principal (Chap. 2, § 6), generated by a function u_F. Hence $g = \alpha \cdot u_F^m$, where α is invertible along F. Then we define $\mathrm{ord}_F(\mathcal{D}) = m$. It is easy to verify that this definition makes sense: it does not depend on the choice of a local equation g, or of a neighbourhood of the general point of F, or of a generator u_F of the ideal $I(F)$. The resulting group homomorphism

$$\mathrm{Div}(X) \to \mathcal{Z}(X)$$

is injective (for X normal). Indeed, on a normal variety, a function is regular if it is regular off a subset of codimension ≥ 2 (cf. Chap. 2, § 7). If X is locally factorial (for instance, smooth) then the groups $\mathrm{Div}(X)$ and $\mathcal{Z}(X)$ are canonically isomorphic; so we shall make no distinction between Cartier and Weil divisors. In the general case they differ, but this is not the whole point. In fact, the Cartier divisors are contravariant, while the Weil divisors are covariant. We shall devote a separate section to Weil divisors and their generalization to higher codimensions. Meanwhile, we proceed with the study of Cartier divisors.

3.3. Divisors and Invertible Sheaves. Let $\mathcal{K}_X$ denote the sheaf of rational functions on X; for $U \subset X$ open, $\mathcal{K}_X(U) = K(U)$. To every Cartier divisor $\mathcal{D} = (U_i, g_i)_{i \in I}$ we shall attach a subsheaf $\mathcal{O}_X(\mathcal{D})$ of $\mathcal{K}_X$. On U_i it is defined as $g_i^{-1}\mathcal{O}_{U_i}$. On the intersections, $g_i^{-1}\mathcal{O}_{U_i}$ and $g_j^{-1}\mathcal{O}_{U_j}$ coincide, since g_i/g_j is invertible. Hence these sheaves can be pasted together into a sheaf $\mathcal{O}_X(\mathcal{D}) \subset \mathcal{K}_X$. For instance, $\mathcal{O}_X(0) = \mathcal{O}_X$ and $\mathcal{O}_X(\mathcal{D} + \mathcal{D}') = \mathcal{O}_X(\mathcal{D}) \cdot \mathcal{O}_X(\mathcal{D}')$.

A nonzero section of $\mathcal{O}_X(\mathcal{D})$ is a rational function f on X such that the functions $f \cdot g_i$ are regular on the U_i, that is to say, such that the divisor $\operatorname{div}(f) + \mathcal{D}$ is effective. If $\mathcal{D}$ itself is effective, the sheaf $\mathcal{O}_X(\mathcal{D})$ has a canonical section $s_\mathcal{D}$, which corresponds to the constant function 1. By contrast, the sheaf $\mathcal{O}_X(-\mathcal{D})$, for $\mathcal{D}$ effective, is an ideal sheaf of $\mathcal{O}_X$. The subscheme it defines will also be denoted by $\mathcal{D}$.

The sheaves $\mathcal{O}_X(\mathcal{D})$ are invertible. In fact, multiplication by g_i defines an isomorphism $\mathcal{O}_X(\mathcal{D})|_{U_i} \xrightarrow{\sim} \mathcal{O}_{U_i}$. Further, the addition of divisors corresponds to the tensor product of these invertible sheaves. This yields a homomorphism

$$\delta\colon \operatorname{Div}(X) \to \operatorname{Pic}(X).$$

The kernel of δ consists of those divisors $\mathcal{D}$ whose associated sheaf $\mathcal{O}_X(\mathcal{D})$ is isomorphic to $\mathcal{O}_X$, that is, principal divisors. Further the homomorphism δ is surjective. Indeed, let $\mathcal{L}$ be an invertible sheaf on X, and U an open dense subset such that $\mathcal{L}|_U \xrightarrow{\sim} \mathcal{O}_U$. Then this isomorphism extends to an inclusion $\mathcal{L} \hookrightarrow \mathcal{K}_X$. This fact is specific to algebraic varieties. On complex analytic varieties there can be substantially fewer divisors than invertible sheaves (cf. Shafarevich [1972], Chapter 8).

The sections of invertible sheaves define some divisors. Let $s \in H^0(X, \mathcal{L})$ be a global section of an invertible sheaf $\mathcal{L}$ which is not identically zero on the components of X. After choosing some trivializations $\varphi_i\colon \mathcal{L}|_{U_i} \xrightarrow{\sim} \mathcal{O}_{U_i}$ on a covering (U_i), we obtain an effective divisor $(U_i, \varphi_i(s_i))$, which we denote by $\operatorname{div}(s, \mathcal{L})$. For instance, if $\mathcal{D}$ is effective, the canonical section $s_\mathcal{D}$ of the sheaf $\mathcal{O}_X(\mathcal{D})$ defines $\mathcal{D}$. Thus we have established the possibility to define any effective divisor globally by one equation $s = 0$, keeping in mind that s is not a function, but a section of an invertible sheaf. If s' is another nonzero section of $\mathcal{L}$ then the divisors $\operatorname{div}(s', \mathcal{L})$ and $\operatorname{div}(s, \mathcal{L})$ differ by the divisor of a rational function, namely s'/s. One also says that they are *linearly equivalent*. We shall discuss this in more detail in the next section.

3.4. Functoriality. Let $f\colon X \to Y$ be a morphism of varieties, and $\mathcal{D}$ a divisor on Y. We suppose that no component of $f(X)$ is contained in the support of $\mathcal{D}$. Then there is on X a well determined divisor $f^*(\mathcal{D})$, called the *inverse image* of $\mathcal{D}$ under f, which is defined in the obvious way: if the g_i are local equations for $\mathcal{D}$ on a covering by subsets V_i, then $f^*(\mathcal{D})$ is given by the equations $f^*(g_i)$ (which are defined and nonzero, since $f(X) \not\subset \operatorname{Supp} \mathcal{D}$) on the covering by the $f^{-1}(V_i)$. In particular, if $f\colon X \to Y$ is dominant then $f^*(\mathcal{D})$ is defined for any $\mathcal{D}$ and we get a homomorphism $f^*\colon \operatorname{Div}(Y) \to \operatorname{Div}(X)$, which commutes with the inverse image on the Picard groups Pic. We notice that, on the level of Pic, the inverse image is always defined; this is also one of the advantages of invertible sheaves as opposed to divisors.

3.5. Excision Theorem. The following simple proposition is useful for computing Picard groups.

Proposition. *Suppose Y is closed in X and X is factorial at the points of Y. Then the following sequence is exact:*

$$\mathbb{Z}^k \to \mathrm{Pic}(X) \to \mathrm{Pic}(X - Y) \to 0,$$

where $\mathbb{Z}^k$ is generated by the irreducible components of Y that have codimension 1 in X.

Example 1. The Picard group of $\mathbb{A}^n$ is equal to 0. This is just a reformulation of the fact that the polynomial ring $K[T_1, \ldots, T_n]$ is factorial. Later we shall give a more geometric explanation.

Example 2. Let us determine the Picard group of projective space $\mathbb{P} = \mathbb{P}(V)$. Let H be a hyperplane in $\mathbb{P}$. Since $\mathbb{P} - H \simeq \mathbb{A}^n$, we see that the group $\mathrm{Pic}(\mathbb{P})$ is generated by the class of the divisor H. Further, the homomorphism $\mathbb{Z} \to \mathrm{Pic}(\mathbb{P})$ is injective. This follows from the theory of the degree or from the following general observation: on a complete variety, 0 is the only divisor that is both effective and principal. Indeed, if $\mathrm{div}(g) \geq 0$ then the function g is regular, and hence constant.

For any hyperplane $H \subset \mathbb{P}$, the invertible sheaf $\mathcal{O}_{\mathbb{P}}(H)$ is isomorphic to the tautological sheaf $\mathcal{O}_{\mathbb{P}}(1)$ of Chap. 1, § 7. The sections of $\mathcal{O}_{\mathbb{P}}(H)$ are of the form l/l_0, where $l \in V^*$ and $l_0 = 0$ is the homogeneous equation of the hyperplane H. Thus the sections of $\mathcal{O}_{\mathbb{P}}(1)$ identify with V^*. More generally, for all $m \in \mathbb{Z}$, we have

$$H^0(\mathbb{P}, \mathcal{O}_{\mathbb{P}}(m)) = \begin{cases} 0 & \text{if } m < 0, \\ \mathrm{Sym}^m(V^*) & \text{if } m \geq 0. \end{cases}$$

As a consequence of this computation we once more obtain that the automorphisms of $\mathbb{P}$ are induced by those of V. Indeed, an automorphism of $\mathbb{P}$ induces an automorphism of $\mathrm{Pic}(\mathbb{P}) \simeq \mathbb{Z}$. Hence the generator $\mathcal{O}_{\mathbb{P}}(1)$ goes into $\mathcal{O}_{\mathbb{P}}(1)$ (for $\mathcal{O}_{\mathbb{P}}(-1)$ has no sections), and hyperplanes into hyperplanes.

Example 3. It is easy to prove, in a similar way, that the Picard group of a product $\mathbb{P}(V) \times \mathbb{P}(W)$ is isomorphic to $\mathbb{Z} \oplus \mathbb{Z}$, and that every invertible sheaf on this product is of the form $\mathcal{O}(m_1, m_2) = p^*\mathcal{O}_{\mathbb{P}(V)}(m_1) \otimes q^*\mathcal{O}_{\mathbb{P}(W)}(m_2)$. The pair of integers (m_1, m_2) is called the *type* of the sheaf. For instance, if $s\colon \mathbb{P}^n \times \mathbb{P}^m \to \mathbb{P}^N$ is the Segre embedding, then $s^*\mathcal{O}(1)$ has type $(1, 1)$. The diagonal in $\mathbb{P}^1 \times \mathbb{P}^1$ is also of type $(1, 1)$.

Example 4. We consider the Grassmann variety $G = G(k, n)$. Let us fix a decomposition $K^n = V \oplus W$, with $\dim V = k$. We shall denote by Σ the set of all linear spaces $L \in G$ such that $L \cap W \neq (0)$. The complement of Σ in G is a 'cell' $U(V, W)$ (cf. Chap. 1, Sect. 3.4), isomorphic to $V^* \otimes W \simeq \mathbb{A}^{k(n-k)}$. It is easy to show that Σ is birationally equivalent to $\mathbb{P}(W) \times G(k-1, n-1)$. Hence it is irreducible and of codimension 1 in G. From the excision exact sequence we deduce that $\mathrm{Pic}(G)$ is isomorphic to $\mathbb{Z}$ and is generated by the class of Σ. Note that $\mathcal{O}_G(\Sigma) \simeq p^*\mathcal{O}(1)$, where $p\colon G \to \mathbb{P}^N$ is the Plücker embedding.

3.6. Divisors on Curves. Obviously, computing the Picard group by means of excision is possible only for simple varieties like $\mathbb{P}^n$ or the Grassmannians. For varieties which are less rational, this group looks more complicated and may contain a 'continuous' part. We will show this on the example of an elliptic curve. First of all, let C be any complete, smooth, irreducible curve. A divisor on C is a linear combination of points, $\Sigma\, n_i\, [x_i]$, where the n_i are integers. The *degree* of this divisor is $\Sigma\, n_i$. It follows readily from the theorem of conservation of number (Chap. 2, Sect. 5.5) that linearly equivalent divisors have the same degree. Thus the degree homomorphism $\mathrm{Div}(C) \to \mathbb{Z}$ factorizes through $\mathrm{Pic}(C)$. We denote by $\mathrm{Pic}^0(C)$ the kernel of $\mathrm{Pic}(C) \to \mathbb{Z}$, that is, the group of divisors of degree 0 modulo principal divisors.

Suppose now that C is a smooth cubic in $\mathbb{P}^2$. We fix a point on it, say p, and associate with every point $x \in C$ the divisor $[x] - [p]$, of degree 0. This yields a mapping $\varphi\colon C \to \mathrm{Pic}^0(C)$. We claim that φ is bijective, a fact which can be reinterpreted in two different manners.

First of all, on the right hand side we have $\mathrm{Pic}^0(C)$, which is a group. Hence C, too, is equipped with a group structure. This fact is specific to curves of degree 3, though it also occurs – in a weakened form – for cubic hypersurfaces.

Secondly, on the left hand side we have the algebraic variety C. Therefore the group $\mathrm{Pic}^0(C)$ is endowed with a structure of algebraic variety. As a matter of fact, this assertion holds for every curve. (Historically, algebraic geometry essentially originated from the discovery of this phenomenon; cf. Shafarevich [1972].) This leads up to the concept of the *Jacobian variety* of a curve. The dimension of $\mathrm{Pic}^0(C)$ as an algebraic variety is nothing else than the genus of C (cf. Chap. 2, Sect. 7.3). However, the Jacobian is a much finer invariant of the curve than the genus (cf. Mumford [1975] and Serre [1959]).

But let us revert to the mapping $\varphi\colon C \to \mathrm{Pic}^0(C)$. We begin with the explicit addition law on the points of C. Let $x, y \in C$; then the line $\overline{xy} \subset \mathbb{P}^2$ meets C in x, y, and one more point, say w. We now draw the straight line $\overline{pw}$ and define z to be its third intersection point with C. This point z will be called the sum of x and y. That this operation agrees with addition in $\mathrm{Pic}^0(C)$ can be seen from the fact that the divisor $[x] + [y] + [w]$ is linearly equivalent to $[p] + [w] + [z]$, for both of them are cut out on C by lines in $\mathbb{P}^2$. It follows that $([x] - [p]) + ([y] - [p]) \sim ([z] - [p])$. One can also write down explicit formulae for the addition operation (cf. Shafarevich [1972], Chap. 3, Sect. 3.1), from which it becomes clear that the addition law is regular.

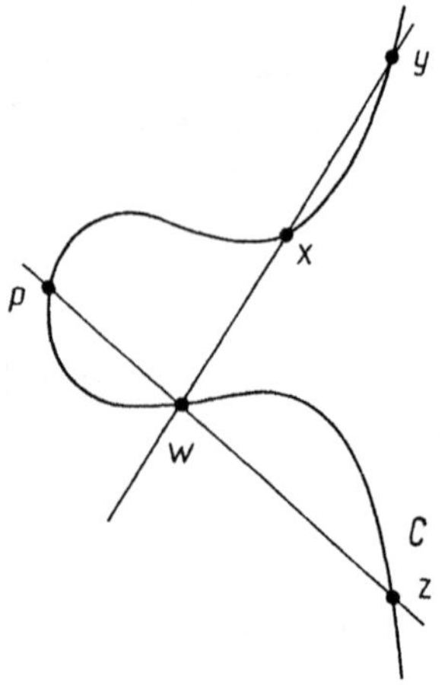

Fig. 17

Thus, $\varphi\colon C \to \mathrm{Pic}^0(C)$ is a group homomorphism, and it is evidently surjective. It is also injective because C is nonrational (Chap. 2, § 7). Indeed, if two points, x and y, have the same image then the divisor $[x] - [y]$ is principal. Hence there exists a rational function f on C with a zero at x and a pole at y. But any such f defines an isomorphism $C \xrightarrow{\sim} \mathbb{P}^1$.

By proceeding in a similar fashion with a singular cubic C, one can define a group law on the set of smooth points $C - \mathrm{Sing}(C)$.

§ 4. Linear Systems of Divisors

4.1. Families of Divisors. In the previous section a divisor was regarded as an individual object. We shall now deal with 'continuous' families of divisors.

Definition. A *family of Cartier divisors* on a variety X, with base S (where S is also a variety), is a Cartier divisor $\mathcal{D}$ on $X \times S$ whose support does not contain any fibre of the projection $X \times S \to S$.

By restricting $\mathcal{D}$ to each fibre $X \times \{s\}$, we obtain a divisor $\mathcal{D}_s$ on that fibre, or on X. In this way, $\mathcal{D}$ gives rise to a family $(\mathcal{D}_s)$, $s \in S$. In future we shall mainly be interested in effective families, when $\mathcal{D}$, or equivalently all of the $\mathcal{D}_s$, are effective.

With families of divisors we can do two things. One is to restrict them to a subvariety $Y \subset X$. This works if Y is not contained in any $\mathcal{D}_s$. The other one is a base extension. If $\varphi\colon T \to S$ is a morphism then one can form the induced family $(\mathcal{D}_{\varphi(t)})$, $t \in T$, which is defined as the divisor $(id \times \varphi)^*\mathcal{D}$ on $X \times T$.

Example. Here is the most important – and simplest – example of a family of divisors. Let V be a vector space, and V^* its dual. Let $\mathcal{H} \subset \mathbb{P}(V) \times \mathbb{P}(V^*)$ be the divisor of zeros of the natural pairing $V \times V^* \to K$, $(v, l) \mapsto l(v)$. For each point $l \in \mathbb{P}(V^*)$, the equation $l = 0$ defines a hyperplane H_l in $\mathbb{P}(V)$. Hence $\mathcal{H}$ is the family of hyperplanes in $\mathbb{P}(V)$, parametrized by the points of $\mathbb{P}(V^*)$ (see Sect. 1.2 for comparison). This family is universal in the sense that if $\mathcal{D} = (\mathcal{D}_s)$, $s \in S$, is any family of hyperplanes in $\mathbb{P}(V)$ then there exists a unique morphism $\varphi\colon S \to \mathbb{P}(V^*)$ such that $\mathcal{D} = \varphi^*(\mathcal{H})$ (that is, $\mathcal{D}_s = H_{\varphi(s)}$). Besides, $\mathcal{H}$ can also be viewed as the family of hyperplanes in $\mathbb{P}(V^*)$, parametrized by $\mathbb{P}(V)$.

4.2. Linear Systems of Divisors. Let $\mathcal{L}$ be an invertible sheaf on an irreducible variety X. As we explained in § 3, a nonzero section, say s, of the sheaf $\mathcal{L}$ defines an effective divisor $\mathrm{div}(s, \mathcal{L})$ on X. Thus, if $V \subset H^0(X, \mathcal{L})$ is any finite-dimensional subspace of sections of $\mathcal{L}$, we get a 'family' of divisors $(\mathrm{div}(s, \mathcal{L}))$, for $s \in V - \{0\}$. As $\mathrm{div}(s, \mathcal{L})$ is not modified when multiplying s by a constant, we may consider that our 'family' is parametrized by the points of $\mathbb{P}(V)$.

The family we have just constructed is also a family in the sense of Sect. 4.1. It is defined by a divisor $\mathcal{D}$ on $X \times \mathbb{P}(V)$, which we exhibit. It is enough to choose a basis $s_0, \ldots, s_n$ of the space V and to define $\mathcal{D}$ by the equation $\sum_i s_i T_i = 0$, where $T_0, \ldots, T_n$ are homogeneous coordinates on $\mathbb{P}(V)$. Such families of divisors are called *linear systems.* In other words, a linear system has for base a projective space $\mathbb{P}^n$, but this is not enough: it must also be defined by a section of some invertible sheaf $\mathcal{L} \otimes q^*\mathcal{O}(1)$ on $X \times \mathbb{P}^n$.

When the variety X is complete, we can take for V the whole space $H^0(X, \mathcal{L})$, thus obtaining what is called a *complete linear system.* This is denoted by $|\mathcal{L}|$, or $|\mathcal{D}|$ if $\mathcal{L} \cong \mathcal{O}(\mathcal{D})$. Every effective divisor that is linearly equivalent to $\mathcal{D}$ appears in the complete system $|\mathcal{D}|$, and precisely once.

The system of hyperplanes $\mathcal{H}$ introduced in the Example of Sect. 4.1 is a complete linear system both in $\mathbb{P}(V)$ and in $\mathbb{P}(V^*)$. If X is a subvariety of $\mathbb{P}(V^*)$ then, by restricting the hyperplanes to X, we get a linear system on X. It turns out that, in some sense, every linear system is built like that.

4.3. Linear Systems without Base Points. Let $\mathcal{D}$ be a linear system on X, corresponding to a space $V \subset H^0(X, \mathcal{L})$. The set (or subscheme) of *base points* of $\mathcal{D}$ is the intersection $B(\mathcal{D}) = \bigcap_{s \in \mathbb{P}(V)} \mathcal{D}_s$ of all divisors in the family $(\mathcal{D}_s)$, $s \in \mathbb{P}(V)$. In other words, x is a base point if $s(x) = 0$ for all $s \in V$. If the set $B(\mathcal{D})$ is empty, one says that the linear system $\mathcal{D}$ is *base-point free,* or without base points. This is equivalent to saying that $\mathcal{L}$ is generated by V. For instance, in $\mathbb{P}$ the system $|mH|$ has no base points if $m \geq 0$.

The set of base points $B(\mathcal{D})$ can also be described as the set of points $x \in X$ with the property that the fibre $\{x\} \times \mathbb{P}(V)$ is contained in the divisor $\mathcal{D} \subset X \times \mathbb{P}(V)$. Thus, if $B(\mathcal{D})$ is empty, $\mathcal{D}$ can be viewed as a family of hyperplanes in $\mathbb{P}(V)$ with base X. Then this family is induced from the universal family $\mathcal{H} \subset \mathbb{P}(V) \times \mathbb{P}(V^*)$ by some morphism $\varphi_{\mathcal{D}}\colon X \to \mathbb{P}(V^*)$. So we have the following

Proposition. *If a linear system $\mathcal{D} = (\mathcal{D}_s)$, $s \in \mathbb{P}(V)$, has no base points then there exists a unique morphism $\varphi_{\mathcal{D}}\colon X \to \mathbb{P}(V^*)$ with the property that the system $(\mathcal{D}_s)$ is induced by the hyperplanes in $\mathbb{P}(V^*)$.*

Here lies the main significance of linear systems: they provide mappings into some projective spaces. We give yet another description of $\varphi_{\mathcal{D}}$. On identifying the fibre $\mathcal{L}(x)$ with K, we obtain a linear mapping $V \to K$, $s \mapsto s(x)$, that is, a point of V^*. This point is nonzero, since $\mathcal{D}$ is base-point free, and it is defined up to multiplication by a constant. Therefore there is a well-defined mapping $X \to (V^* - \{0\})/K^* = \mathbb{P}(V^*)$, which is in fact $\varphi_{\mathcal{D}}$. If $s_0, \dots, s_n$ is a basis for V then $\varphi_{\mathcal{D}}(x) = (s_0(x), \dots, s_n(x))$.

Examples. a) The complete linear system $|mH|$ of hypersurfaces (more precisely, of divisors) of degree m in projective space $\mathbb{P}(V)$ is parametrized by the points of the space $\mathbb{P}(\mathrm{Sym}^m V^*)$. The map $\varphi_{|mH|}\colon \mathbb{P}(V) \to \mathbb{P}(\mathrm{Sym}^m V)$ is nothing but the Veronese embedding (cf. § 5 of Chap. 1).

b) The Segre embedding is given by the complete linear system of divisors of type $(1,1)$ in $\mathbb{P}^n \times \mathbb{P}^m$.

c) The Plücker embedding $G(k,V) \to \mathbb{P}(\Lambda^k V)$ is given by the complete linear system $|\Sigma|$ (see Example 4 in Sect. 3.5).

4.4. Ample Systems. A linear system $\mathcal{D}$ is said to be *very ample* if it is base-point free and defines an embedding of X in $\mathbb{P}^n$. An invertible sheaf $\mathcal{L}$ is said to be *very ample* if the complete linear system $|\mathcal{L}|$ is very ample. Thus, a very ample system consists of all hyperplane sections of X with respect to some embedding $X \hookrightarrow \mathbb{P}^n$. Now, hyperplanes are in some sense quite numerous. At any rate, they separate points (including infinitely near ones). Conversely, if a linear system $\mathcal{D} = (\mathcal{D}_s)$ separates points then $\varphi_{\mathcal{D}}$ is an embedding. Indeed, $\varphi^{-1}(\varphi(x))$ is nothing else than the subscheme of base points of the linear system $\mathcal{D} - x$ (that is, of the subsystem of all $\mathcal{D}_s$ such that $x \in \mathcal{D}_s$), which is just $\{x\}$, by assumption. Thus we can apply Chap. 2, Sect. 5.2.

A divisor $\mathcal{D}$ is said to be *ample* if, for some $m > 0$, the system $|m\mathcal{D}|$ is very ample. We quote the *Nakai–Moĭshezon criterion for ampleness:*

Theorem. *A divisor $\mathcal{D}$ on a complete variety X is ample if and only if for every irreducible subvariety $Y \subset X$ we have $(\mathcal{D}^r.Y) > 0$, where $r = \dim Y$.*

The definition of the intersection numbers $(\mathcal{D}^r.Y)$ will be given in § 6. Proofs of the theorem can be found in Hartshorne [1977], Iitaka [1982], and Kleiman [1966]. In particular, a divisor on a curve is ample if it has positive degree. On a complete smooth surface S, a divisor $\mathcal{D}$ is ample if $(\mathcal{D}^2) > 0$ and $(\mathcal{D}.C) > 0$ for all curves $C \subset S$.

4.5. Linear Systems and Rational Maps. Even if a linear system $\mathcal{D}$ has some base points, it defines a mapping of $X - B(\mathcal{D})$ into $\mathbb{P}(V^*)$, which we can regard as a rational map from X to $\mathbb{P}(V^*)$ (provided, of course, that $B(\mathcal{D}) \neq X$, that is, that the system $\mathcal{D}$ is nonempty).

Example 1. Let $\mathcal{D} = |H - p|$ be the linear system of all hyperplanes in $\mathbb{P}^n$ that pass through a point $p \in \mathbb{P}^n$. Then the mapping $\varphi_{\mathcal{D}}\colon \mathbb{P}^n - \{p\} \to \mathbb{P}^{n-1}$ is nothing but the linear projection with centre p.

Example 2. We consider linear systems of conics in $\mathbb{P}^2$.

a) Let $\mathcal{D} = |2H - a|$ be the system of all conics that pass through the point a. Its dimension is equal to 4. (The family of all conics is five-dimensional, and passing through a point imposes one condition.) Hence $\varphi_{\mathcal{D}}$ maps $\mathbb{P}^2 - \{a\}$ (or the blow-up $\widetilde{\mathbb{P}}^2_a$) into $\mathbb{P}^4$. It is easy to see that this is an embedding and that the degree of the image is equal to $2 \cdot 2 - 1 = 3$. This map can be looked upon as the composite of the Veronese embedding $v\colon \mathbb{P}^2 \to \mathbb{P}^5$ and the linear projection $\mathbb{P}^5 \dashrightarrow \mathbb{P}^4$ centred at the point $v(a)$. This shows once again that the variety of secants of $v(\mathbb{P}^2)$ is not $\mathbb{P}^5$ (cf. Sect. 2.3).

b) Let $\mathcal{D} = |2H - a - b|$ be the system of all conics that pass through the points a and b. It maps $\mathbb{P}^2 - \{a, b\}$ into $\mathbb{P}^3$ and is generically injective. Indeed, if a point $x \in \mathbb{P}^2$ does not lie on the line $\overline{ab}$ then the set (or the subscheme) of base points of the system $|2H - a - b - x|$ is $\{a, b, x\}$. So $\varphi_{\mathcal{D}}$ is injective at x. But if x lies on $\overline{ab}$ then the set of base points of the system $|2H - a - b - x|$ is the whole line $\overline{ab}$, which is therefore mapped into one point p. It is easy to see that the image of $\varphi_{\mathcal{D}}$ is a smooth quadric in $\mathbb{P}^3$, which contains two mutually transversal families of straight lines, coming from the pencils of lines $|H - a|$ and $|H - b|$ in $\mathbb{P}^2$ (see Fig. 18). It can be shown that the inverse mapping to $\varphi_{\mathcal{D}}$ is the linear projection from p, and that every smooth quadric in $\mathbb{P}^3$ can be obtained in this way.

c) Consider the system $\mathcal{D} = |2H - a - b - c|$. It has dimension 2 and degree 1. Hence it yields a birational map from $\mathbb{P}^2$ to $\mathbb{P}^2$, which is undefined only at the points a, b, and c.

Example 3. We now consider systems of cubics in $\mathbb{P}^2$. The dimension of the system $|3H|$ of all cubics is equal to 9. If we take the system $|3H - a_1 - \ldots - a_6|$ of all cubics passing through a set of six points $a_1, \ldots, a_6$ in reasonably general position, it defines an embedding of $\widetilde{\mathbb{P}}^2_{a_1,\ldots,a_6}$ (the blow-up of $\mathbb{P}^2$ at these points) into $\mathbb{P}^3$ as a surface of the third degree. A more detailed analysis reveals that every smooth surface of degree 3 in $\mathbb{P}^3$ is obtained in this way.

Let C be a cubic curve in $\mathbb{P}^2$, and fix eight points on it. The linear system of all cubics passing through these points has dimension $9 - 8 = 1$. Hence all these cubics meet again in a ninth point. Thus we have arrived at a classical

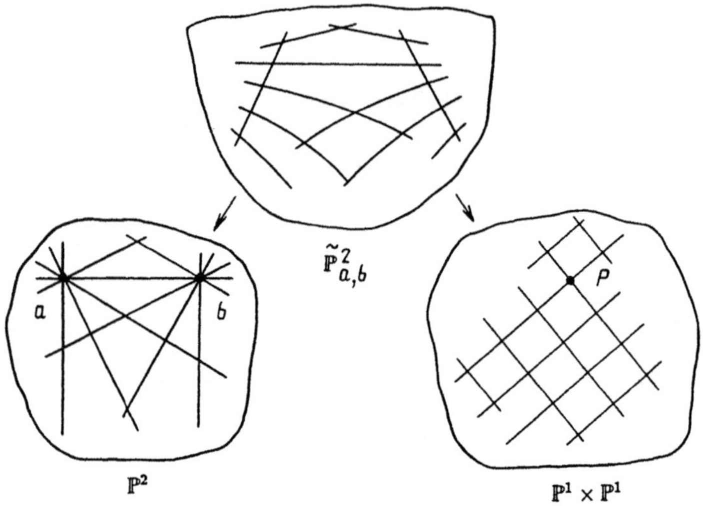

Fig. 18

assertion: if a cubic C'' passes through eight intersection points of two cubics, C and C', then it passes also through the ninth intersection point of C and C'.

Corollary (Pascal's theorem). *The pairs of opposite sides of a hexagon inscribed in a conic meet in three collinear points.*

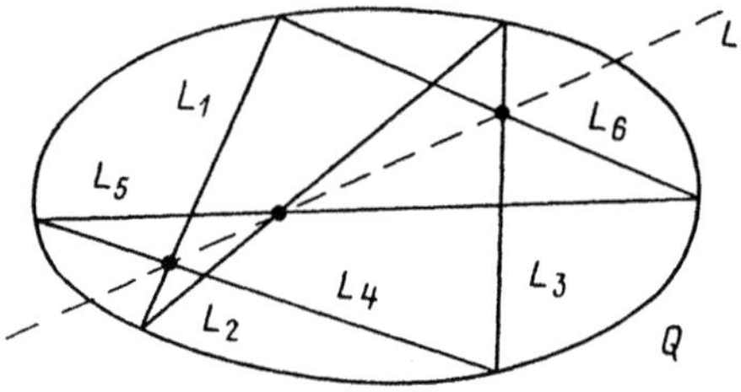

Fig. 19

Indeed, let Q be the conic, and $L_1, \ldots, L_6$ the sides of the inscribed hexagon. Define $C = L_1 + L_3 + L_5$, $C' = L_2 + L_4 + L_6$, and $C'' = Q + L$, where L is the line passing through the points p_{14} and p_{36}, with $p_{ij} = L_i \cap L_j$. Since C'' passes through 8 intersection points of C and C', it must also pass through the ninth point p_{25}.

As a further application, one can derive the associativity of the group law on a cubic (cf. Sect. 3.6). Numerous examples of linear systems will be found in Chapters 4 and 5 of Griffiths-Harris [1978]. A classical procedure for studying algebraic varieties and their classification depends on using the so-called pluricanonical (rational) maps $X \dashrightarrow \mathbb{P}^N$, which are defined by the multiples of the canonical linear system $|\omega_X^{\otimes n}|$ (cf. Iitaka [1982]). From the spectrum of possible situations, let us mention only two extremes: the varieties of general type (when ω_X is ample) and the Fano varieties (when ω_X^{-1} is ample).

4.6. Pencils. The *dimension of a linear system* $(\mathcal{D}_s)$, $s \in \mathbb{P}(V)$, is the dimension of its base $\mathbb{P}(V)$. A linear system of dimension one is called a *pencil;* it is given by a mapping into $\mathbb{P}^1$. The fibres of this map can be identified with the divisors $\mathcal{D}_s$.

Pencils have long been used to fibre varieties into subvarieties of dimension one less, so as to study them by induction. Here is the simplest means of constructing a pencil on a projective variety $X \subset \mathbb{P}$. Take a straight line l in the dual space $\mathbb{P}^*$; it defines a pencil of hyperplane sections on X. Usually one tries to select the pencil in such a way that its members $X \cap H_s$, $s \in l$, are somewhat simpler.

Suppose, for example, that X is smooth, and let $X^* \subset \mathbb{P}^*$ be its dual variety (cf. § 1). If the line l does not intersect X^* then all the members $X \cap H_s$ are smooth. As we know from § 1, such an ideal situation rarely occurs: as a rule, X^* is a hypersurface, l intersects X^*, and among all fibres of the pencil some are singular. Nevertheless, if the line l meets X^* transversally then the singularities of $X \cap H_s$ are especially simple (cf. Sect. 1.3). Such pencils are called *Lefschetz pencils.* They have been employed for quite a long time as a powerful tool for analysing varieties in an inductive way (cf. Deligne [1974] and Grothendieck, Deligne & Katz [1973]). One may say that this represents the analogue of Morse theory.

4.7. Linear and Projective Normality. Let $X \subset \mathbb{P}$ be a projective variety. On restricting to X the linear system $|mH| = |\mathcal{O}_{\mathbb{P}}(m)|$, we obtain on X a linear system $L_X(m)$. More precisely, this system is defined by the image of the restriction homomorphism

$$H^0(\mathbb{P}, \mathcal{O}_{\mathbb{P}}(m)) \to H^0(X, \mathcal{O}_X(m)).$$

In general this homomorphism is not surjective, that is, the system $L_X(m)$ is not complete. But it is complete if m is large enough (see, for example, Hartshorne [1977], Mumford [1966], or Mumford [1976]).

Notice that the proposition of Sect. 2.3 provides an upper estimate for the dimension of $H^0(X, \mathcal{O}_X(1))$: it does not exceed $\dim X + \deg X$. Similarly, one can estimate the dimension of $H^0(X, \mathcal{O}_X(m))$, thereby obtaining another proof that this space is finite-dimensional.

Definition. A projective variety $X \subset \mathbb{P}$ is said to be *linearly normal* if the system $L_X(1)$ is complete.

For instance, every hypersurface in $\mathbb{P}^n$ (where $n \geq 2$) is linearly normal. The rational normal curve $v_n(\mathbb{P}^1) \subset \mathbb{P}^n$ is also linearly normal (which explains why it is called normal). On projecting a variety into a projective space of smaller dimension, we get a variety which is not linearly normal. Thus, linear normality means that the variety lies in $\mathbb{P}$ in the freest way. F.L. Zak [1981], in answer to a question of Hartshorne [1974], has obtained the following result:

Theorem. *Let $X \subset \mathbb{P}^n$ be a smooth variety and suppose* $3 \dim X > 2(n-1)$. *Then X is linearly normal.*

Projective normality places a stronger restriction. Let $X = \mathbb{P}(C)$, where C is a cone in K^{n+1}. The variety X is said to be *projectively normal* if the cone C is a normal variety (that is, if the ring $K[C]$ is normal). In case the variety X is normal, this is equivalent to the condition that all systems $L_X(m)$ are complete (or that all Veronese images $v_m(X) \subset \mathbb{P}^{N_m}$ are linearly normal). Again, a smooth hypersurface in $\mathbb{P}^n$ is projectively normal. Here is another interesting example.

Example. Let G be a semisimple algebraic group, V an irreducible linear representation of the group G with highest weight Λ, and let $v \in V$ be a maximal vector. Then the set $Gv \cup \{0\} = C$ – the closure of the orbit of the maximal vector – is a cone. It is proved in Vinberg-Popov [1972] that the cone C is normal; so the variety $\mathbb{P}(C) \subset \mathbb{P}(V)$ is projectively normal. In particular, the Grassmann variety is projectively normal under the Plücker embedding.

§ 5. Algebraic Cycles

5.1. Definitions. We now move on to the figures of codimension greater than 1. Already on the example of divisors, we saw that it is often convenient to deal with subvarieties equipped with some 'multiplicities'. The classic authors used to call these objects virtual varieties. Nowadays, one talks about algebraic cycles, thus emphasizing the analogy with homology theory.

An *algebraic cycle* of dimension k (or a k-cycle) on a variety X is a finite sum $\alpha = \Sigma n_i [V_i]$, where the $n_i \in \mathbb{Z}$, and the V_i are irreducible k-dimensional subvarieties of X. The cycle α is said to be *effective* if all $n_i \geq 0$. The *support* of α is the union of all the V_i such that n_i is nonzero. Cycles can be added together, and the group of k-cycles on X is denoted by $\mathcal{Z}_k(X)$.

As a rule, we shall assume that X is irreducible. Then, if $n = \dim X$, the group $\mathcal{Z}_n(X)$ is isomorphic to $\mathbb{Z}$ and is generated by $[X]$. Further, $\mathcal{Z}_{n-1}(X)$ is identical with the group $\mathcal{Z}(X)$ of Weil divisors.

5.2. Direct Image of a Cycle. Cycles are covariant objects. More precisely, if $f: X \to Y$ is a proper morphism then one can define a direct-image homomorphism (*push-forward*)

$$f_*: \mathcal{Z}_k(X) \to \mathcal{Z}_k(Y).$$

By the additiveness, it is enough to define $f_*[V]$ for an irreducible k-dimensional subvariety $V \subset X$. If $\dim f(V) < k$ then we set $f_*[V] = 0$. And if $\dim f(V) = k$, we set $f_*[V] = d \cdot [f(V)]$, where $d = [K(V) : K(f(V))]$ is the degree of V over $f(V)$. Of course, if $g\colon Y \to Z$ is also proper, we have $(g \circ f)_* = g_* \circ f_*$.

We recall that in Sect. 3.2, with each Cartier divisor $\mathcal{D}$ on a normal n-dimensional variety X, we associated an $(n-1)$-cycle $[\mathcal{D}]$. This can now be done for any variety X, by setting $[\mathcal{D}] = \pi_* [\pi^*(\mathcal{D})]$, where $\pi\colon X^\nu \to X$ is the normalization morphism.

5.3. Rational Equivalence of Cycles. Let S be a smooth irreducible curve. By a *family of k-cycles* on X with base S, we mean a $(k+1)$-cycle α on $X \times S$ whose support projects dominantly to S.

Such a cycle α on $X \times S$ is called a family because, for every point $s \in S$, it defines a k-cycle α_s on X (called the *specialization* of α at the point s), which depends on s 'in a continuous way'. For the definition of specialization we can assume that $\alpha = [V]$, where V is an irreducible subvariety of $X \times S$ which dominates S. We consider the point $s \in S$ as a Cartier divisor on S and set $[V]_s = p_* [q^*(s)]$, where p and q are the projections of V to X and to S. Note that, if α is effective then so is its specialization α_s.

Two cycles are said to be *algebraically equivalent* if there is a family of cycles containing both of them. If, moreover, the base curve S is rational (usually it is $\mathbb{A}^1$ or $\mathbb{P}^1$) then we say that the cycles are *rationally equivalent.* It is easy to convince oneself that rational equivalence (represented by the symbol $\sim$) is indeed an equivalence relation, and that it is compatible with addition in $\mathcal{Z}_k(X)$. The factor group $\mathcal{Z}_k(X)/\sim$ is called the *group of k-cycle classes* on X and is denoted by $A_k(X)$.

Example. We show that $A_k(\mathbb{A}^n) = 0$ for $k < n$ (cf. Example 1 in Sect. 3.5). Let V be a subvariety of $\mathbb{A}^n$ with dimension $< n$. By moving V if necessary, we may assume that $0 \notin V$. Then, as $t \to 0$, scalar multiplication by t shifts $V_t = t^{-1}V$ to infinity, so that $[V] \sim 0$.

Proposition. *The push-forward homomorphism is compatible with rational equivalence and induces a homomorphism $f_*\colon A_k(X) \to A_k(Y)$.*

Indeed, let $f\colon X \to Y$ be a proper morphism, and α a family of cycles on X with base S. Then $\beta = (f \times id)_*(\alpha)$ is a family of cycles on Y with the same base. Thus it remains to check that $\beta_s = f_*(\alpha_s)$ for every point $s \in S$. Now, if we remember the definition of the specialization of a cycle, everything reduces to the so-called projection formula.

Projection formula. *Let $g\colon V \to W$ be a dominant morphism of two varieties with the same dimension, and let $\mathcal{D}$ be a Cartier divisor on W. Then*

$$g_* [g^*(\mathcal{D})] = \deg(g) \cdot [\mathcal{D}] .$$

For the proof we can throw away from W the subvarieties of codimension ≥ 2. So we may assume that g is finite. Now, provided we normalize V and W, we may assume that they are normal, and even smooth. Then g is a locally free morphism (see the proposition of Chap. 2, Sect. 6.6) and the projection formula derives from the Principle of conservation of number.

The degree of a 0-cycle $\alpha = \Sigma n_i[x_i]$ is the integer $\deg(\alpha) = \Sigma n_i$ (cf. Sect. 3.6). As a corollary of the above proposition – or of the theorem of conservation of number – we see that, on a complete variety, the degree of a 0-cycle is preserved by rational and by algebraic equivalence.

5.4. Excision Theorem. Let Y be a subvariety of X. Then we have an exact sequence

$$A_k(Y) \xrightarrow{i_*} A_k(X) \xrightarrow{j^*} A_k(X-Y) \to 0,$$

where i is the inclusion map $Y \hookrightarrow X$, and j^* the restriction of cycles to $X - Y$.

From this sequence (cf. Sect. 3.5), one can determine A_* for some simple varieties. For instance, it is easy to prove by induction that $A_k(\mathbb{P}^n)$ is the free abelian group generated by the class of a k-plane L^k. That $m\left[L^k\right]$ is nontrivial follows from the theory of the degree.

Similarly, one can compute A_* for products of projective spaces and, more generally, for varieties admitting a 'cellular decomposition'. This means that there is a filtration

$$X = X_n \supset X_{n-1} \supset \ldots \supset X_0 = \varnothing$$

by closed subsets, where each $X_i - X_{i-1}$ is the union of several copies of $\mathbb{A}^i$ (the 'cells'). Then $A_k(X)$ is generated by the closures of all the k-dimensional cells. The most important example is as follows.

Example. Let $G = G(k, V)$ be the Grassmann variety of k-dimensional vector subspaces of an n-dimensional space V. To construct a cellular decomposition of G, we start from a fixed flag of subspaces

$$0 = V_0 \subset V_1 \subset \ldots \subset V_n = V,$$

with $\dim V_i = i$. For every sequence of integers $a = (a_1, \ldots, a_k)$ such that $n - k \geq a_1 \geq \ldots \geq a_k \geq 0$, we set

$$W_a = \{L \in G, \quad \dim(L \cap V_{n-k+i-a_i}) = i\}.$$

One can check that W_a is isomorphic to an affine space of dimension $k(n-k) - (a_1 + \ldots + a_k)$. Its closure in G, namely

$$\overline{W}_a = \{L \in G, \quad \dim(L \cap V_{n-k+i-a_i}) \geq i\},$$

is known as a *Schubert variety,* of type a. The cells W_a cover G, and $A_*(G)$ is generated (freely) by the *Schubert cycles* $\sigma_a = [\overline{W}_a]$ (cf. Fulton [1984] and Griffiths-Harris [1978]).

For instance, the cycle $\overline{W}_{(1,0,\ldots,0)}$ has codimension 1 in G, and it consists of those L for which $L \cap V_{n-k} \neq 0$. Hence it is simply the divisor Σ of Example 4 in § 3.

5.5. Intersecting Cycles with Divisors. A structure of fundamental importance on cycles is the operation of intersecting with divisors, which induces an action of $\mathrm{Pic}(X)$ on $A_*(X)$. Let $\mathcal{L}$ be an invertible sheaf, and [V] a prime k-cycle, on X. The restriction $\mathcal{L}|_V$ of $\mathcal{L}$ to V is an invertible sheaf on V. So it defines a class $[\mathcal{L}|_V]$ of Weil divisors on V, which is denoted by $\mathcal{L} \cap [V]$.

Extending this definition by linearity, we obtain a bilinear action

$$\cap\colon \mathrm{Pic}(X) \times \mathcal{Z}_k(X) \to A_{k-1}(X).$$

For a divisor $\mathcal{D}$ on X we also write $\mathcal{D} \cap [V]$ instead of $\mathcal{O}_X(\mathcal{D}) \cap [V]$.

To be more precise, if the subvariety V is not contained in the support of $\mathcal{D}$, one can exhibit a well-determined cycle $\mathcal{D} \cdot [V] = [\mathcal{D}|_V]$ in the group $\mathcal{Z}_{k-1}(V \cap \mathrm{Supp}\,\mathcal{D})$. But if $V \subset \mathrm{Supp}\,\mathcal{D}$ then $\mathcal{D} \cdot [V]$ is defined only up to rational equivalence.

We quote the most important properties of the action of $\mathrm{Pic}(X)$ on cycles:

a) if $\alpha \sim 0$ then $\mathcal{L} \cap \alpha = 0$; hence there is an induced action

$$\mathrm{Pic}(X) \times A_k(X) \to A_{k-1}(X);$$

b) if $f\colon X \to Y$ is a proper morphism, and $\mathcal{L} \in \mathrm{Pic}(Y)$, then

$$f_*(f^*\mathcal{L} \cap \alpha) = \mathcal{L} \cap f_*(\alpha) \qquad \text{(projection formula)} ;$$

c) $\mathcal{L} \cap (\mathcal{L}' \cap \alpha) = \mathcal{L}' \cap (\mathcal{L} \cap \alpha)$ for $\mathcal{L}, \mathcal{L}' \in \mathrm{Pic}(X)$.

The main point in the proof of these properties is that $\mathcal{D} \cdot [\mathcal{D}'] = \mathcal{D}' \cdot [\mathcal{D}]$ for any two divisors $\mathcal{D}$ and $\mathcal{D}'$ on X. This formula is immediate when $\mathcal{D}$ and $\mathcal{D}'$ have simple intersection, and it is obvious if $\mathcal{D} = \mathcal{D}'$. The general case reduces to these two by blowing up $\mathcal{D} \cap \mathcal{D}'$ on X.

5.6. Segre Classes of Vector Bundles. A bunch of operations on $A_*(X)$ are induced not only by line bundles, but by any locally free vector bundle. Let $p\colon E \to X$ be a vector bundle of rank $e+1$, and let $q\colon \mathbb{P}_X(E) \to X$ be the corresponding projective bundle, with tautological sheaf $\mathcal{O}(1)$ on $\mathbb{P}_X(E)$. The morphism q is locally trivial, with fibre $\mathbb{P}^e$, and one can define a pull-back homomorphism

$$q^*\colon A_k(X) \to A_{k+e}(\mathbb{P}_X(E))$$

by setting $q^*[V] = [q^{-1}(V)]$ for any subvariety $V \subset X$. For every integer $i \geq -e$, we can now define an operation

$$s_i(E)\colon A_k(X) \to A_{k-i}(X)$$

by the formula $s_i(E) \cap \alpha = q_*(\mathcal{O}(1)^{e+i} \cap q^*(\alpha))$, where $\mathcal{O}(1)^{e+i}$ denotes the iterated action of the invertible sheaf $\mathcal{O}(1)$ on $\mathbb{P}_X(E)$. The $s_i(E)$ are known as the *Segre classes* of E. For further details on their properties and on how to construct the Chern classes from them, see Fulton [1984]. For our own part, we shall need the following simple remarks.

By dimension considerations, $s_i(E) = 0$ if $i < 0$. Almost as evidently, $s_0(E)$ is the identity on $A_*(X)$. Indeed, we must check that $s_0(E) \cap [V] = [V]$. By restricting E to V, we may assume that $X = V$. Now, replacing X by an open subset, we may assume that the bundle E is trivial. It remains only to observe that the e-fold intersection of a hyperplane in $\mathbb{P}^e$ is a point.

Corollary. *The homomorphism* $q^*\colon A_k(X) \to A_{k+e}(\mathbb{P}_X(E))$ *is injective.*

5.7. The Splitting Principle. This principle allows us to reduce certain questions about vector bundles to the case of line bundles. We know from Chap. 1, § 5 that there exists on $\mathbb{P}_X(E)$ an exact sequence of vector bundles $0 \to S \to q^*E \to Q \to 0$. On going over to $\mathbb{P}_{\mathbb{P}_X(E)}(Q)$, we get a line subbundle in the pull-back of Q, and so forth. In the end we obtain a morphism $f\colon X' \to X$ with the following properties:

a) the homomorphism $f^*\colon A_*X \to A_*X'$ is injective (see the previous corollary);

b) the vector bundle f^*E has a flag of subbundles

$$f^*E = E_{e+1} \supset E_e \supset \ldots \supset E_1 \supset E_0 = 0$$

with line bundle quotients E_i/E_{i-1}.

Theorem. *Let* $p\colon E \to X$ *be a vector bundle of rank* $e+1$. *Then the homomorphism* $p^*\colon A_k(X) \to A_{k+e+1}(E)$ *is a bijection.*

That p^* is injective is proved in three steps. If E is a line bundle then intersecting $p^*(\alpha)$ with the zero-section (which is a divisor on E) gives back α. If E has a flag of subbundles then we argue by induction. Finally, in the general case we use the splitting principle.

We establish now that p^* is surjective, that is, that every cycle $[V]$ on E is equivalent to some $p^*(\alpha)$. Using excision (and restricting X if necessary), we may assume that V does not meet some section of E. Then we argue as in the case of $\mathbb{A}^n$ (cf. Sect. 5.3) and push V away to infinity.

The inverse homomorphism to p^* can be naturally interpreted as sending a cycle β on E to its intersection with the zero-section of E. This is a special case of the Gysin homomorphism.

§ 6. Intersection Theory

6.1. Intersection of Cycles. Suppose, for simplicity, that X is a smooth n-dimensional variety, and let Y and Z be two subvarieties of X. As was shown in Chap. 2, § 6, we have

$$\dim(Y \cap Z) \geq \dim Y + \dim Z - n.$$

If equality holds, that is, if the varieties Y and Z meet properly then, to each component W of $Y \cap Z$, intersection theory assigns some multiplicity $i(W;\ Y, Z)$, and defines the intersection of Y and Z to be the cycle

$$Y \cdot Z = \sum_W i(W;\ Y, Z) \cdot [W].$$

In the general case, when Y and Z do not intersect properly, it attempts to 'move them a bit', replacing them by rationally equivalent cycles, Y' and Z', which do intersect properly. Then $Y \cdot Z$ is defined only as a cycle class. This definition makes $A_*(X)$ into a ring, which is called the Chow ring.

Until recently, these two steps were carried out separately and appeared rather cumbersome. Fulton [1984] succeeded in bringing a remarkable simplification to the foundations of intersection theory. We have followed his approach in this and the preceding section. His major idea is that a variety X looks, in some 'neighbourhood' of a subvariety Y, like the normal cone $C_{Y|X}$ (see § 7 of Chap. 1).

6.2. Deformation to the Normal Cone. Suppose first that Y is a point and that X lies in affine space $\mathbb{A}^n$ in such a way that Y is situated at the origin. For each $t \in K^*$ we set $X_t = t^{-1} \cdot X$, that is, X_t is X distended by a factor t^{-1}. In this way we obtain a family of subvarieties X_t, with $t \in K - \{0\}$. It turns out that, as $t \to 0$, this family has a limit X_0, which is precisely the tangent cone C_0X.

This construction can be made global and yields a family of inclusions $(Y_t \subset X_t)$, $t \in \mathbb{A}^1$, with the property that the pair (Y_t, X_t) is isomorphic to (Y, X) if $t \neq 0$, and to the embedding of the zero-section in the normal cone $C_{Y|X}$ if $t = 0$. This is utterly simple to do. Let $X \times \mathbb{A}^1 \to \mathbb{A}^1$ be the trivial family, and $\widetilde{X \times \mathbb{A}^1}$ the blow-up of $X \times \mathbb{A}^1$ along the subvariety $Y \times \{0\}$. The fibre of $\widetilde{X \times \mathbb{A}^1}$ above $t = 0$ consists of two components: $\widetilde{X}$, the blow-up of X along Y, and the closure $\overline{C}_{Y|X}$ of the normal cone. Define now $\mathcal{X} = \widetilde{X \times \mathbb{A}^1} - \widetilde{X}$ and let $\rho\colon \mathcal{X} \to \mathbb{A}^1$ be the projection onto $\mathbb{A}^1$. The fibres of ρ over $t \neq 0$ are isomorphic to X, while the fibre $\mathcal{X}_0 \simeq C_{Y|X}$. The subvariety Y embeds in each fibre $\mathcal{X}_t$, and the inclusion of Y in $\mathcal{X}_0 = C_{Y|X}$ is the embedding of the zero-section (see Fig. 20).

More generally, let V be a subvariety of X. If we perform the same operation on the pair $(V \cap Y, V)$, we obtain a subvariety $\mathcal{V} \subset \mathcal{X}$. If we prefer, it

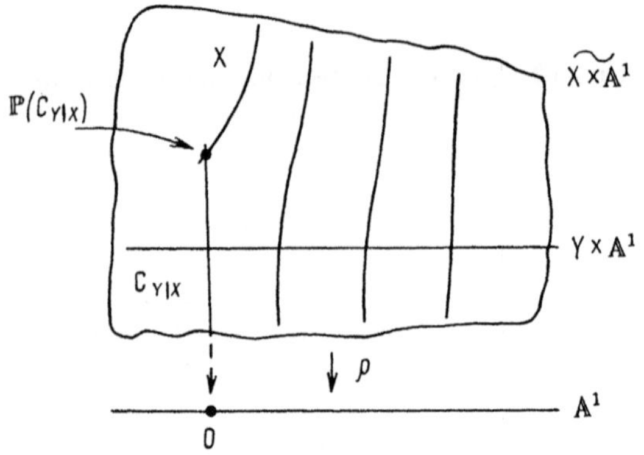

Fig. 20

is the closure of $V \times (\mathbb{A}^1 - \{0\})$ in $\mathcal{X}$. Let $\rho_V : \mathcal{V} \to \mathbb{A}^1$ be the induced morphism. The fibre of ρ_V over 0 is isomorphic to $C_{V \cap Y|V}$. What is important for us, is that it can be construed as a cycle $[\rho_V^*(0)]$ on $\mathcal{X}_0 \cap \mathcal{V}$. This cycle is called the *specialization* of V to the normal cone $C_{Y|X}$ and is denoted by $\sigma[V]$. We note that $\sigma[V]$ is an effective cycle.

On extending by linearity, we get a specialization homomorphism $\sigma\colon \mathcal{Z}_k(X) \to \mathcal{Z}_k(C_{Y|X})$. Specialization is compatible with rational equivalence and induces $\sigma\colon A_k(X) \to A_k(C_{Y|X})$.

6.3. Gysin Homomorphism. Suppose now that Y is a smooth subvariety of the smooth variety X. Then the normal cone bundle $C_{Y|X}$ reduces to the normal bundle $N = N_{Y|X}$. If we remember the isomorphism $p^*\colon A_k(Y) \to A_{k+r}(N)$ of Sect. 5.7, where r is the codimension of Y in X, we can define a homomorphism

$$A_k(X) \xrightarrow{\sigma} A_k(N_{Y|X}) \xrightarrow{(p^*)^{-1}} A_{k-r}(Y).$$

This is called the *Gysin homomorphism* of the embedding $i\colon Y \to X$ and is denoted by $i^*\colon A_k(X) \to A_{k-r}(Y)$. We can interpret its action as sending a cycle on X to its intersection with Y.

We point out that, even though the cycle $\sigma[V]$ is effective on N, its intersection with the zero-section of N may fail to be effective (see the proof of the theorem of Sect. 5.7). However, if V meets Y properly then the intersection $Y \cdot [V]$ is defined as an effective cycle on $Y \cap V$. This means that $Y \cdot [V]$ is of the shape $\sum m_W [W]$, where W runs through the irreducible components of $Y \cap V$. And the positive integers m_W are called the *intersection multiplicities* of Y and V along W.

6.4. The Chow Ring. We turn now to the intersection of arbitrary cycles, but still on a smooth n-dimensional variety X. We shall restrict attention to the intersection of two cycles, say, α and β. Here again we use reduction to the diagonal. Let $\delta\colon X \to X \times X$ be the diagonal embedding, where both the diagonal and $X \times X$ are smooth. Thus we have the Gysin homomorphism δ^* and we can define $\alpha \cdot \beta = \delta^*(\alpha \times \beta)$.

It is convenient here to use codimension and to write $A^p = A_{n-p}$ and $A^*(X) = \bigoplus_p A^p(X)$. When intersecting cycles, their codimensions add up, and $A^*(X)$ is a graded commutative and associative ring, which is called the *Chow ring*. In some sense it is the analogue of a cohomology ring and it has similar functorial properties.

Let $f\colon X \to Y$ be a morphism of smooth varieties. Given a cycle class $\alpha \in A^p(Y)$, we can intersect $X \times \alpha$ with the graph of f in $X \times Y$. The resulting cycle class $f^*(\alpha) \in A^p(X)$ is called the *inverse image* (or *pull-back*) of α under f. The pull-back commutes with taking intersections, that is, it induces a homomorphism of graded rings

$$f^*\colon A^*(Y) \to A^*(X).$$

The projection formula establishes a connection between f^* and f_*: if f is proper then $f_*(f^*(\alpha) \cdot \beta) = \alpha \cdot f_*(\beta)$. In particular, if the variety X is complete and smooth, and the cycles α and β have complementary dimensions, then the degree of the 0-cycle $\alpha \cdot \beta$ is called the *intersection number* of α and β on X and is denoted by $(\alpha \cdot \beta)_X$ or $(\alpha \cdot \beta)$.

An important special case is the intersection of divisors. Given n effective divisors, say, $\mathcal{D}_1, \ldots, \mathcal{D}_n$ on X, and a point P which is isolated in $\mathcal{D}_1 \cap \ldots \cap \mathcal{D}_n$, the intersection multiplicity of $\mathcal{D}_1, \ldots, \mathcal{D}_n$ at P is equal to $\dim \mathcal{O}_{X,P}/(f_1, \ldots, f_n)$, where the f_i are local equations for the $\mathcal{D}_i$ at the point P. It must be said that the intersection of n divisors can be defined on any variety X (not necessarily smooth) as the intersection of the diagonal in X^n with the product $\mathcal{D}_1 \times \ldots \times \mathcal{D}_n$. Indeed, $\prod \mathcal{D}_i$ is locally a complete intersection in X^n and once more the normal cone comes down to the normal bundle, so that one can argue as in Sect. 6.3. However, we note that if X is not Cohen-Macaulay then the intersection multiplicity of $\mathcal{D}_1, \ldots, \mathcal{D}_n$ at P is no longer equal to the dimension of $\mathcal{O}_{X,P}/(f_1, \ldots, f_n)$.

We introduce two examples of computation of the Chow ring.

6.5. The Chow Ring of Projective Space. By the *degree* of a k-cycle α in $\mathbb{P}^n$ we mean its intersection number with $[H]^k$. This definition is consistent with the definition of degree given in § 2. One sees from Sect. 5.4 that $\alpha \sim \deg(\alpha)\,[L^k]$; so the Chow ring $A^*(\mathbb{P}^n)$ is isomorphic to $\mathbb{Z}[\zeta]/(\zeta^{n+1})$, where $\zeta \in A^1(\mathbb{P}^n)$ is the hyperplane class $[H]$. In particular, we obtain the general formulation of the *Bézout theorem:* if $\alpha_1, \ldots, \alpha_r$ are cycles of codimension $p_1, \ldots, p_r$, and $\Sigma p_i \le n$, then

$$\deg(\alpha_1 \cdot \ldots \cdot \alpha_n) = \deg(\alpha_1) \cdot \ldots \cdot \deg(\alpha_r).$$

Similarly, it is easy to show that

$$A^*(\mathbb{P}^n \times \mathbb{P}^m) = \mathbb{Z}[\zeta, \xi]/(\zeta^{n+1}, \xi^{m+1}),$$

where ζ and ξ are the hyperplane classes in $\mathbb{P}^n$ and $\mathbb{P}^m$. For instance, the diagonal Δ in $\mathbb{P}^n \times \mathbb{P}^n$ represents the class $\zeta^n + \zeta^{n-1}\xi + \ldots + \xi^n$, whence $(\Delta.\Delta) = n+1$. As a further example, let $I \subset \mathbb{P}^n \times \mathbb{P}^{n*}$ be the incidence divisor $I = \{(x, H),\ x \in H\}$. Then, in the Chow ring of $\mathbb{P} \times \mathbb{P}^*$, we have $[I] \sim \zeta + \xi$, so that $(I^{2n}) = \binom{2n}{n}$.

For the computation of the Chow rings of projectivized bundles and of the blow-ups of smooth varieties, see Manin [1971] and Fulton [1984].

6.6. The Chow Ring of a Grassmannian. We consider the Grassmann variety $G = G(k,n)$. As we saw in Sect. 5.4, $A^*(G)$ is generated by the classes of the Schubert cycles σ_a. Now, there are some elegant multiplication formulae for the cycles σ_a (see Fulton [1984], Griffiths-Harris [1978] or Hodge-Pedoe [1952]). In particular, the σ_a can be expressed as polynomials in the so-called 'special cycles' $\sigma_{(a,0,\ldots,0)}$. We indicate here only the simplest formula, which applies to the intersection of cycles of complementary dimension. For any sequence $a = (a_1, \ldots, a_k)$ as in Sect. 5.4, we denote by a^* the sequence $(n-k-a_k, \ldots, n-k-a_1)$. Then the cycle σ_{a^*} has complementary dimension to that of σ_a, and for two cycles of complementary dimension we have $(\sigma_a \cdot \sigma_b) = \delta_{a,b^*}$.

We shall dwell in more detail on the simplest Grassmannian, namely $G(2,4)$ – the Grassmann variety of lines in $\mathbb{P}^3$. In this case we have the following Schubert cycles:

$\sigma_{1,0} = \sigma_{1,0}(L) = \{l,\ l \cap L \neq \varnothing\}$ consists of the lines which meet a fixed line L; $\dim \sigma_{1,0} = 3$.

$\sigma_{2,0} = \sigma_{2,0}(P) = \{l,\ P \in l\}$ consists of the lines that pass through a fixed point P; $\dim \sigma_{2,0} = 2$.

$\sigma_{1,1} = \sigma_{1,1}(H) = \{l,\ l \subset H\}$ consists of the lines that lie in a fixed plane H; $\dim \sigma_{1,1} = 2$.

$\sigma_{2,1} = \sigma_{2,1}(P,H) = \{l,\ P \in l \subset H\}$ consists of the lines that lie in the plane H and pass through the point $P \in H$; $\dim \sigma_{2,1} = 1$.

The intersection products are easily found. For instance, if H and H' are two distinct planes then the cycles $\sigma_{1,1}(H)$ and $\sigma_{1,1}(H')$ meet transversally in one point, corresponding to the line $H \cap H'$; hence $(\sigma_{1,1}^2) = 1$. Similarly, $(\sigma_{2,0}^2) = 1$ and $(\sigma_{2,0} \cdot \sigma_{1,1}) = 0$. We shall now determine $\sigma_{1,0}^2$. To this end we take two lines, L and L', which meet in a point P. Then $\sigma_{1,0}(L) \cap \sigma_{1,0}(L')$ consists of those lines l which intersect both L and L'. Such a line either lies in the plane $\overline{LL'}$, or it goes through P. Hence $\sigma_{1,0}^2 = \sigma_{2,0} + \sigma_{1,1}$ and therefore $(\sigma_{1,0}^4) = 2$. This last result can also be deduced from the fact that G is a quadric in $\mathbb{P}^5$ (cf. Chap. 1, Sect. 5.7), and $\sigma_{1,0} = \Sigma$ a hyperplane section of G.

6.7. Intersections on Surfaces. Let S be a smooth projective surface. In this case, the intersection product $A^1(S) \times A^1(S) \to \mathbb{Z}$ is a symmetric bilinear pairing on $A^1(S)$.

Let H be the class of a hyperplane section of S; then $(H \cdot H) = (H^2) > 0$. If a curve C on S can be sufficiently moved (that is, if there exists an effective cycle $C' \sim C$ which intersects C properly) then $(C^2) \geq 0$. However, for some curves one may also have $(C^2) < 0$. This happens, for example, with the exceptional curves of a blowing-up. We proceed to explain this in more detail.

Let $\sigma\colon \widetilde{S} \to S$ be the blowing-up of the surface S at a point $p \in S$. We know that $\widetilde{S}$ is again a smooth surface and that the curve $E = \sigma^{-1}(p)$ is isomorphic to $\mathbb{P}^1$. We shall determine the self-intersection number $(E^2)_{\widetilde{S}}$. To this effect we take a curve $C \subset S$ that passes through p and is smooth at p. Clearly, $\sigma^*(C) = \widetilde{C} + E$, where $\widetilde{C}$ is the proper transform of C, that is, the closure in $\widetilde{S}$ of the curve $\sigma^{-1}(C - \{p\})$. The curve $\widetilde{C}$ meets E transversally in one point (which corresponds to the tangent direction $T_pC \subset T_pS$). Now, since $\sigma_*(E) = 0$, it follows from the projection formula that

$$0 = (\sigma_*(E) \cdot C)_S = (E \cdot \sigma^*(C))_{\widetilde{S}} = (E^2)_{\widetilde{S}} + (E \cdot \widetilde{C})_{\widetilde{S}},$$

whence $(E^2) = -1$. In particular, it is not possible to move the curve E (or any multiple of it) into an effective curve on $\widetilde{S}$. This is also the reason why it is called exceptional. Incidentally, we have obtained that $(\widetilde{C}^2)_{\widetilde{S}} = (C^2)_S - 1$. Thus, when blowing up a point of a curve, the self-intersection number of this curve drops by one (provided one blows up a smooth point of C; otherwise, it drops by $\operatorname{mult}_p C$). Any curve can therefore be made exceptional by blowing up sufficiently many points lying on it.

It can be shown that, if a curve E on a surface S' is isomorphic to $\mathbb{P}^1$ and $(E^2) = -1$, then E can be contracted (blown down) into a smooth point (theorem of Castelnuovo). More precisely, there exists a surface S, with a smooth point p, such that S' is isomorphic to $\widetilde{S}$ (the blow-up of S at p) and $E \simeq \sigma^{-1}(p)$. For instance, if we blow up two points of a line L in $\mathbb{P}^2$, we get a curve $\widetilde{L}$ on $\widetilde{\mathbb{P}}_{xy}$ with self-intersection -1. Contracting $\widetilde{L}$ into a point p, we get a surface S which is isomorphic to a quadric $Q \subset \mathbb{P}^3$. The mapping $\mathbb{P}^2 \dashrightarrow Q$ we have constructed is the inverse of the linear projection $Q \dashrightarrow \mathbb{P}^2$ from the point $p \in Q$.

A very deep fact concerning intersections on surfaces is the *Hodge Index Theorem*. Let $\mathcal{D}$ be a divisor on S, whose intersection number with a hyperplane section of S is equal to 0. Then $(\mathcal{D} \cdot \mathcal{D}) \leq 0$; further, $(\mathcal{D} \cdot \mathcal{D}) = 0$ only if $\mathcal{D}$ belongs to the kernel of the pairing. A proof can be found in Griffiths-Harris [1978], Hartshorne [1977], Iitaka [1982] or Mumford [1966]. Various consequences of this theorem, including a proof of the Weil conjecture for curves, will be found in any textbook on algebraic surfaces. See also the survey on cohomology in this collection (volume 35).

§ 7. The Chow Variety

7.1. Cycles in $\mathbb{P}^n$. In this section we will show that the effective cycles in $\mathbb{P}^n$, of given dimension and degree, are parametrized by the points of some algebraic variety. This we have already seen for varieties of degree 1, in which case we defined the Grassmann varieties. Basically, we have also seen it for cycles of codimension 1, which are parametrized by the complete linear systems. We will discuss this in more detail.

The effective divisors of degree m in projective space $\mathbb{P}(V)$ form a linear system $S_m = |mH|$, parametrized by the points of the projective space $\mathbb{P}(\mathrm{Sym}^m V^*)$. The corresponding divisor $S_m \subset \mathbb{P}(V) \times \mathbb{P}(\mathrm{Sym}^m V^*)$ is given as the divisor of zeros of the function $\Phi\colon V \times \mathrm{Sym}^m V^* \to K$ which maps a point $v \in V$ and a form $F \in \mathrm{Sym}^m V^*$ into $F(v)$. This system is universal in the sense that, for every family $(\mathcal{D}_t)$, $t \in T$, of effective divisors of degree m in $\mathbb{P}(V)$, there exists a unique morphism $\varphi\colon T \to \mathbb{P}(\mathrm{Sym}^m V^*)$ such that $\mathcal{D}_t = (S_m)_{\varphi(t)}$. Of course, a 'general' divisor in our system is simple, and even smooth; but, for $m \geq 2$, there also occur divisors with multiple components.

Example 1. The divisors of degree 2 on the line $\mathbb{P}^1$ are parametrized by the points of the plane $\mathbb{P}^2$. If T_0, T_1 are coordinates on $\mathbb{P}^1$, a form of degree 2 is of shape $aT_0^2 + bT_0T_1 + cT_1^2$, where one should think of a, b, and c as being homogeneous coordinates on $\mathbb{P}^2$. Merging (i.e., double) points correspond to forms with vanishing discriminant $b^2 - 4ac$. So they define a curve C in $\mathbb{P}^2$ of degree 2, which is dual to the Veronese curve. The universal family $S_2 \to \mathbb{P}^2$ is a two-sheeted covering ramified over the curve C. It is not hard to convince oneself that $S_2 \simeq \mathbb{P}^1 \times \mathbb{P}^1$, whose diagonal lies precisely over C.

Note that, in characteristic 2, the equation of C takes the form $b^2 = 0$; hence C is a double line. This has two geometric consequences. First, all the tangents to a conic meet in one point. And next, there exists a nontrivial étale covering of the affine plane, namely, $(\mathbb{P}^1 \times \mathbb{P}^1 - \Delta) \to (\mathbb{P}^2 - C) = \mathbb{A}^2$ (compare with Example 1 in Sect. 1.3).

Example 2. The divisors of degree 3 in $\mathbb{P}^1$ are parametrized by the space $\mathbb{P}^3$ of forms $aT_0^3 + bT_0^2T_1 + cT_0T_1^2 + dT_1^3$. The degree-three covering $S_3 \to \mathbb{P}^3$ ramifies over a surface $W \subset \mathbb{P}^3$, which corresponds to divisors of the form $2P + Q$. The explicit equation of W is as follows: $b^2c^2 - 4ac^3 - 4b^3d - 27a^2d^2 + 18abcd = 0$; hence $\deg W = 4$. Divisors of type $3P$ form a curve $C \subset W$ of degree 3. It can be shown that W is spanned by the tangents to C.

Example 3. The divisors of degree 2 in $\mathbb{P}^2$, that is, the plane conics, are parametrized by the points of $\mathbb{P}^5$. A degenerate conic is a pair of lines. Such conics are parametrized by some 4-dimensional subvariety $W \subset \mathbb{P}^5$. Indeed, W is the quotient variety of $\mathbb{P}^2 \times \mathbb{P}^2$ by the action of the group of order 2 exchanging the factors, for a pair of lines consists of two indistinguishable members. We will determine the degree of W. To that purpose, we pick a 'general' pencil of conics $\overline{QQ'} \subset \mathbb{P}^5$. The conics Q and Q' intersect in four points, say p, q, r, and s; and the pencil $\overline{QQ'}$ consists of those conics which pass through these four points. The degree of W is the number of degenerate conics in this pencil. Now, there are exactly three of them: $\overline{pq}+\overline{rs}$, $\overline{pr}+\overline{qs}$, and $\overline{ps}+\overline{qr}$. Hence $\deg W = 3$.

The non-reduced conics, that is, the double lines in $\mathbb{P}^2$, are parametrized by the Veronese surface $S \subset \mathbb{P}^5$; of course, $S \subset W$. Furthermore, as is readily seen, every chord of S is entirely contained in W. This agrees with Example 2 of Sect. 2.3.

7.2. From Cycles to Divisors. We now move on to cycles in $\mathbb{P}^n$ whose degree and codimension are greater than 1. It so happens that they can be converted into divisors on a Grassmann variety, which in turn can be parametrized by a subset of the corresponding complete linear system.

Let $G = G(k, n+1)$ be the Grassmann variety of $(k-1)$-planes in $\mathbb{P}^n$. We denote by $\Psi = \{(x, L),\ x \in L\}$ the incidence set in $\mathbb{P}^n \times G$. We shall regard Ψ as a correspondence from $\mathbb{P}^n$ to G, which takes a point $x \in \mathbb{P}^n$ to the set $\Psi(x) = \{L,\ x \in L\} \cong G(k-1, n)$. In particular, to every subset $Z \subset \mathbb{P}^n$ one can attach the set

$$\Psi(Z) = \bigcup_{x \in Z} \Psi(x) = \{L \in G, \quad L \cap Z \neq \varnothing\}.$$

$\Psi(Z)$ is closed if Z is closed, irreducible if Z is, and its codimension in G is equal to 1 if $\operatorname{codim}(Z, \mathbb{P}^n) = k$. Finally, we observe that even the 'degree' of $\Psi(Z)$ in G is equal to $d = \deg Z$. By 'degree' we mean here the intersection number of $\Psi(Z)$ with the 1-dimensional Schubert cycle $\sigma = \sigma_{(n+1-k, \ldots, n+1-k, n-k)}$. And, in our case, $\Psi(Z)$ is linear equivalent to $d\,[\Sigma]$ (cf. Sect. 6.6). Indeed, suppose M is a k-dimensional linear variety in $\mathbb{P}^n$ which intersects Z transversally in $\deg Z$ points. Then, if σ is a pencil of subspaces $L \subset M$, precisely $\deg Z$ of them meet Z, whence $(\Psi(Z) \cdot \sigma)_G = d$.

Extending by linearity, we get a mapping

$$\Psi\colon \mathcal{Z}^k_+(\mathbb{P}^n) \to \operatorname{Div}_+(G),$$

which preserves the degree. We claim that Ψ is injective, that is, that Z can be retrieved from the divisor $\Psi(Z)$. For this we have an explicit formula: $Z = \{x \in \mathbb{P}^n,\ \Psi(x) \subset \Psi(Z)\}$. One inclusion is trivial; the other one comes from the fact that for each point $x \notin Z$ there exists an $L \in G$ which contains x and does not meet Z.

7.3. From Divisors to Cycles. It remains for us to describe the image of Ψ and to show that it is closed in the space $\mathbb{P}\big(H^0(G, \mathcal{O}_G(d))\big)$ that parametrizes all effective divisors of degree d on G. To this purpose, with every effective divisor $\mathcal{D} \subset G$, we associate the set $Z_{\mathcal{D}} = \{x \in \mathbb{P}^n,\ \Psi(x) \subset \mathcal{D}\}$. We claim that it is closed in $\mathbb{P}^n$. For this we consider the set $\Phi = \Psi \cap (\mathbb{P}^n \times \mathcal{D})$ in $\mathbb{P}^n \times G$.

The fibre of Φ above a point $x \in \mathbb{P}^n$ is isomorphic to $\Psi(x) \cap \mathcal{D}$, and its dimension is $\leq \dim \Psi(x) = (k-1)(n+1-k)$. Further, equality holds precisely when $\Psi(x) \subset \mathcal{D}$. In view of the theorem on the dimension of the fibres, applied to the morphism $\Phi \to \mathbb{P}^n$, the set $Z_\mathcal{D} = \{x, \dim \Phi(x) \geq (k-1)(n+1-k)\}$ is closed.

Clearly, $\mathcal{D} \supset \Psi(Z_\mathcal{D})$; and $\mathcal{D}$ belongs to the image of Ψ if and only if $\mathcal{D} = \Psi(Z_\mathcal{D})$, for in that case $Z_\mathcal{D}$ has codimension k. Thus, belonging to the image of Ψ can be expressed as the identity of two correspondences, and a condition of this nature designates a closed subset in the space of all divisors. We omit details. The resulting variety S_d^k, which parametrizes the cycles of degree d and codimension k in $\mathbb{P}^n$, is called the *Chow variety*. Its computation in some very simple cases is done in Shafarevich [1972].

7.4. Cycles on Arbitrary Varieties. With the foregoing construction one can give a parametrization for the effective cycles on any projective variety $X \subset \mathbb{P}^n$. Indeed, a k-dimensional cycle Z on X can be regarded as a cycle in $\mathbb{P}^n$. Conversely, if $Z \subset X$, such a cycle can be regarded as a cycle on X. Now the condition $Z \subset X$ can be rephrased as $\dim(Z \cap X) \geq k$. Therefore the cycles that lie on X are parametrized by a closed subset of the Chow variety of cycles in $\mathbb{P}^n$.

This is a very important qualitative property. As a result, all subvarieties of X of a given degree are parametrized by a finite number of algebraic varieties. This finiteness theorem plays a dominant role in the arithmetic theory of varieties.

A more satisfactory and modern construction of the Chow variety has been given by Angéniol [1981]. A closely related concept is the Hilbert scheme (see Dolgachev [1972]).

7.5. Enumerative Geometry. Enumerative geometers are interested in finding the number of geometric figures that satisfy imposed geometric conditions (Manin [1969]). Usually the task reduces to computing some intersection numbers on the variety that parametrizes those figures. We shall illustrate the problems and methods of this branch of geometry with three examples. An innumerable quantity of further examples can be found in Griffiths-Harris [1978], Hodge-Pedoe [1952], Kleiman [1977], and Semple-Roth [1949].

Our first example is quite simple: given four lines in $\mathbb{P}^3$, situated in 'general' position, how many lines intersect all four of them? The lines in $\mathbb{P}^3$ are parametrized by the Grassmann variety $G(2,4)$. Those meeting a given line L form a Schubert cycle $\sigma_1(L)$. Hence the set we are concerned with, namely $\bigcap_{i=1}^{4} \sigma_1(L_i)$, has cardinality $(\sigma_1^4) = 2$. More generally, four curves $C_1, \ldots, C_4$, 'in general position' in $\mathbb{P}^3$, are met by $2 \cdot \deg(C_1) \cdot \ldots \cdot \deg(C_4)$ straight lines.

7.6. Lines on a Cubic. How many straight lines lie on a cubic surface in $\mathbb{P}^3$? If $X \subset \mathbb{P}^3$ is our cubic, let $H_1, \ldots, H_4$ be four 'general' planes in $\mathbb{P}^3$, and let $C_i = X \cap H_i$. The C_i have degree 3; so there exist $2 \cdot 3^4$ lines in $\mathbb{P}^3$ that meet $C_1, \ldots, C_4$. Let us determine the number of lines that meet $\bigcup C_i$ in four points. As these four points belong to X, such a line is obliged to lie on the cubic X. We observe that C_1 and C_2 meet in three points, and the same holds for C_3 and C_4. Now, connecting the first three points with the second set of three, we obtain altogether 9 lines. Similarly, on connecting $C_1 \cap C_3$ with $C_2 \cap C_4$, and $C_1 \cap C_4$ with $C_2 \cap C_3$, we find that 27 of our lines intersect $\bigcup C_i$ in two points. Taking once more a point in the intersection $C_1 \cap C_2$, and projecting from it into $\mathbb{P}^2$, we get two curves, $\pi(C_3)$ and $\pi(C_4)$, which intersect in 9 points. Three of these are uninteresting, because they spring from $C_3 \cap C_4$; the remaining ones yield 6 lines. Now, there are six different ways of selecting a pair of curves C_i, C_j, and three ways of choosing a point in their intersection. Hence we have $6 \cdot 3 \cdot 6$ lines intersecting $\bigcup C_i$ in three points. And if we subtract $27 + 4 \cdot 27$ from $2 \cdot 3^4$, we are left with 27 lines, which meet X in four points. For further details, see Griffiths-Harris [1978] and Mumford [1976].

7.7. The Five Conics Problem. Given five conics $C_1, \ldots, C_5$ in 'general' position in $\mathbb{P}^2$, how many conics are tangent to all five of them?

In the space $\mathbb{P}^5$ of all conics, those which are tangent to a given conic C form a hypersurface W_C. It is fairly easy to see that $\deg W_C = 6$. It would seem pretty clear now that 6^5 conics are tangent to $C_1, \ldots, C_5$. However, this conclusion is incorrect, because the hypersurfaces W_{C_i} do not intersect properly. In fact, every W_C contains the Veronese surface of double lines, $S \subset \mathbb{P}^5$. A double line, which geometrically is in no way tangent to C, meets C in two double points and is therefore counted as being 'tangent'. We have to rule out the double lines.

The correct answer is obtained by going over to $\widetilde{\mathbb{P}^5}$, the blow-up of $\mathbb{P}^5$ along S. Geometrically this means that, instead of a double line $2l$, we consider a finer object, namely: the double line $2l$, together with a pair of points on it, say p and p'. It is agreed that this object is tangent to a conic C if either l is tangent to C, or if one of p and p' lies on C. Then, if $\widetilde{W_{C_i}}$ denotes the proper transform of W_{C_i}, it remains to find the intersection number of $\widetilde{W_{C_1}}$, $\ldots$, $\widetilde{W_{C_5}}$ on $\widetilde{\mathbb{P}^5}$.

Here we use another trick. Imagine that the conic C degenerates into a pair of lines, l and l', meeting in one point P. Then W_C degenerates into the hypersurface $W_l + W_{l'} + 2W_P$, where W_l is the set of conics tangent to the line l, and W_P the set of conics passing through P. Of course, W_P is a hyperplane in $\mathbb{P}^5$, and W_l a hypersurface of degree 2. Further, on $\widetilde{\mathbb{P}^5}$ the divisor $\widetilde{W_C}$ is equivalent to $2\widetilde{W_l} + 2\widetilde{W_P}$. So we may start computing the intersection number

$$(2W_l + 2W_P)^5 = 32(\widetilde{W}_P^5 + 5\widetilde{W}_P^4\widetilde{W}_l + 10\widetilde{W}_P^3\widetilde{W}_l^2 + 10\widetilde{W}_P^2\widetilde{W}_l^3 + 5\widetilde{W}_P\widetilde{W}_l^4 + \widetilde{W}_l^5).$$

Each intersection number $\widetilde{W}_P^i \widetilde{W}_l^{5-i}$ can be interpreted as being the number of conics passing through i 'general' points and tangent to $5-i$ 'general' lines. It is clear that $\widetilde{W}_P^5 = 1$ and $\widetilde{W}_P^4 \widetilde{W}_l = 2$. Further, $\widetilde{W}_P^3 \widetilde{W}_l^2 = W_P^3 W_l^2 = 4$. The remaining numbers are obtained by duality: $\widetilde{W}_P^2 \widetilde{W}_l^3 = \widetilde{W}_P^3 \widetilde{W}_l^2 = 4$, $\widetilde{W}_P \widetilde{W}_l^4 = 2$, and $\widetilde{W}_l^5 = 1$. The definitive answer

$$32\,(1 + 5 \cdot 2 + 10 \cdot 4 + 10 \cdot 4 + 5 \cdot 2 + 1) = 32 \cdot 102 = 3264$$

was found by Chasles as early as 1864. We have assumed here that the characteristic of K is not equal to 2. For further details, see Fulton [1984] and Griffiths-Harris [1978].

Having completed our presentation of the general theory of algebraic varieties, we should like to point out something rather astonishing. We have learnt many profound and delicate facts about varieties over a field K, without practically knowing anything of the field itself. In the forthcoming chapter, devoted to schemes, we shall see that for many questions one does not even need to have a field K, to say nothing about its being algebraically closed.

Chapter 4
Schemes

This chapter is concerned with extending the geometric language to algebraic equations over arbitrary fields and rings. The corresponding geometric object, which generalizes algebraic varieties, is called a scheme. Like varieties, schemes are glued together from local pieces, called affine schemes. That is why we shall devote most of our attention to local questions, since otherwise everything goes on (almost) as in the case of varieties.

The theory of schemes unites Geometry with Arithmetic and achieves the goal pursued by Kronecker; cf. Klein [1926]:

"... *The problem at issue can be stated as follows: for configurations of the most general type which simultaneously pertain to arithmetic and to the theory of functions – that is, which depend on any given algebraic numbers and algebraic functions of whatsoever parameters –, try to attain, in the general case, the same degree of achievement and completion as characterizes – in some measure or other – the simplest results brought to light.*

There opens up here before us a vast expanse in the domain of pure theory. The universal regularity and harmony which rules here confers on it a supreme degree of fascination and beauty. It must be observed, besides, that this domain is still far from practical applications, though in future the situation may change."

§ 1. Algebraic Equations

This section has introductory character. Its task is to prepare and give motivations for the notion of affine scheme.

1.1. Real Equations. We started Chapter 1 with algebraic equations over an algebraically closed field K. From now on we shall consider equations over an arbitrary field K, starting from the simplest (or more usual) one, namely the field $\mathbb{R}$ of real numbers. Let $X = (F_j = 0,\ j \in J)$ be a system of algebraic equations over $\mathbb{R}$, that is, the F_j are polynomials in $T_1, \ldots, T_n$ with real coefficients.

We denote by $X(\mathbb{R})$ the set of real solutions of our system, that is, the set of all n-tuples $(x_1, \ldots, x_n) \in \mathbb{R}^n$ such that $F_j(x_1, \ldots, x_n) = 0$ for all $j \in J$. But even if we are interested only in real solutions, it is useful to know that there exist some complex ones and to figure out the structure of the set $X(\mathbb{C})$ of all complex solutions of the system X. Not least because it is the variety $X(\mathbb{C})$ which gives the correct picture of what goes on. The fact of realizing that the roots of the equation $T^2 + a = 0$ do not disappear when $a > 0$, but pass to the complex domain, was one of the greatest achievements of mathematics.

The real points (or solutions) are singled out from among all complex solutions by their being invariant under complex conjugation. In particular, the non-real solutions occur in pairs. With this trivial observation one can sometimes already establish the existence of real solutions. For instance, by the Bézout theorem, the number of solutions of a system of homogeneous equations with odd degree is odd. Hence there must be a real solution. From this, one can deduce that the dimension of a division algebra over $\mathbb{R}$ is a power of two (cf. Shafarevich [1972], Chap. 4, § 2). A further, related example is Harnack's theorem on the maximum number of ovals of a real curve (see Shafarevich [1972], Chap. 7, § 4).

1.2. Equations over a Field. The situation is quite analogous over any field K. If $X = (F_j = 0)$ is a system of algebraic equations over K then a principal role is played by the set $X(\overline{K})$ of all solutions to this system with values in $\overline{K}$, where $\overline{K}$ denotes the algebraic closure of K. This set is an algebraic variety over $\overline{K}$, that is, the kind of object studied in the previous chapters. But, because the equations are defined over K, there is some additional structure: for each intermediate field $K \subset K' \subset \overline{K}$, one can distinguish the subset $X(K') \subset X(\overline{K})$ of solutions with values in K'. To use an image, the points of the variety $X(\overline{K})$ are characterized by their relative 'closeness' to the ground field K, just like the stars in the sky stand out by their brightness. To express it in another way, the additional structure is given by the action of the Galois group $\mathrm{Gal}(\overline{K}/K)$ on the set $X(\overline{K})$ (not on the $\overline{K}$-variety $X(\overline{K})$); again, the points differ by the size of their stabilizer. Such objects will temporarily be called K-varieties.

1.3. Equations over Rings. Let now K be a ring, commutative and unitary, like the ring $\mathbb{Z}$ of integers. Given any K-algebra L, that is, any commutative ring L endowed with a ring homomorphism $\varphi\colon K \to L$, we denote by $X(L)$ the set of solutions of the system $X = (F_j = 0,\ j \in J)$ with values in L, that is, the set of n-tuples $(x_1, \ldots, x_n) \in L^n$ such that $F_j(x_1, \ldots, x_n) = 0$ for all $j \in J$. We note that every K-algebra homomorphism $\psi\colon L \to L'$ induces a mapping $X(L) \to X(L')$.

As the ring $\mathbb{Z}$ embeds in the field $\mathbb{R}$ we see that if $X(\mathbb{R}) = \varnothing$ then $X(\mathbb{Z}) = \varnothing$. In fact, for such purposes one can use any homomorphism $\mathbb{Z} \to L$. Mostly, one takes for L the ring $\mathbb{Z}/m\mathbb{Z}$ of residues modulo m. For instance, the equation $y^2 = 3x^2 + 3$ has no solution modulo 4, and hence no integer solution either.

Thus it is natural, also in the general case, to test equations over a ring K by using all possible homomorphisms $\varphi\colon K \to L$, where L is a field. On replacing the coefficients of the equations F_j by their images under φ, we get a system of algebraic equations $X_\varphi = (\varphi(F_j) = 0)$ over the field L. And there corresponds an L-variety, which we shall also denote by X_φ. So, for each homomorphism $\varphi\colon K \to L$ into a field L (which will be called a *point* of the ring K), we obtain an algebraic variety X_φ. This suggests that we might attach to the system X a geometric object whose fibre over a point φ is the variety X_φ. Then the solutions in $X(K)$ would emerge as sections of this fibration.

We make a preliminary remark. Not all points of the ring K are equally interesting. If a point $\varphi'\colon K \to L'$ factorizes through a point $\varphi\colon K \to L$, that is, if it is induced by a field inclusion $L \subset L'$, then the variety $X_{\varphi'}$ carries no new information, compared to X_φ. The only essential points of K are those which are minimal, or prime in the sense we have indicated. Saying that a point $\varphi\colon K \to L$ is *prime* means that L is generated as a field by $\varphi(K)$. Hence we see that every point factorizes through a unique prime point. In terms of the ring K, a prime point $\varphi\colon K \to L$ is given by its kernel $\varphi^{-1}(0)$, which is a prime ideal. Conversely, every prime ideal $\mathfrak{p} \subset K$ (not equal to K) defines a prime point $K \to k(\mathfrak{p})$, where $k(\mathfrak{p})$ is the quotient field of the integral domain $K/\mathfrak{p}$ (which is called the residue field of the point $\mathfrak{p}$).

1.4. The Prime Spectrum. The *prime spectrum* of a ring K is the set of prime points, or of prime ideals, of K; it is denoted by $\operatorname{Spec}(K)$. A justification of this terminology, and the connection with the spectrum of an operator, can be found in Bourbaki [1967]. For instance, $\operatorname{Spec}(\mathbb{Z})$ consists of the point $\mathbb{Z} \hookrightarrow \mathbb{Q}$ (the ideal (0)) and the points $\mathbb{Z} \to \mathbb{Z}/p\mathbb{Z}$ (the ideals (p)) for all prime numbers p. Of course, the spectrum of a field K consists of the single point $K \xrightarrow{id} K$.

Spec is a functorial construct: for every ring homomorphism $\varphi\colon A \to B$ there is a mapping in the reverse direction $\mathrm{Spec}(\varphi)\colon \mathrm{Spec}(B) \to \mathrm{Spec}(A)$, given as follows. If $\psi\colon B \to L$ is a prime point of B then $\mathrm{Spec}(\varphi)(\psi)$ is the unique prime point of A through which the point $A \xrightarrow{\varphi} B \xrightarrow{\psi} L$ factorizes. In terms of prime ideals, everything appears even simpler: $\mathrm{Spec}(\varphi)$ carries a prime ideal $\mathfrak{q} \subset B$ into the prime ideal $\varphi^{-1}(\mathfrak{q}) \subset A$.

It is now already quite easy to associate a geometric object with a system of equations $X = (F_j = 0,\ j \in J)$. It suffices to take $\mathrm{Spec}(A_X)$, where $A_X = K[T_1, \ldots, T_n]/(F_j)_{j \in J}$. Then the structure homomorphism $K \to A_X = K[T]/(F)$ yields a mapping $\mathrm{Spec}(A_X) \to \mathrm{Spec}(K)$, whose fibres are in some sense the varieties X_φ. We shall come back to this later.

Clearly, the fact that there is a finite number of variables $T_1, \ldots, T_n$ is of no relevance. Neither are the variables themselves! The only significant point is that we can attach a set $\mathrm{Spec}\,A$ to each commutative ring A. In the next section we shall provide it with a topology and a sheaf of rings, as was done in Chapter 1 for varieties. But, before anything, it is worth examining the relationship between Spec and the variety when K is an algebraically closed field.

1.5. Comparison with Varieties. Let K be an algebraically closed field, and X a system of equations over K, which we shall identify with the corresponding affine variety. How is the set X related to $\mathrm{Spec}(K[X])$?

Of course, every point $x \in X$ determines a prime point $\varphi_x\colon K[X] \to K$, $\varphi_x(f) = f(x)$. However, $\mathrm{Spec}\,K[X]$ contains many other points. Namely, every irreducible subvariety $Y \subset X$ gives a point of $K[X]$, either as the prime ideal $I(Y)$ or in the guise of a homomorphism $K[X] \to K[Y] \to K(Y)$. In this way one already obtains all points of $\mathrm{Spec}\,K[X]$. Indeed, a prime ideal $\mathfrak{p} \subset K[X]$ is of the form $I(Y)$, where $Y = V(\mathfrak{p})$ is irreducible. The points of X thereby correspond to the maximal ideals of the ring $K[X]$.

This raises another question: why is it not possible in general to get along using only the maximal ideals, instead of all primes? There are several reasons for that. First of all, when dealing with equations over a ring K, why should one ignore points of the type $\mathbb{Z} \hookrightarrow \mathbb{Q}$? The next reason has to do with functoriality: the inverse image of a maximal ideal may fail to be maximal, even in such a simple instance like the inclusion $\mathbb{Z} \hookrightarrow \mathbb{Q}$. Thirdly, maximal ideals are in general so 'scarce' that we cannot decide the behaviour of functions from the values they take on those ideals. In the case of algebraic varieties there were enough of them, owing to Hilbert's Nullstellensatz, a fact which is specific to rings of finite type over a field. Finally, even for algebraic varieties the new points have a natural interpretation as generic points of irreducible subvarieties, and this is often convenient.

§ 2. Affine Schemes

2.1. Functions on the Spectrum. In the preceding section we associated with each ring A the set $\mathrm{Spec}(A)$ of its prime points $\varphi\colon A \to k(\varphi)$. The elements of the ring A are thereby interpreted as functions on $\mathrm{Spec}(A)$. The only novelty here, is that the value of a function $a \in A$ at a point φ is not taken in a fixed field K, but in a field $k(\varphi)$ which is peculiar to each point φ. Further the value of a at the point φ is simply $\varphi(a)$. Fig. 21 illustrates the numbers 3, 5, and 6 as 'functions' on $\mathrm{Spec}\,\mathbb{Z}$.

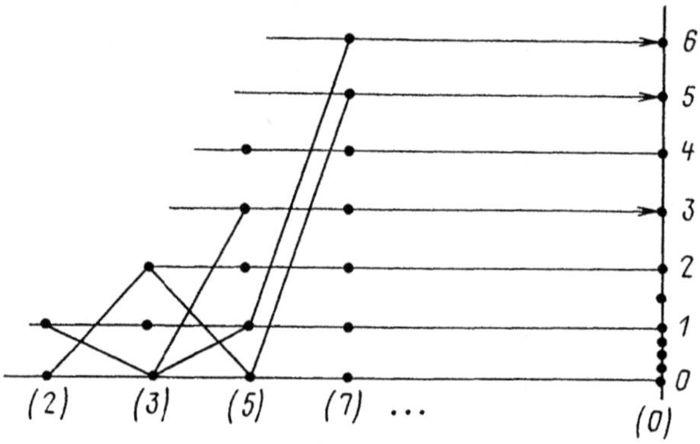

Fig. 21

Besides, as in Sect. 2.3 of Chap. 1, the representation of the elements of A by functions on $\mathrm{Spec}\,A$ has one defect: some nonzero elements of the ring A can lead to the zero-function. However, as with varieties, this can happen only for nilpotent elements. Furthermore, if in the case of varieties we derived that from Hilbert's Nullstellensatz, it is now the almost tautological fact that the nilradical of a ring is the intersection of all its prime ideals (cf. Atiyah-Macdonald [1969]). All of this shows that it is more accurate to envisage the elements of A as being sections of some sheaf of rings on $\mathrm{Spec}\,A$. To this purpose we need a topology.

2.2. Topology on the Spectrum. As in the case of varieties, the closed subsets of $\mathrm{Spec}\,A$ are defined as the 'zeros of the functions' $a \in A$. In other words, the closed subsets in the *Zariski topology* on $\mathrm{Spec}\,A$ are of the form $V(I)$, where I is a subset of A, and $V(I)$ consists of all points φ with the property that $\varphi(a) = 0$ for all $a \in I$.

As in the case of varieties, the topology on $\mathrm{Spec}\,A$ is not Hausdorff in general. Moreover, there may be some non-closed points. For instance, if a point corresponds to an irreducible subvariety Y of a variety X then its closure contains all points of Y. For this reason it is called the generic point of the subvariety Y. More generally, a point ξ that is everywhere dense in $\mathrm{Spec}\,A$

is said to be *generic*. For instance, the point $\operatorname{Spec}\mathbb{Z}$, which corresponds to the zero ideal of $\mathbb{Z}$, is generic. Closed points, on the other hand, are associated with the maximal ideals of A. More generally, if a point x belongs to the closure of a point y, we say that x is a *specialization* of y. This is depicted by a short arrow $y \to x$.

2.3. Structure Sheaf. This is built essentially as in Chap. 1, Sect. 2.8 for varieties. For an open subset U of $\operatorname{Spec} A$ we set

$$S[U] = \{a \in A, \quad \varphi(a) \neq 0 \ \forall \varphi \in U\}.$$

Clearly, $S[U]$ is a multiplicative subset of A. Define $\widetilde{A}(U)$ to be the ring of fractions of A with respect to $S[U]$. In other words, $\widetilde{A}(U)$ is the set of ratios of the form a/s with $a \in A$ and $s \in S[U]$. If $U \subset U'$ then $S[U'] \subset S[U]$; so we get a ring homomorphism $\widetilde{A}(U') \to \widetilde{A}(U)$. In this way we get on $\operatorname{Spec} A$ a presheaf of rings $\widetilde{A}$. One shows, as in the proposition of Chap. 1, Sect. 2.8, that $\widetilde{A}$ is actually a sheaf, which is denoted by $\mathcal{O}_{\operatorname{Spec}(A)}$, or simply $\mathcal{O}$, and is called the *structure sheaf* on $\operatorname{Spec}(A)$.

The ringed space $(\operatorname{Spec}(A), \widetilde{A})$ is called the *affine scheme* corresponding to the ring A; it is also denoted by $\operatorname{Spec}(A)$. This scheme now carries all the information on the ring A, since $A = \widetilde{A}(\operatorname{Spec} A)$. Thus, in place of a ring, we may work with a more geometric object: its affine scheme.

A module M over the ring A leads, as in Sect. 6.4 of Chap. 1, to a sheaf of modules $\widetilde{M}$ on $\operatorname{Spec} A$. In this way the theory of modules over rings reduces to that of quasi-coherent sheaves. We note, further, that the sheaf of rings $\widetilde{A}$ has the property that its stalk $\widetilde{A}_\varphi$ at any point $\varphi \in \operatorname{Spec} A$ is a local ring. In fact, it is the ring of fractions of the form a/b, where $b(\varphi) \neq 0$. Its unique maximal ideal $\mathfrak{m}_\varphi$ consists of the fractions such that $a(\varphi) = 0$. The residue field of the local ring $\widetilde{A}_\varphi$ is just $k(\varphi)$.

2.4. Functoriality. Let $\psi\colon A \to B$ be a ring homomorphism. An easy verification shows that the mapping $\operatorname{Spec}(\psi)\colon \operatorname{Spec} B \to \operatorname{Spec} A$ is continuous. Furthermore, ψ induces a natural homomorphism of sheaves of rings $\psi^\sharp\colon \widetilde{A} \to (\operatorname{Spec}\psi)_*\widetilde{B}$, so that $f = \operatorname{Spec}(\psi)$ is a morphism of ringed spaces, and even of locally ringed spaces. This latter property means that the homomorphisms of local rings $\widetilde{A}_{f(\varphi)} \to \widetilde{B}_\varphi$ are local, that is, carry maximal ideals into maximal ideals. It turns out that the converse is also true: every morphism $f\colon \operatorname{Spec} B \to \operatorname{Spec} A$ of locally ringed spaces, is of the shape $\operatorname{Spec}(\psi)$ for some ring homomorphism $\psi\colon A \to B$. Thus, the category of commutative rings is dual to that of affine schemes; and every statement about rings and modules can be rephrased geometrically in terms of affine schemes and sheaves of modules on them.

2.5. Example: the Affine Line. Given any ring R, the spectrum of the polynomial ring $R[T]$ is called the *affine line* over R and is denoted by $\mathbb{A}^1_R$.

In any case, this terminology agrees with the earlier one when $R = K$ is an algebraically closed field. Now suppose K is an arbitrary field; we shall examine what the line $\mathbb{A}^1_K$ looks like as a set. First of all, it contains the generic point, which corresponds to the zero ideal in $K[T]$. Further, $K[T]$ being a principal ideal domain, the remaining prime ideals are maximal, of the form (f), where f is an irreducible polynomial. By looking at the roots of f in the algebraic closure $\overline{K}$ of the field K, we see that the closed points of $\mathbb{A}^1_K$ identify with the orbits for the action of the Galois group $\mathrm{Gal}(\overline{K}/K)$ on $\overline{K}$. See also Sect. 4.2.

In the case of a general ring R we shall make use of the structure morphism $p\colon \mathbb{A}^1_R \to \operatorname{Spec} R$, which corresponds to the inclusion of rings $R \to R[T]$. For every prime point $\mathfrak{p} \in \operatorname{Spec} R$, the fibre of p over it is isomorphic to the affine line $\mathbb{A}^1_{k(\mathfrak{p})}$ over the field $k(\mathfrak{p})$. Thus, $\mathbb{A}^1_R$ represents the family of lines $\mathbb{A}^1_{k(\mathfrak{p})}$, as $\mathfrak{p}$ runs through $\operatorname{Spec} R$. The main point is that all these lines do not form a heterogeneous collection, but are grouped into one whole. Consider, for instance, the line over $\mathbb{Z}$ (cf. Fig. 21). Above each prime number p one finds the line $\mathbb{A}^1_{\mathbb{Z}/p\mathbb{Z}}$. However, there is also the line $\mathbb{A}^1_{\mathbb{Q}}$ above the generic point of $\operatorname{Spec} \mathbb{Z}$. The points situated on this line bind all the preceding fibres together. Indeed, the closure in $\mathbb{A}^1_{\mathbb{Z}}$ of a point on the line $\mathbb{A}^1_{\mathbb{Q}}$ is seen to be a subset of $\mathbb{A}^1_{\mathbb{Z}}$ which is a finite covering of the base $\operatorname{Spec} \mathbb{Z}$. We recommend the reader to draw for himself the closure of the points $\frac{1}{2}$, $\sqrt{2}$, and $\sqrt{-1}$.

The spectrum of the polynomial ring $R[T_1, \dots, T_n]$ is called the *n-dimensional affine space* over R and is denoted by $\mathbb{A}^n_R$.

2.6. Example: the Abstract Vector. Even if $\operatorname{Spec} A$ consists of only one point, the presence of nilpotent elements in the ring A (or in the structure sheaf, which in our case comes essentially to the same thing) has an effect on the morphisms of this scheme into other schemes. We consider the simplest, but nonetheless very important, example.

The simplest algebra with nilpotents is the algebra of dual numbers, $K[\varepsilon]$, where K is a field and $\varepsilon^2 = 0$. In other words, its elements are of the form $a + b\varepsilon$, with $a, b \in K$. Set-theoretically, $\operatorname{Spec} K[\varepsilon]$ is made up of one point (call it $*$), which corresponds to the prime ideal (ε) or to the ring homomorphism $\delta\colon K[\varepsilon] \to K$, where $\delta(a + b\varepsilon) = a$. Let now X be an affine variety over K. Giving a morphism of schemes $f\colon \operatorname{Spec} K[\varepsilon] \to X$ (more precisely, a morphism of K-schemes; cf. Sect. 3.3) amounts to giving a K-algebra homomorphism $\psi\colon K[X] \to K[\varepsilon]$. Now, such a homomorphism is determined by the maximal ideal $\mathfrak{m} = \ker(\delta \circ \psi)$ (that is, by the point $f(*) = x \in X$) and by a K-linear map of vector spaces $\mathfrak{m}/\mathfrak{m}^2 \to K\varepsilon \cong K$ (that is, by a tangent vector to X at the point x; cf. Chap. 1, Sect. 7.2). Thus, $\operatorname{Spec} K[\varepsilon]$ should be viewed as a point together with a vector protruding from it.

The ring $\mathbb{Z}/4\mathbb{Z}$ furnishes another type of nilpotent elements, having an 'arithmetic bias'.

§ 3. Schemes

3.1. Definitions. A *scheme* is a ringed space $X = (\operatorname{sp} X, \mathcal{O}_X)$ which is locally isomorphic to an affine scheme.

In other words, a scheme X consists of a topological space $\operatorname{sp} X$ (often denoted simply by X), together with a sheaf of commutative rings $\mathcal{O}_X$ on it. Further, each point of X has an open neighbourhood U such that the pair $(U, \mathcal{O}_X|_U)$ is isomorphic to an affine scheme (namely, $\operatorname{Spec} \mathcal{O}_X(U)$). The sheaf $\mathcal{O}_X$ is called the *structure sheaf* of the scheme; from Sect. 2.3 we see that its stalk $\mathcal{O}_{X,x}$ at any point $x \in X$ is a local ring.

A *morphism of schemes* is a morphism of the underlying locally ringed spaces. In other words, a morphism $f: X \to Y$ of a scheme X into a scheme Y consists of a continuous mapping $\operatorname{sp} f: \operatorname{sp} X \to \operatorname{sp} Y$, together with a homomorphism of sheaves of rings $f^\sharp: \mathcal{O}_Y \to (\operatorname{sp} f)_* \mathcal{O}_X$. Further, for every point $x \in X$ and every section s of the sheaf $\mathcal{O}_Y$ in a neighbourhood of $f(x)$, the relation $s(f(x)) = 0$ implies $(f^\sharp s)(x) = 0$. Every morphism of schemes is locally of the form $\operatorname{Spec}(\varphi): \operatorname{Spec} B \to \operatorname{Spec} A$ for some ring homomorphism $\varphi: A \to B$. More generally, for every scheme X and every ring A, the morphisms $X \to \operatorname{Spec} A$ are in one-to-one correspondence with the ring homomorphisms of A into $\mathcal{O}_X(X)$.

3.2. Examples.

a) Every affine scheme is a scheme.

b) If U is an open subset of a scheme X then the pair $(U, \mathcal{O}_X|_U)$ is also a scheme; we say it is an *open subscheme* of the scheme X. A morphism of schemes $Y \to X$ is an *open immersion* if it induces an isomorphism of Y with an open subscheme of X. For instance, for any element $a \in A$ the scheme $\operatorname{Spec} A[a^{-1}]$ can be identified with an open subscheme of $\operatorname{Spec} A$, which is induced on the complement of the closed subset $V(a)$.

c) A morphism $f: Y \to X$ is a *closed immersion* if $\operatorname{sp} f$ identifies Y with a closed subset of X and, furthermore, the homomorphism of sheaves $\mathcal{O}_X \to f_* \mathcal{O}_Y$ is surjective. For instance, if $\mathfrak{a}$ is an ideal of the ring A then the surjection $A \to A/\mathfrak{a}$ defines a closed immersion $\operatorname{Spec}(A/\mathfrak{a}) \to \operatorname{Spec}(A)$.

A quasi-coherent sheaf of ideals $\mathcal{J} \subset \mathcal{O}_X$ defines a closed subscheme $(\operatorname{Supp} \mathcal{O}_X/\mathcal{J}, \mathcal{O}_X/\mathcal{J})$, and this correspondence induces a bijection between the ideals of $\mathcal{O}_X$ and the closed subschemes of X. In particular, the descending chain of ideal sheaves $\mathcal{J} \supset \mathcal{J}^2 \supset \ldots \supset \mathcal{J}^n$ leads to an ascending chain of closed subschemes $Y \subset Y^{(2)} \subset \ldots \subset Y^{(n)}$. The spaces $\operatorname{sp} Y^{(n)}$ are identical for all these subschemes. Nevertheless, as n increases, the subscheme $Y^{(n)}$ reflects more and more accurately the structure of X in an infinitesimal neighbourhood of Y. The direct limit $Y^{(\infty)} = \varinjlim Y^{(n)}$ plays the same role as the Taylor series of a differentiable function. Strictly speaking, $Y^{(\infty)}$ is no longer a scheme, but what is called a formal scheme. In some sense, this is a substitute for the concept of a tubular neighbourhood of Y in X.

d) One usually builds more complicated schemes by gluing together some simpler ones, in the spirit of Chap. 1, Sect. 3.3. So one can obtain the relative affine space $\mathbb{A}^n_X$ over any scheme X, or the relative projective space $\mathbb{P}^n_X$. In the spirit of Chap. 1, Sect. 6.7, for every quasi-coherent sheaf $\mathcal{A}$ of $\mathcal{O}_X$-algebras, one can build a scheme $\operatorname{Spec}\mathcal{A} \to X$. Similarly, for a graded sheaf $\mathcal{A} = \bigoplus_{i\geq 0} \mathcal{A}_i$ of $\mathcal{O}_X$-algebras, one can construct the projective spectrum $\operatorname{Proj}(\mathcal{A}) \to X$. In particular, one can talk about the blowing-up of a scheme along a subscheme.

3.3. Relative Schemes. The theory of schemes is characterized, like that of algebraic varieties, by a heavy use of relative notions. We fix a scheme S; by an *S-scheme,* or a scheme over S, we mean a scheme X together with a morphism $f\colon X \to S$ (which is called the structure morphism). A morphism of an S-scheme X into an S-scheme Y is a morphism $X \to Y$ that commutes with the structure morphisms. Such a morphism of S-schemes is also called an X-valued point of the S-scheme Y. The set of all such points is denoted by $Y(X)$ (cf. Sect. 1.3).

For schemes, as for varieties, there exist fibre products. Thus the product of two S-schemes, X and Y, is an S-scheme $X \times_S Y$, equipped with projections onto X and Y and such that, for every S-scheme Z,

$$(X \times_S Y)(Z) = X(Z) \times Y(Z).$$

In the case of affine schemes, this product is dual to the tensor product of rings, that is,

$$\operatorname{Spec} A \times_{\operatorname{Spec} C} \operatorname{Spec} B = \operatorname{Spec}(A \otimes_C B).$$

As with varieties, the fibre product allows to do base extensions. In particular, for a point $s \in S$, the fibre product $X \times_S \operatorname{Spec}(k(s)) = X_s$ is called the *fibre* of the morphism $f\colon X \to S$ above s. The reason for this name is that, set-theoretically, X_s coincides with $f^{-1}(s)$. Thus a relative scheme $X \to S$ can be viewed as a family of $k(s)$-schemes X_s, parametrized by the points $s \in S$.

Further, any scheme X can be regarded as a scheme over $\operatorname{Spec}\mathbb{Z}$. This makes it possible to consider the family of varieties (or schemes) X_p ($p = 0, 2, 3, 5, \dots$), which are defined over fields of varying characteristics p, and to switch from positive characteristic to characteristic zero, and conversely. This is where schemes have proved most useful.

3.4. Properties of Schemes. Basically it is the same properties as for varieties. Thus, a scheme X is said to be *irreducible* if the topological space $\operatorname{sp} X$ is irreducible. And a scheme X is *reduced* if the structure sheaf $\mathcal{O}_X$ does not contain any nilpotent elements. For every scheme X there exists a reduced closed subscheme with the same underlying space $\operatorname{sp} X$. It is denoted by X_{red} and is defined by the ideal sheaf $\mathcal{J} \subset \mathcal{O}_X$ consisting of all sections that vanish at all points of X.

The notions of *normal scheme* and of *normalization* of a reduced scheme are introduced as in the case of varieties. For instance, the scheme $\operatorname{Spec}\mathbb{Z}\left[\sqrt{-3}\right]$ is not normal, and its normalization is $\operatorname{Spec}\mathbb{Z}\left[\frac{1+\sqrt{-3}}{2}\right]$.

The *dimension* of a scheme X at a point $x \in X$ is the maximal length of a chain $x = x_0 \leftarrow x_1 \leftarrow \cdots \leftarrow x_n$ of distinct points. It is denoted by $\dim_x X$. This is essentially the definition we used for varieties. For instance, the dimension of $\operatorname{Spec}\mathbb{Z}$ is equal to 1. Hence $\operatorname{Spec}\mathbb{Z}$ should be regarded as a 'curve'. The affine line $\mathbb{A}^1_{\mathbb{Z}}$ over $\mathbb{Z}$ should be regarded as an 'arithmetic surface', for its dimension at closed points is equal to 2. One must point out that for general schemes the dimension does not behave in such an ideal way as in the case of varieties.

For varieties we often used Hilbert's Basis Theorem, to the effect that rings of finite type over a field are noetherian. In the case of schemes this must be specifically demanded. A scheme X is said to be *noetherian* if there exists a *finite* open covering of X by spectra of noetherian rings. The topological space $\operatorname{sp} X$ of a noetherian scheme X is noetherian, so that a noetherian scheme is quasi-compact and can be decomposed into a finite number of irreducible components. However, there exists a finer, purely scheme-theoretic decomposition of a noetherian scheme into primary components, which takes nilpotents into account (cf. Manin [1970]).

In what follows we shall limit ourselves mainly to noetherian schemes. Apart from varieties, they include the spectrum of $\mathbb{Z}$ and, more generally, of any ring of algebraic integers. For every point x of a noetherian scheme X, the dimension $\dim_x X$ is finite. If a function f is not a zero-divisor in the ring $\mathcal{O}_{X,x}$ then $\dim_x V(f) = \dim_x X - 1$, as in the case of varieties (Chap. 2, Sect. 4.3).

3.5. Properties of Morphisms. As a rule, each property of schemes has a relative analogue, which is a property of the corresponding morphism. Roughly speaking, this property must be shared by all fibres of the morphism.

For instance, a morphism $f\colon X \to S$ is said to be *affine* if, for every affine chart $U \subset S$, the scheme $f^{-1}(U)$ is affine. Every such morphism is of the form $\operatorname{Spec}(\mathcal{A}) \to S$, where $\mathcal{A} = f_*(\mathcal{O}_X)$ is a quasi-coherent sheaf of $\mathcal{O}_S$-algebras. An affine morphism $f\colon X \to S$ is said to be *finite* if the sheaf $f_*(\mathcal{O}_X)$ is coherent over $\mathcal{O}_S$. All that was said about finite morphisms in Chap. 2, § 2 carries over to schemes.

Of greatest importance is the concept of a morphism of finite type. Suppose first that the base $S = \operatorname{Spec} A$ is affine. In this case an S-scheme X is of finite type if there exists a *finite* open covering of X by affine charts $\operatorname{Spec} B_i$, such that the A-algebras B_i are finitely generated. In the general case, a morphism $f\colon X \to S$ (or an S-scheme X) is of *finite type* if, for every affine chart $U \subset S$, the morphism $f^{-1}(U) \to U$ has finite type. For instance, the following S-schemes are of finite type: affine space $\mathbb{A}^n_S$, projective space $\mathbb{P}^n_S$, as well as their closed subschemes. In algebraic geometry, the action almost always takes place in the framework of schemes of finite type over a noetherian base.

We say that a morphism of finite type is *proper* if it is separated and universally closed. Everything that was said about proper morphisms in Chap. 2, § 3 carries over to schemes.

3.6. Regular Schemes. The concept of a smooth variety generalizes in two ways: to schemes and to morphisms. We shall study here the former case; the latter will be considered in the next section. A noetherian scheme X is said to be *regular at a point* x if the dimension of the vector space $\mathfrak{m}_x/\mathfrak{m}_x^2$ over the residue field $k(x)$ is equal to $\dim_x X$. (Here $\mathfrak{m}_x$ denotes the maximal ideal of the local ring $\mathcal{O}_{X,x}$.) We say that a noetherian scheme is *regular* if it is regular at all points (or merely at all closed points). For further details on the properties of regular schemes and rings, see Grothendieck-Dieudonné [1964–1967], Serre [1965], and Zariski-Samuel [1958,1960].

For varieties over an algebraically closed field, regularity and smoothness coincide. As with varieties, the regularity of a scheme implies that it is reduced, normal, factorial, and Cohen-Macaulay. For one-dimensional schemes, regularity is the same thing as normality, once more as in the case of varieties. Nevertheless, there are also some points of difference. Thus a scheme may be non-regular at all points, like $\operatorname{Spec} K[\varepsilon]$; this, of course, has to do with nilpotent elements. Furthermore, the set of regular points of a noetherian scheme is not always open.

3.7. Flat Morphisms. This is for us a new concept, which we have not yet encountered though it is very important even for varieties. In spite of their rather algebraic definition, flat morphisms play an important role in algebraic geometry, as they permit to formalize our intuitive idea of a 'continuous' algebraic family of schemes and varieties (cf. Sect. 4.4).

An algebra B over a ring A is said to be *flat* if, for every exact sequence of A-modules $0 \to M' \to M \to M'' \to 0$, the sequence

$$0 \to M' \otimes_A B \to M \otimes_A B \to M'' \otimes_A B \to 0$$

is exact. Besides, it suffices to require that the map $\mathfrak{a} \otimes_A B \to B$ be injective for every ideal $\mathfrak{a} \subset A$. Further details on flat rings and modules can be found in Bourbaki [1961–1965].

A morphism of schemes $f \colon X \to S$ is said to be *flat at a point* $x \in X$ if the ring $\mathcal{O}_{X,x}$ is flat over $\mathcal{O}_{S,f(x)}$. We say that f is *flat* if it is flat at all points of X. For instance, a morphism $\operatorname{Spec} B \to \operatorname{Spec} A$ is flat if and only if the A-algebra B is flat. In particular, open immersions are flat morphisms. By contrast, closed immersions are very rarely flat (only when they are also open). The structure morphism $\mathbb{A}_S^n \to S$ is also flat. Flatness is preserved under composition and base extension.

A finite morphism $X \to S$ is flat if and only if it is locally free (cf. Chap. 2, Sect. 5.7). The principle of conservation of number remains true for such morphisms.

§ 4. Algebraic Schemes and Families of Algebraic Schemes

4.1. Algebraic Schemes. By an *algebraic scheme* we mean a scheme of finite type over a field. These schemes are those which resemble algebraic varieties most. One can even define an algebraic variety as a reduced algebraic scheme over an algebraically closed field. Systems of algebraic equations over K with finitely many unknowns lead to algebraic schemes. Hence it is worth looking into them in more detail.

It is worth pointing out right away that the topological space $\operatorname{sp} X$ does not always convey an accurate idea of the algebraic scheme X. For instance, consider all finite field extensions $K \subset K'$. The K-schemes $\operatorname{Spec} K'$ each consist of only one point, but how different they are! To help us understand what kind of object we are dealing with, we can use the process of *geometrization.*

4.2. Geometrization. By this we mean going over from an algebraic K-scheme X to the $\overline{K}$-scheme $\overline{X} = X \times_{\operatorname{Spec} K} \operatorname{Spec} \overline{K}$, where $\overline{K}$ is the algebraic closure of the field K. If we forget about the nilpotent elements of $\overline{X}$, what we obtain is a variety over $\overline{K}$, that is, an object we know pretty well from the previous chapters. The schemes X and $\overline{X}$ are connected by the projection morphism $\pi \colon \overline{X} \to X$. Further, for each point $x \in X$, the fibre $\pi^{-1}(x) = \operatorname{Spec}(k(x) \otimes_K \overline{K})$ is a nonempty zero-dimensional scheme. In particular, π is surjective and X should be viewed as a quotient space of $\overline{X}$ (compare with Sect. 1.2).

Moreover, the morphism $\pi \colon \overline{X} \to X$ is flat and essentially finite. Hence X and $\overline{X}$ have the same dimension, and the theory of dimension for algebraic schemes looks like that for varieties. Thus, if a scheme X is irreducible, its dimension is the same at all *closed* points and is equal to the transcendence degree of the field $k(\xi)$ over K, where ξ is the generic point of X.

A word of caution: we talked about the algebraic closure $\overline{K}$ as if it were a canonical construction. Now, all closures of the field K are of course isomorphic, but not canonically! For instance, there is no way of pinpointing either of the roots $\sqrt{-1}$ in $\mathbb{C}$. This has as a consequence that non-isomorphic schemes may become isomorphic under geometrization. (They are said to be *forms* of one another.) For instance, the $\mathbb{R}$-schemes defined by the equations $T^2 = 1$ and $T^2 = -1$ are different, but they become isomorphic over $\mathbb{C}$. The same thing happens with the real conics $X^2 + Y^2 + Z^2 = 0$ and $X^2 + Y^2 - Z^2 = 0$.

4.3. Geometric Properties of Algebraic Schemes. Under geometrization a scheme which is reduced and irreducible may become reducible and/or non-reduced. Here is a very simple example. We consider the $\mathbb{R}$-scheme $X = \operatorname{Spec}(\mathbb{C})$; it is a point. However, $\overline{X} = \operatorname{Spec}(\mathbb{C} \otimes_{\mathbb{R}} \mathbb{C})$ is made up of two points. Besides, the scheme $\overline{X}$ here is reduced; in characteristic $p > 0$ there may appear nilpotent elements as well. For suppose $\alpha \in K$ is not a p-th power, and let $K' = K(\alpha^{1/p})$ and $X = \operatorname{Spec} K'$. Then $\overline{X}$ is the spectrum of the ring

$$K' \otimes_K \overline{K} = \overline{K}[T]/(T^p - \alpha) = \overline{K}[T]/(T - \alpha^{1/p})^p \simeq \overline{K}[S]/(S^p),$$

where $S = T - \alpha^{1/p}$.

If, nevertheless, some property of schemes holds for the geometrization $\overline{X}$, then one says that the K-scheme X has that property *geometrically*. Thus X is said to be *geometrically irreducible* (respectively, *geometrically reduced, normal,* or *regular*) according as $\overline{X} = X \otimes_K \overline{K}$ is irreducible (respectively, reduced, normal, or regular). In these cases X, too, has these properties. The converse, as we have seen, is not always true. However, if K is a perfect field and the K-scheme X is reduced (respectively, normal or regular), then so is its geometrization $\overline{X}$. We note, further, that a geometrically regular K-scheme X is also said to be *smooth,* since this is equivalent to the variety $\overline{X}$ being smooth.

4.4. Families of Algebraic Schemes. Let $f\colon X \to S$ be a morphism of finite type. Then, for each point $s \in S$, the $k(s)$-scheme $X_s = f^{-1}(s) = X \times_S \operatorname{Spec} k(s)$ is algebraic. Hence a morphism of finite type can be thought of as a family of algebraic schemes, parametrized by the points of S. In this connection, the special members of such a family may rather strongly differ from the 'general' ones. We observe that it is now possible to give a meaning to the notion of 'generic' fibre, as the fibre over the generic point of the base S (provided S is irreducible, which is often tacitly understood). As a rule, the properties of the generic fibre are inherited by the fibres which lie over the points of some neighbourhood of the generic point of S, especially if we are talking about geometric properties. For instance, if the generic fibre is geometrically irreducible then so are the 'neighbouring' fibres, and they have the same dimension. As for the dimension of 'remote' fibres, it may jump, as we saw in Chapter 2.

There are no such jumps in the dimension if the morphism $f\colon X \to S$ is flat. In this case all fibres X_s have the same dimension, which is denoted by $\dim(f)$ and is called the relative dimension of f. If, in addition, the morphism f is projective then the degree of X_s is independent of $s \in S$. The same holds for some other numerical invariants like the Hilbert polynomial, the arithmetic genus, and so forth. All of this gives grounds for regarding the fibres of a flat morphism of finite type as varying 'continuously'. It is important to note that this conception of 'continuity' makes sense – and works well – also when the base S has nilpotent elements (for instance when S is the spectrum of an Artin local ring, which enables one to build a theory of deformation for schemes and for varieties).

Even for studying one particular variety, it is sometimes useful to include it in a family. Thus, by including a singularity in a versal family of deformations, we straighten this singularity, as it were, which enables us to analyse more thoroughly its genesis and its nature (see Arnol'd, Varchenko & Guseĭn-Zade [1982]). Here is another example: suppose we are given a variety $X_{\mathbb{Q}}$ over the field $\mathbb{Q}$. Then there exists a scheme X, of finite type over $\mathbb{Z}$, whose fibre over the generic point is isomorphic to $X_{\mathbb{Q}}$. The variety $X_{\mathbb{Q}}$ unfolds here in an arithmetic direction. On applying to such an arithmetic scheme X the concepts and methods of algebraic geometry, and particularly intersection theory and the theory of cohomology, various interesting number-theoretic consequences are obtained.

4.5. Smooth Families. A morphism $f\colon X \to S$ of finite type is said to be *smooth* if it is flat and all its fibres are smooth algebraic schemes. A smooth morphism of relative dimension 0 is said to be *étale*. If S is a regular scheme, and the morphism $X \to S$ is smooth, then the scheme X is also regular.

Smooth morphisms have the following infinitesimal lifting property. Suppose we are given an S-scheme Y and a closed subscheme Y_0 with the same underlying topological space as Y. Then, if $f\colon X \to S$ is a smooth morphism, every S-morphism $Y_0 \to X$ extends locally to an S-morphism $Y \to X$. (More precisely, if Y is an affine scheme then every morphism $Y_0 \to X$ extends to a morphism $Y \to X$.) Further, the extension is unique if $f\colon X \to S$ is étale.

As in the case of varieties, the smoothness of a morphism is closely related to its differentiability properties. It is apparently impossible to tell what we should mean by a differential form (or a tangent bundle) on an arbitrary scheme. Nevertheless, for relative schemes these notions make sense and are introduced exactly as for varieties (see Chap. 1, Sect. 7.7). Let X be an S-scheme of finite type; the *sheaf of relative differentials* of X over S is the conormal sheaf to the diagonal in $X \times_S X$. It is denoted by $\Omega^1_{X/S}$; under the natural identification of the diagonal with X, this sheaf $\Omega^1_{X/S}$ is a coherent sheaf on X. The functor Ω^1 commutes with base extension and is contravariant. The sheaves $\Omega^p_{X/S}$ of higher-order differentials are constructed, as usual, by means of exterior powers.

For an algebraic K-scheme X, smoothness over $\operatorname{Spec} K$ means that, as an $\mathcal{O}_X$-module, the sheaf $\Omega^1_{X/K}$ is locally free of rank $\dim X$. Hence, saying that a morphism $f\colon X \to S$ is smooth, is equivalent to saying that two properties hold: f is flat and $\Omega^1_{X/S}$ is a locally free $\mathcal{O}_X$-module of rank $\dim(f)$.

References*

At the present time, several modern presentations of the foundations of algebraic geometry are available, and each of them is good in its own way. First of all, one must mention the ambitious work of Grothendieck, *Éléments de géométrie algébrique* (usually referred to as *EGA*), conceived as a masterpiece, but which has remained unfinished. From the 13 chapters originally planned, only four have appeared, and they already fill eight books. Besides, many subjects which were not covered in *EGA* have been taken up in *SGA* (*Séminaire de géométrie algébrique;* see also the bibliography in Hartshorne [1977]). The exposition in *EGA* is conducted in the language of schemes and with maximal generality. More accessible introductions to the theory of schemes will be found in Manin [1970], Dieudonné [1969], Iitaka [1982], and Mumford [1966].

For a presentation of algebraic geometry which lays greater emphasis on algebraic varieties and classical problems, we refer to Shafarevich [1972] and Hartshorne [1977]. Algebraic varieties over the field of complex numbers, where – in addition to algebraic methods – one can also use analytic and transcendental techniques, are investigated in the books by Griffiths and Harris [1978], and by Mumford [1976]. A similar point of view is taken by Chirka [1985], who discusses complex analytic sets. As specimens of books published in the period before Grothendieck we may mention Baldassarri [1956], Hodge-Pedoe [1952], Lang [1958], Samuel [1955], and Semple-Roth [1949].

Some of the books mentioned in the list of references deal with more specialized questions. On intersection theory there is Fulton [1984]; the associated K-theory is presented in Manin [1971]. Iitaka [1982] leans towards birational geometry. Algebraic groups are discussed in Borel [1969], Humphreys [1975], and Serre [1959]; abelian varieties, in Mumford [1970]. All the required notions from commutative algebra are contained in Atiyah-Macdonald [1969], Bourbaki [1961–1965], and Zariski-Samuel [1958,1960]. Differential calculus on varieties is presented in Bourbaki [1967–1971], Cartan [1967], Lang [1962], and Wells [1973].

Abhyankar, S.S. [1968] On the Problem of Resolution of Singularities; in : Tr. Mezhdunarod. Kongr. Mat. Moskva 1966, Mir, Moscow, 1968, 469–481, Zbl. 215, 371.

Angéniol, B. [1981] Familles de cycles algébriques—Schéma de Chow. Springer (Lect. Notes Math. 896), Berlin, Heidelberg, New York, 1981; 140 pp., Zbl. 496.14004.

Arnol'd, V.I., Varchenko, A.N., Guseĭn-Zade S.M. [1982] Singularities of Differentiable Mappings. Nauka, Moscow, 1982; 304 pp. English transl.: Birkhäuser, Boston, 1985, Zbl. 513.58001.

Artin, M. [1971] Algebraic Spaces. Yale Math. Monographs 3. Yale Univ. Press, New Haven, 1971; 39 pp., Zbl. 226.14001.

Atiyah, M., Macdonald, I. [1969] Introduction to Commutative Algebra. Addison-Wesley, Reading, Mass., 1969; 128 pp., Zbl. 175, 36.

Baldassarri, M. [1956] Algebraic Varieties. Springer, Berlin, Heidelberg, New York, 1956; 195 pp., Zbl. 75,159.

Bass, H., Connell, E.H., Wright, D. [1982] The Jacobian conjecture: reduction of degree and formal expansion of the inverse. Bull. Am. Math. Soc., New Ser. 7 (1982), 287–330, Zbl. 539.13012.

Borel, A. [1969] Linear Algebraic Groups. Benjamin, New York, 1969; 398 pp., Zbl. 186, 332.

Bourbaki, N. [1961–1965] Algèbre commutative. Hermann, Paris, 1961–1965, Zbl. 108, 40; Zbl. 119, 36; Zbl. 141, 35.

* For the convenience of the reader, references to reviews in Zentralblatt für Mathematik (Zbl.), compiled using the MATH database, and Jahrbuch über die Fortschritte der Mathematik (Jbuch) have, as far as possible, been included in this bibliography.

Bourbaki, N. [1967] Théories spectrales. Hermann, Paris, 1967; 166 pp., Zbl. 152, 326.
Bourbaki, N. [1967–1971] Variétés différentielles et analytiques. Hermann, Paris, 1967–1971, Zbl. 171, 220; Zbl. 217, 204.
Cartan, H. [1967] *a*) Calcul différentiel; *b*) Formes différentielles. Hermann, Paris, 1967; 178 pp.; 186 pp., Zbl. 156, 361; Zbl. 184, 127.
Chirka, E.M. [1985] Complex Analytic Sets. Nauka, Moscow, 1985; 272 pp. English transl.: Kluwer, Dordrecht, 1989, Zbl. 586.32013.
Danilov, V.I. [1978] The geometry of torus embeddings. Usp. Mat. Nauk 33 (1978), 85–135. English transl.: Russ. Math. Surv. 33, No. 2, 97–154 (1978), Zbl. 425.14013.
Deligne, P. [1974] La conjecture de Weil I. Publ. Math. Inst. Hautes Etud. Sci. 43 (1974), 273–307, Zbl. 287.14001.
Dieudonné, J. [1969] Algebraic geometry. Adv. Math. 3 (1969), 233–321, Zbl. 185, 491.
Dolgachev, I.V. [1972] Abstract algebraic geometry; in: *Itogi Nauki Tekh., Ser. Algebra, Topologiya, Geometriya* 10, 1972, 47–112. English transl.: J. Sov. Math. 2, 264–303 (1974), Zbl. 277.14001.
Fulton, W. [1984] Intersection Theory. Springer, Berlin, Heidelberg, New York, 1984; 470 pp., Zbl. 541.14005.
Fulton, W., Lazarsfeld, R. [1981] Connectivity and its applications in algebraic geometry; in: Algebraic geometry, Proc. Conf. Chicago 1980, Lect. Notes Math. 862, 1981, 26–92, Zbl. 484.14005.
Godement, R. [1958] Topologie algébrique et théorie des faisceaux. Hermann, Paris, 1958; 283 pp., Zbl. 80, 162.
Griffiths, Ph., Harris, J. [1978] Principles of Algebraic Geometry. Wiley, New York, 1978; 813 pp., Zbl. 408.14001.
Grothendieck, A., Dieudonné, J. [1971] Éléments de géométrie algébrique I. 2nd ed. Springer, Berlin, Heidelberg, New York, 1971; 466 pp., Zbl. 204, 233. (1st ed. Publ. Math. Inst. Hautes Etud. Sci. 4 (1960), Zbl. 118, 362)
Grothendieck, A., Dieudonné, J. [1961] Éléments de géométrie algébrique II. Publ. Math. Inst. Hautes Etud. Sci. 8 (1961), 222 pp., Zbl. 118, 362.
Grothendieck, A., Dieudonné, J. [1961–1963] Éléments de géométrie algébrique III. Publ. Math. Inst. Hautes Etud. Sci. 11, 17 (1961–1963), Zbl. 118, 362; Zbl. 122, 161.
Grothendieck, A., Dieudonné, J. [1964–1967] Éléments de géométrie algébrique IV. Publ. Math. Inst. Hautes Etud. Sci. 20, 24, 28, 32 (1964–1967), Zbl. 136, 159; Zbl. 135, 397; Zbl. 144, 199; Zbl. 153, 223.
Grothendieck, A. [1971] *SGA 1,* Revêtements étales et groupe fondamental. Springer, Berlin, Heidelberg, New York, 1971; 447 pp., Zbl. 234.14002.
Grothendieck, A. [1968] *SGA 2,* Cohomologie locale des faisceaux cohérents et théorèmes de Lefschetz locaux et globaux. North-Holland, Amsterdam, 1968; 287 pp., Zbl. 197, 472.
Grothendieck, A., Demazure, M. [1970] *SGA 3,* Schémas en groupes I. Springer, Lect. Notes Math. 151, Berlin, Heidelberg, New York, 1970; 564 pp., Zbl. 207, 514.
Grothendieck, A., Deligne, P., Katz, M. [1973] *SGA 7,* Groupes de monodromie en géométrie algébrique II. Springer, Berlin, Heidelberg, New York, 1973, Zbl. 258.00005.
Hartshorne, R. [1974] Varieties of small codimension in projective space. Bull. Amer. Math. Soc. 80 (1974), 1017–1032, Zbl. 304.14005.
Hartshorne, R. [1977] Algebraic Geometry. Springer, New York, Berlin, Heidelberg, 1977; 496 pp., Zbl. 367.14001.
Hodge, W.V.D., Pedoe, D. [1952] Methods of Algebraic Geometry; vol. 2. Cambridge Univ. Press, Cambridge, 1952; 394 pp., Zbl. 48, 145.
Humphreys, J.E. [1975] Linear Algebraic Groups. Springer, New York, Berlin, Heidelberg, 1975; 248 pp., Zbl. 325.20039.
Iitaka, S. [1982] Algebraic Geometry. An Introduction to Birational Geometry of Algebraic Varieties. Springer, New York, Berlin, Heidelberg, 1982; 357 pp., Zbl. 491.14006.

Iskovskikh, V.A. [1979] Anticanonical models of algebraic threefolds; in: *Itogi Nauki Tekh., Ser. Sovrem. Probl. Mat.* 12, 1979, 59–157. English transl.: J. Sov. Math. 13, 745–814 (1980), Zbl. 415.14024.

Kleiman, S.L. [1966] Towards a numerical theory of ampleness. Ann. Math., II. Ser. 84 (1966), 293–344, Zbl. 146, 170.

Kleiman, S.L. [1977] The enumerative theory of singularities; in: Real and complex singularities, Proc. Nord. Summer Sch., Symp. Math., Oslo 1976, 297–396 (1977), Zbl. 385.14018.

Klein, F. [1926] Vorlesungen über die Entwicklung der Mathematik im 19. Jahrhundert; Part 1. Springer, Berlin, Heidelberg, 1926; 385 pp., Jbuch 52, 22.

Lang, S. [1958] Introduction to Algebraic Geometry. Interscience Publ., New York, 1958; 260 pp., Zbl. 95, 153.

Lang, S. [1962] Introduction to Differentiable Manifolds. Interscience Publ., New York, 1962; 126 pp., Zbl. 103, 151.

Manin, Yu.I. [1969] On Hilbert's 15th Problem; in: Hilbert's Problems. Nauka, Moscow, 1969, 175–181, Zbl. 213, 288.

Manin, Yu.I. [1970] Lectures on Algebraic Geometry. Part 1: Affine Schemes. Moscow State Univ., Moscow, 1970; 133 pp. (Russian).

Manin, Yu.I. [1971] Lectures on Algebraic Geometry. Part 2: The K-Functor in Algebraic Geometry. Moscow State Univ., Moscow, 1971; 86 pp.; also published in Usp. Mat. Nauk 24, No. 5, 3–86. English transl.: Russ. Math. Surv. 24, No. 5, 1–89 (1969), Zbl. 199, 555.

Manin, Yu.I. [1984] New directions in geometry. Usp. Mat. Nauk 39 (1984), No. 6, 47–73. English transl.: Russ. Math. Surv. 39, No. 6, 51–83 (1984), Zbl. 577.14002.

Miyanishi, M. [1981] Non-Complete Algebraic Surfaces. Springer, Berlin, Heidelberg, New York, 1981; 244 pp., Zbl. 456.14018.

Mumford, D. [1965] Geometric Invariant Theory. Springer, Berlin, Heidelberg, New York, 1965; 146 pp., Zbl. 147, 393.

Mumford, D. [1966] Lectures on Curves on an Algebraic Surface. Princeton Univ. Press, Princeton, 1966; 200 pp., Zbl. 187, 427.

Mumford, D. [1970] Abelian Varieties. Oxford Univ. Press, Oxford, 1970; 242 pp., Zbl. 223.14022.

Mumford, D. [1975] Curves and their Jacobians. Univ. of Michigan Press, Ann Arbor, 1975; 104 pp., Zbl. 316.14010.

Mumford, D. [1976] Algebraic Geometry I. Springer, Berlin, Heidelberg, New York, 1976; 186 pp., Zbl. 356.14002.

Quillen, D. [1976] Projective modules over polynomial rings. Invent. Math. 36 (1976), 167–171, Zbl. 337.13011.

Samuel, P. [1955] Méthodes d'algèbre abstraite en géométrie algébrique. Springer, Berlin, Heidelberg, New York, 1955; 133 pp., Zbl. 67, 389.

Semple, J.G., Roth, L. [1949] Introduction to Algebraic Geometry. Oxford Univ. Press, Oxford, 1949; 446 pp., Zbl. 41, 279.

Serre, J-P. [1955] Faisceaux algébriques cohérents. Ann. Math., II. Ser. 61 (1955), 197–278, Zbl. 67, 162.

Serre, J-P. [1956] Géométrie algébrique et géométrie analytique. Ann. Inst. Fourier 6 (1956), 1–42, Zbl. 75, 304.

Serre, J-P. [1959] Groupes algébriques et corps de classes. Hermann, Paris, 1959; 202 pp., Zbl. 97, 356.

Serre, J-P. [1965] Algèbre locale. Multiplicités. Springer, Berlin, Heidelberg, New York, 1965; 190 pp., Zbl. 142, 286.

Shafarevich, I.R. [1972] Basic Algebraic Geometry. Nauka, Moscow, 1972, 567 pp., Zbl. 258.14001. English transl.: Grundlehren der mathematischen Wissenschaften 213. Springer, New York, Berlin, Heidelberg, 1974, Zbl. 284.14001; 2nd ed. 1977 (Zbl. 318.14001 = VEB Deutscher Verlag d. Wiss., Berlin 1972).

Shafarevich, I.R. [1986] Basic notions of algebra; in: *Itogi Nauki Tekh., Ser. Sovrem. Probl. Mat. Fundam. Napravleniya* 11. VINITI, Moscow, 1986, 290 pp. English transl.: Encycl. Math. Sci. 11: Algebra I, Springer, Heidelberg, 1990, Zbl. 655.00002, Zbl. 711.16001.

Suslin, A.A. [1976] Projective modules over a polynomial ring are free. Dokl. Akad. Nauk SSSR 229 (1976), 1063–1066. English transl.: Sov. Math., Dokl. 17, 1160–1164 (1977), Zbl. 354.13010.

Tyurin, A.N. [1979] The intermediate Jacobian of a threefold; in: *Itogi Nauki Tekh., Ser. Sovrem. Probl. Mat.* 12, 1979, 5–57. English transl.: J. Sov. Math. 13, 707–745 (1980), Zbl. 428.14015.

Vinberg, È.B., Popov, V.L. [1972] On a class of quasi-homogeneous affine varieties. Izv. Akad. Nauk SSSR 36 (1972), 749–764. English transl. Math. USSR, Izv. 6, 743–758 (1973), Zbl. 248.14014.

Wells, R.O. [1973] Differential Analysis on Complex Manifolds. Prentice-Hall, New York, 1973; 252 pp., Zbl. 262.32005.

Zak, F.L. [1981] Projections of algebraic varieties. Mat. Sb., Nov. Ser. 116 (1981), 593–602. English transl.: Math. USSR, Sb. 44, 535–544 (1983), Zbl. 484.14016.

Zariski, O., Samuel, P. [1958,1960] Commutative Algebra; 2 vol. Van Nostrand, Princeton, 1958, 1960; 329 pp., 414 pp., Zbl. 81, 265; Zbl. 121, 278. 2nd ed. Springer, Berlin, Heidelberg, New York, 1975, 1976. Zbl. 313.13001; Zbl. 322.13001.

Author Index

Subject Index

Encyclopaedia of Mathematical Sciences

Editor-in-Chief: R. V. Gamkrelidze

Algebraic Geometry IV

Volume 55: **A. N. Parshin, I. R. Shafarevich** (Eds.)

Algebraic Geometry IV

Linear Algebraic Groups. Invariant Theory

1993. ISBN 3-540-54682-0

Lie Groups and Lie Algebras

Volume 20: **A. L. Onishchik** (Ed.)

Lie Groups and Lie Algebras I

Foundations of Lie Theory. Lie Groups of Transformations

1993. ISBN 3-540-18697-2

Volume 21: **A. L. Onishchik, E. B. Vinberg** (Eds.)

Lie Groups and Lie Algebras II

Discrete Subgroups of Lie Groups. Cohomologies of Lie Groups and Lie Algebras

1993. ISBN 3-540-50585-7

Volume 41: **A. L. Onishchik, E. B. Vinberg** (Eds.)

Lie Groups and Lie Algebras III

Structure of Lie Groups and Lie Algebra

1994. ISBN 3-540-54683-9

Number Theory

Volume 49: **A. N. Parshin, I. R. Shafarevich** (Eds.)

Number Theory I

Fundamental Problems, Ideas and Theories

1993. ISBN 3-540-53384-2

Volume 62: **A. N. Parshin, I. R. Shafarevich** (Eds.)

Number Theory II

Algebraic Number Theory

1992. ISBN 3-540-53386-9

Volume 60: **S. Lang**

Number Theory III

Diophantine Geometry

1991. ISBN 3-540-53004-5

Algebra

Volume 11: **A. I. Kostrikin, I. R. Shafarevich** (Eds.)

Algebra I

Basic Notions of Algebra

1990. ISBN 3-540-17006-5

Volume 18: **A. I. Kostrikin, I. R. Shafarevich** (Eds.)

Algebra II

Noncommutative Rings. Identities

1991. ISBN 3-540-18177-6

Volume 37: **A. I. Kostrikin, I. R. Shafarevich** (Eds.)

Algebra IV

Infinite Groups. Linear Groups

1993. ISBN 3-540-53372-9

Volume 57: **A. I. Kostrikin, I. R. Shafarevich** (Eds.)

Algebra VI

Combinatorial and Asymptotic Methods of Algebra

1994. ISBN 3-540-54699-5

Volume 58: **A. N. Parshin, I. R. Shafarevich** (Eds.)

Algebra VII

Combinatorial Group Theory. Applications to Geometry

1993. ISBN 3-540-54700-2

Volume 73: **A. I. Kostrikin, I. R. Shafarevich** (Eds.)

Algebra VIII

Representations of Finite-Dimensional Algebras

1992. ISBN 3-540-53732-5

Springer

Encyclopaedia of Mathematical Sciences

Editor-in-Chief: R. V. Gamkrelidze

Dynamical Systems

Volume 1: **D. V. Anosov, V. I. Arnol'd** (Eds.)

Dynamical Systems I

Ordinary Differential Equations and Smooth Dynamical Systems

1988. ISBN 3-540-17000-6

Volume 2: **Ya. G. Sinai** (Ed.)

Dynamical Systems II

Ergodic Theory with Applications to Dynamical Systems and Statistical Mechanics

1989. ISBN 3-540-17001-4

Volume 3: **V. I. Arnol'd**

Dynamical Systems III

Mathematical Aspects of Classical and Celestial Mechanics

2nd ed. 1994. ISBN 3-540-57241-4

Volume 4: **V. I. Arnol'd, S. P. Novikov** (Eds.)

Dynamical Systems IV

Symplectic Geometry and its Applications

1990. ISBN 3-540-17003-0

Volume 5: **V. I. Arnol'd** (Ed.)

Dynamical Systems V

Bifurcation Theory and Catastrophe Theory

1993. ISBN 3-540-18173-3

Volume 6: **V. I. Arnol'd** (Ed.)

Dynamical Systems VI

Singularity Theory I

By V. I. Arnol'd, V. V. Goryunov, O. V. Lyashko, V. A. Vasil'ev
Translated from the Russian by A. Iacob

1993. ISBN 3-540-50583-0

Volume 16: **V. I. Arnol'd, S. P. Novikov** (Eds.)

Dynamical Systems VII

Integrable Systems. Nonholonomic Dynamical Systems

1993. ISBN 3-540-18176-8

Volume 39: **V. I. Arnol'd** (Ed.)

Dynamical Systems VIII

Singularity Theory II – Applications

1993. ISBN 3-540-53376-1

Volume 66: **D. V. Anosov**

Dynamical Systems IX

Hyperbolic Theory

1994. ISBN 3-540-57043-8

Several Complex Variables

Volume 7: **A. G. Vitushkin** (Ed.)

Several Complex Variables I

Introduction to Complex Analysis

1989. ISBN 3-540-17004-9

Volume 8: **G. M. Khenkin, A. G. Vitushkin** (Eds.)

Several Complex Variables II

Function Theory in Classical Domains. Complex Potential Theory

1993. ISBN 3-540-18175-X

Volume 9: **G. M. Khenkin** (Ed.)

Several Complex Variables III

Geometric Function Theory

1989. ISBN 3-540-17005-7

Volume 10: **S. G. Gindikin, G. M. Khenkin** (Eds.)

Several Complex Variables IV

Algebraic Aspects of Complex Analysis

1990. ISBN 3-540-18174-1

Volume 54: **G. M. Kenkin** (Ed.)

Several Complex Variables V

Convolution Equations. Complex Geometry with Applications to Physics

1993. ISBN 3-540-54451-8

Volume 69: **W. Barth, R. Narasimhan** (Eds.)

Several Complex Variables VI

Complex Manifolds

1990. ISBN 3-540-52788-5

H. Grauert, T. Peternell, R. Remmert (Eds.)

Several Complex Variables VII

Sheaf-Theoretical Methods in Complex Analysis

1993. ISBN 3-540-56259-1

Springer